Atomic Theory Workshop on Relativistic and QED Effects in Heavy Atoms

(National Bureau of Standards, Gaithersburg, MD, 1985)

AIP Conference Proceedings
Series Editor: Rita G. Lerner
Number 136

Atomic Theory Workshop on Relativistic and QED Effects in Heavy Atoms

(National Bureau of Standards, Gaithersburg, MD, 1985)

Edited by
Hugh P. Kelly and Yong-Ki Kim

American Institute of Physics
New York 1985

L.C. Catalog Card No. 85-73790
ISBN 0-88318-335-8
DOE CONF-8505189

Printed in the United States of America

Contents

FOREWORD

Within the community of atomic theorists, recently there has been a strong desire to begin holding workshops on various aspects of atomic theory. During the winter of 1984-85 it was decided that a workshop on relativistic and QED effects in atoms would be particularly appropriate for many reasons. For example, there is increasing interest in atomic calculations and measurements involving both relativistic and QED effects in the interpretation of spectra of highly-charged ions. Another area has been the calculation and measurement of parity nonconservation in heavy atoms, where the tests of the electro-weak theory depend crucially upon accurate calculations. Very recent advances in heavy-ion accelerators have made possible a wide range of new experiments which test a new regime of relativistic and QED effects. An additional reason concerns controversy regarding the foundations of relativistic many-body calculations. It was therefore felt to be important to address these questions and to discuss new and alternative methods in order to extend the range of validity of relativistic atomic many-body theory.

In response to this challenge, the National Bureau of Standards kindly supported the Atomic Theory Workshop on Relativistic and QED Effects in Heavy Atoms which was held on May 23-24, 1985, at the National Bureau of Standards in Gaithersburg, Maryland. Although the meeting was arranged on short notice and was not widely publicized, there were more than fifty participants from the United States and many foreign countries. Discussion at this meeting was lively and wide-ranging, and we felt that a record of the proceedings would bring the discussions (and some afterthoughts) to the entire physics community.

Organization and planning of the meeting was carried out by Lloyd Armstrong, Jr., Hugh P. Kelly, Yong-Ki Kim, and Andrew W. Weiss. We are grateful to the participants of this workshop who provided us with a written record of their presentations and comments. We are also very grateful to the National Bureau of Standards for providing facilities and financial support which made this workshop possible on short notice.

Editors: Hugh P. Kelly
Yong-Ki Kim

Healthy Hamiltonians for Relativistic Atomic Physics

J. Sucher
Dept. of Physics and Astronomy, University of Maryland
College Park, MD 20742

ABSTRACT

It has long been known that the Dirac-Coulomb Hamiltonian H_{DC}, often used in the past as the basis for relativistic calculations of the spectra of heavy atoms, has no bound states. A review is given of a QED-based approach to this problem. It yields a configuration-space Hamiltonian H_+, which is protected from the troubles of H_{DC} by the appearance of positive-energy projection operators surrounding the electron-electron interaction. Several choices for H_+ are considered and compared. The relation between the Dirac-Hartree-Fock type equations obtained from H_{DC} and H_+ is discussed. Applications to concrete physical problems which have been made with H_+-type hamiltonians are described.

I. INTRODUCTION

Suppose we want to calculate the energy levels and the amplitudes for radiative decay of an atom or ion with N electrons and a nucleus of charge Ze. Within the framework of nonrelativistic atomic theory the procedure to be followed has been well known for many years. To the level of accuracy usually sought after (order α^2 Ryd for energy levels and leading order for decay amplitudes) there are "only" practical difficulties- not questions of principle. When it comes to relativistic atomic theory the situation is different: for one-electron atoms it is a highly developed art, firmly grounded in quantum electrodynamics (QED) and well described in modern textbooks.[1] However this is not so for many-electron atoms. Although many practical advances have been made in the past three decades, the theoretical basis for the modern calculations has, until relatively recently, not received much attention.

In the remainder of this section I will first review briefly the nonrelativistic theory of many-electron atoms, in a somewhat non-standard way designed to set the stage for the discussion of the relativistic theory. Next, I will outline the standard approach to the relativistic theory of one-electon atoms. I will then turn to a discussion of a peculiar difficulty encountered in the formulation of a relativistic QED-based approach when more than one electron is present, the problem of continuum dissolution, which plagues the Dirac-Coulomb Hamiltonian H_{DC}. I will sketch the derivation of healthy Hamiltonians from QED in Sec. II, describe some applications in Sec. III and give a summary in Sec. IV.

A. Nonrelativistic atomic theory and the definition of a genuine relativistic effect.

The approach of ordinary nonrelativistic atomic theory to the problem posed above may be thought of as involving the following steps:

a) Solve, as well as possible, the nonrelativistic Schroedinger equation

$$H_{nr}\phi = W\phi\ ,\quad H_{nr} = \sum_i (\vec{p}_i^{\,2}/2m - Z\alpha/r_i) + \sum_{i<j} \alpha/r_{ij}\ , \tag{1.1}$$

for totally-antisymmetric normalizable multi-Pauli spinor wave functions ϕ.

b) Introduce the quantized radiation field $\vec{A}_T(\vec{x})$, its free Hamiltonian H_{rad}, and the interaction

$$H' = :H_{nr}(\vec{\pi}_i, \vec{r}_i): - H_{nr}(\vec{p}_i, \vec{r}_i)\ ,\quad \vec{\pi}_i = \vec{p}_i + e\vec{A}_T(\vec{r}_i)\ . \tag{1.2}$$

Here it is important to reinterpret $\vec{p}_i^{\,2}$ as $(\vec{\sigma}_i \cdot \vec{p}_i)^2$ and $\vec{\pi}_i^{\,2}$ as $(\vec{\sigma}_i \cdot \vec{\pi}_e)^2$. The colons in (1.2) denote normal ordering.

c) Compute decay amplitudes in terms of H', using first order perturbation theory for one-photon emission and first- and second-order perturbation theory for two-photon emission.

d) Compute a level-shift arising from H', using second-order perturbation theory:

$$\Delta W' = \langle\phi|H'(W-H_{nr}-H_{rad})^{-1}H'|\phi\rangle_{int}\ . \tag{1.3}$$

Here the subscript "int" means that only terms in which a photon is absorbed by an electron other than the one emitting it are to be kept. Thus $\Delta W'$ is finite as it stands, since no self-energy effects are included. On neglect of electron recoil energy in the energy denominators appearing in (1.3) and use of completeness of the eigenstates of H_{nr} one finds, after carrying out the integration over the virtual-photon momentum, the result

$$\Delta W' \approx \langle\phi|H'_{fs}|\phi\rangle \tag{1.4a}$$

where

$$H'_{fs} = \Sigma[H_{o-o}(i,j) + H_{s-o-o}(i,j) + H_{s-s}(i,j)]\ . \tag{1.4b}$$

The various H(i,j) denote the familiar orbit-orbit, spin-other-orbit, and spin-spin interaction operators. These are usually described as arising from the reduction of the Breit operator B(i,j) to Schroedinger-Pauli (SP) form. The treatment given here serves to emphasize that in fact nothing need be known about relativistic quantum theory, from which the Breit operator is obtained, to arrive at the operator H'_{fs}, which describes part of the fine structure.

e) Compute another level shift, defined by

$$\Delta W'' = \langle \phi | H''_{fs} | \phi \rangle \,, \tag{1.5a}$$

where

$$H''_{fs} = \Sigma\, H_{s-o}(i) + \Sigma\, H_{s-o}(i,j) + \Sigma\text{"}(-\vec{p}_i^{\,4}/8m^3)\text{"} + H^{cont}_{s.i.} \,. \tag{1.5b}$$

Here $H_{s-o}(i)$ is the spin-orbit interaction operator, coming from the interaction of "i" with the magnetic field produced in its rest frame by the nucleus, and $H_{s-o}(i,j)$ is the analogous term with "j" replacing the nucleus. The $\vec{p}_i^{\,4}$ terms are kinematic corrections to the kinetic-energy operator and have quotes on them because in $\vec{r}$-space they must be evaluated as $\langle \vec{p}_i^{\,2}\phi | \vec{p}_i^{\,2}\phi \rangle$, to avoid inclusion of a spurious contact term. All these terms can be obtained from semiclassical considerations. The last term in (1.5b) is a spin-independent contact operator given by

$$H^{cont}_{s.i.} = (\text{const.})(\sum_i Z\delta(\vec{r}_i) - 2\sum_{i<j} \delta(\vec{r}_{ij})) \,. \tag{1.5c}$$

It can be written down by analogy with the so-called S-state correction found in the one-electron case on reduction of the Dirac equation to SP form. However viewed, the shift defined by (1.5a) is essentially grafted onto the "theory" defined by (1.1) and (1.2), i.e. it is put in by hand.

In summary, apart from $H^{cont}_{s.i.}$, all terms contributing to the leading fine structure can be obtained without use of relativistic field theory or even the Dirac description of the electron.

Of course, the only really hard part of this program is step (a)! Let us imagine that we have found exact solutions to (1.1). Then we can ask: How good is this hybrid theory? How far can it be pushed?

With regard to the first question, the energy levels are correct through terms of order α^2 Ryd. The decay amplitudes are correct in leading order for electric dipole transitions and for most but not all magnetic dipole transitions. Thus, in its most naive form, the hybrid theory fails for as simple a process as $2^2S_{1/2} \rightarrow 1^2S_{1/2} + \gamma$ in hydrogen, an example of a hindered or relativistic magnetic-dipole transition.

With regard to the second question, one can go a bit further: The nonrelativistic parts of the Lamb shift can be obtained from (1.3) by including the self-absorption of emitted photons and adding a mass counter term to H'. But basically that's where it ends. In particular, the operator $H'_{fs} + H''_{fs}$ can only be used in lowest order -- already in the next order spurious divergences appear, because the operators involved are so singular at short distances.

A succinct definition of what is meant by a genuine relativistic or QED effect in atomic spectroscopy would be to say that it's anything not accurately described by the naive hybrid theory. An example is provided by virtual-pair effects arising from the electron-electron Coulomb interaction, which are totally missing from this theory and yield level shifts of order α^3 Ryd.

B. Relativistic atomic theory: N=1 or $1 < N \ll Z$

For $N = 1$ and $Z < 137$ a wonderful starting point is the external-field Dirac equation:

$$h_{D;ext}(1)\psi(\vec{x}_1) = E\psi(\vec{x}_1),\ h_{D;ext}(1) = \vec{\alpha}_1\cdot\vec{p}_1 + \beta_1 m + V_{ext}(1) \qquad (1.6)$$

with $V_{ext}(1) = -Z\alpha/r_1$. From the viewpoint of QED, this equation is really "too exact": E contains terms of all orders in $(\alpha Z)^2$ whereas radiative corrections begin in order $\alpha(\alpha Z)^2\log(\alpha Z)$Ryd . However, it doesn't hurt to have more accuracy than needed in parts of the calculation. The computation of the radiative corrections to E has in the past forty years been elaborated into a highly sophisticated art.[1] From the theoretical point of view it corresponds to the following approach to QED: Write

$$H_{QED} = H_{rad} + H^{ext}_{mat} + H' + c.t. \qquad (1.7)$$

with

$$H_{rad} = \frac{1}{2}\int d\vec{x} : \vec{E}^2(\vec{x}) + \vec{H}^2(\vec{x}) : \qquad (1.8a)$$

$$H^{ext}_{mat} = \int d\vec{x} : \psi^\dagger_D(\vec{x})(\vec{\alpha}\cdot\vec{p} + \beta m + V_{ext}(\vec{x}))\psi_D(\vec{x}) : \qquad (1.8b)$$

and, in Coulomb gauge,

$$H' = H^C_{mat} + H_T, \qquad H^C_{mat} = \frac{1}{2}\int d\vec{x}d\vec{x}'\ \frac{:j^0(\vec{x})j^0(\vec{x}'):}{|\vec{x}-\vec{x}'|},$$

$$H_T = -\int d\vec{x}\,\vec{j}(\vec{x})\cdot\vec{A}_T(\vec{x}) . \qquad (1.8c)$$

Here $\vec{E}(\vec{x})$ and $\vec{H}(\vec{x})$ are the electric and magnetic field operators associated with $\vec{A}_T(\vec{x})$, $\psi_D(\vec{x})$ is the quantized Dirac matter field and $j^\mu(\vec{x}) = -e:\bar{\psi}_D(\vec{x})\gamma^\mu\psi_D(\vec{x}):$ is the electromagnetic current; the symbol "c.t." denotes renormalization counter terms. Next, introduce the equivalent bound-state or Furry picture (within the context of the Schroedinger picture we are using here) by expanding $\psi_D(\vec{x})$ in the positive and negative energy eigenfunctions u and v of $h_{D;ext} = \vec{\alpha}\cdot\vec{p} + \beta m + V_{ext}$,

$$\psi_D(\vec{x}) = \sum_n u_n(\vec{x})a_m + \sum_m v_m(\vec{x})b_m^\dagger \ . \tag{1.9}$$

Then the electron and positron number operators, N_- and N_+ respectively, associated with this decomposition of ψ_D commute with $H_o = H_{rad} + H_{mat}^{ext}$ and one can look for eigenstates of H_o, with eigenvalue E_o, for which $N_- = 1$, $N_+ = 0$, $N_\gamma = 0$. These are necessarily of the form

$$\Psi = \sum_n f_n a_n^\dagger |vac\rangle \ . \tag{1.10}$$

The Dirac spinor function $u(\vec{x})$ defined by

$$u(\vec{x}) = \sum_n f_n u_n(\vec{x}) \tag{1.11}$$

then satisfies (1.6) and we are back to the c-number Dirac equation. Note that any function of the form (1.11) satisfies

$$L_+ u(\vec{x}) = u(\vec{x}) \ , \tag{1.12}$$

where L_+ is the projection operator onto the subspace spanned by the u_n's.

The calculation of QED effects then normally proceeds by treating <u>all</u> of the remaining interaction of the Dirac matter field as a perturbation. Since $N_- = 1$ it is actually most convenient to use the formally covariant quantization of the radiation field (Gupta-Bleuler method) which allows use of the Feynman propagator for virtual photons and facilitates implementation of the renormalization program.

For $N_- = N > 1$ but $Z >> N$ one can also treat all of H' as a perturbation. With the help of a symmetrized version of the Gell-Mann-Low level shift formula, one can then use formally covariant methods of calculation.[2] This approach is currently being exploited, for N=2, by P. Mohr.[3]

I now turn to the case of dominant interest for this talk:

C. Z large and $N \sim Z$, the difficulties of H_{DC}.

What's the problem? On the one hand, with Z large one doesn't want to be committed to an expansion in powers of v/c from the outset. This implies that, at a minimum, we want to use relativistic kinematics for the electrons. On the other hand, with $N \sim Z$, we know from the nonrelativistic theory that at least the static part of the electron-electron interaction cannot be treated by ordinary perturbation theory. Thus one would like to have a practical starting point which both treats the electrons relativistically and the most important part of their interaction nonperturbatively.

By analogy with the nonrelativistic case, an obvious candidate is given by the equation

$$H_{DC}\phi = E\phi \tag{1.13a}$$

where H_{DC} is the Dirac-Coulomb hamiltonian, defined by:

$$H_{DC} = \sum_{i=1}^{N} h_{D;ext}(i) + \sum_{i<j} \alpha/r_{ij} \; . \tag{1.13b}$$

Indeed for N>1 this hamiltonian is related to $h_{D;ext}(1)$ in the same way that H_{nr} is related to the one-electron Schroedinger hamiltonian. It incorporates relativistic kinematics and the electron-electron Coulomb interaction in what is seemingly the simplest possible way.

Unfortunately, H_{DC} is sick -- it has no normalizable eigenstates. Although this has been known since the work of Brown and Ravenhall in 1951,[4] it did not draw much attention until about five years ago.[5,6] Some of the reasons for this have been discussed elsewhere.[7] Here I just want to review, for the benefit of newcomers to the field of relativistic atomic physics, the disease which afflicts (1.13).

It suffices to consider N=2. Turn off the Coulomb interaction and examine the spectrum of $H_D(1,2) = h_{D;ext}(1) + h_{D;ext}(2)$. Although $H_D(1,2)$ has normalizable eigenstates, of the product form $u_{n_1}(1)u_{n_2}(2)$, it does not have a discrete spectrum because degenerate in energy with each of these states is a continuous family of improper (non-normalizable) states, with one electron in a continuum positive-energy state and the other in a continuum negative-energy state. ($H_D(1,2)$ doesn't know about the filled negative sea and the exclusion principle.) On turning on an ordinary local interaction such as α/r_{12} the nominal bound states of $H_D(1,2)$ will mix with these non-normalizable states and dissolve into the continuum. I will refer to this phenomenon as "continuum dissolution" (CD). It is analogous in its mathematical aspects to the process of autoionization, but unlike the latter it has no physical counterpart.

What is the source of the disease? What is to be done about it? With regard to the first question, the answer is simple: the failure of naive intuition. Eq. (1.13) is not a reasonable approximation to what one obtains from quantum field theory: It corresponds to an extension of Dirac's one-electron theory rather than to an extension of Dirac's hole theory to many electrons. The answer to the second is equally simple (to announce): Return to field theory, i.e. to the basic equations of QED, and consider the problem from scratch.

II. FIELD THEORY AND HEALTHY HAMILTONIANS.

A. Guidelines

Of course, the injunction to "return to field theory" is by itself perhaps on a par to suggesting to someone with a personal problem to "read the bible". There are various ways to go and one needs some guidelines to pick a definite path. In any approach one will have to make some approximations; this is O.K. as long as one can indicate how the calculation can be improved upon, if the approximations turn out to be inadequate in a given context. The approach I will describe is based on the following criteria, patterned after the nonrelativistic framework as much as feasible:

a) The formulation is to involve, in zeroth order, an N-particle configuration-space Hamiltonian, call it H_+, which has bona-fide normalizable eigenfunctions ψ and eigenvalues which may be associated with atomic bound states and energy levels.

b) In the nonrelativistic limit ($\alpha Z \ll 1$) the quantities ψ and $E-Nm$ are to approach nonrelativistic counterparts ϕ and W, associated with eq. (1.1).

c) The connection of the hamiltonian H_+ to field theory and the field-theoretic interpretation of the eigenfunctions ψ is to be transparent; level shifts and decay amplitudes should be calculable in a straightforward way, at least in principle, without fear of double counting, etc.

These are minimum requirements. In addition one can impose the following:

d) The eigenvalues of H_+ should be correct to all orders in αZ ; thus if all e-e interaction is neglected, the bound state spectrum is to reduce to the sum of one-electron Dirac energies.

e) The eigenvalues of H_+ should be correct at least to order α^2 Ryd.

Criterion (d) implies that no approximation is made with regard to the magnitude of Z. Criterion (e) implies that all the familiar fine structure effects are included in E, without further computation.

B. Healthy hamiltonians[8]

To arrive at a class of Hamiltonians which satisfy properties (a), (b) and (c), consider a potential U which may be zero or V_{ext} or something in between, and expand the matter field operator $\psi_D(\vec{x})$ in terms of the positive and negative energy eigenfunctions $u(\vec{x})$ and $v(\vec{x})$ of

$$h_{D;U} = \vec{\alpha}\cdot\vec{p} + \beta m + U \; . \tag{2.1}$$

Working in Coulomb gauge, separate H_{QED} into a part $H^{(o)}_{QED}$ involving no creation or destruction of electron-positron pairs or of photons and a remainder H_{QED}^{rem}. Look for eigenstates of $H_{QED}^{(o)}$ of the form

$$\Psi = \frac{1}{\sqrt{N!}} \sum f_{n_1 n_2 \ldots n_N} a^\dagger(n_1) a^\dagger(n_2) \ldots a^\dagger(n_N) |vac\rangle , \quad (2.2)$$

corresponding to $N_+ = N_\gamma = 0$ and $N_- = N$.

On defining a configuration-space wave function ϕ by

$$\phi = \sum f_{n_1 n_2 \ldots n_N} u_{n_1}(1) \ldots u_{n_N}(N) , \quad (2.3)$$

one then finds, without further approximation, that ϕ satisfies

$$H_+^U \phi = E\phi \quad (2.4a)$$

and

$$L_+^U(i)\phi = \phi , \quad (2.4b)$$

where

$$H_+^U = \sum_i h_{D;U}(i) + L_+^U [V_{ext}^{tot} - U^{tot} + \sum_{i<j} \alpha/r_{ij}] L_+^U \quad (2.5)$$

and L_+^U is a product of positive-energy projection operators

$$L_+^U = L_+^U(1) \ldots L_+^U(N) \quad (2.6a)$$

with

$$L_+^U(i) = \sum_n |u_n(i)\rangle\langle u_n(i)| . \quad (2.6b)$$

The condition (2.4b), which follows directly from the definition (2.3), is compatible with (2.4a) because the $L_+^U(i)$ commute with H_+^U :

$$[L_+^U(i), H_+^U] = 0 . \quad (2.7)$$

Since $L_+^U(i)$ is a projection operator we have

$$(L_+^U(i))^2 = L_+^U(i) . \quad (2.8)$$

Use of (2.7), (2.8) and the constraint (2.4b) allows (2.4a) to be written in a variety of other ways. In particular, on multiplying (2.4a) on the left by L_+^U one sees that it is equivalent to

$$\hat{H}_+^U \phi = E\phi , \quad (2.9)$$

where

$$\hat{H}_{+}^{U} = L_{+}^{U} H_{DC} L_{+}^{U} , \tag{2.10}$$

and (2.4b) is of course still imposed.

While the mathematical relation between $\hat{H}_{+}^{U}$ and H_{DC} is simpler than that between H_{+}^{U} and H_{DC} the operator H_{+}^{U} has a more transparent physical interpretation than the hatted operator. The first sum in (2.5) represents the propagation of "electrons", as defined by U, moving without direct interaction with each other, the term $L_{+}^{U}(V_{ext}^{tot}-U^{tot})L_{+}^{U}$ generates virtual scattering of these electrons (without pair production) arising from the residual potential $V_{ext}-U$, while the term $L_{+}^{U}(\Sigma\alpha/r_{ij})L_{+}^{U}$ describes such scattering arising from the total electron-electron Coulomb interaction V_{ee}^{tot} . The physical difference between H_{+}^{U} and H_{DC} is most transparent for the choice $U = V_{ext}$ (Furry picture) where

$$H_{+}^{U} \rightarrow H_{+} = \Sigma\, h_{D;ext}(i) + L_{+}(\Sigma\alpha/r_{ij})L_{+} \tag{2.11}$$

with L_{+} the product of the external-field projection operators. The operator H_{+} differs from H_{DC} precisely in the presence of factors L_{+} surrounding V_{ee}^{tot} . The right-hand factor is purely cosmetic --- since $L_{+}\psi = \psi$ it could be removed --- but it serves to make the hermiticity of H_{+} manifest. It is the left-hand factor L_{+} which plays the key role of preventing V_{ee}^{tot} from generating unphysical virtual transitions to negative-energy states and thereby makes H_{+} healthy. Its presence is a direct consequence of the fact that QED is the mathematical formulation of the physical ideas of Dirac's hole theory rather than Dirac's one-electron theory. The left-hand factor L_{+} is not put in by hand --- it comes out of the theory, which "knows" that the negative-energy sea is filled.

C. Choice of U

The most important of the criteria listed above for a satisfactory configuration-space Hamiltonian, viz. (a), (b) and (c), are satisfied for any reasonable choice of U. Three natural possibilities are (i) $U = 0$, (ii) $U = V_{ext}$ and (iii) $U = V_{int}$, where V_{int} denotes some potential "intermediate" between 0 and V_{ext}.

(i) $\underline{U = 0}$. This is by far the simplest choice, both from a conceptual and a mathematical point of view. It is conceptually the simplest because it involves a description of the atomic electrons in terms of free or "physical" electrons (apart from self-field effects), pictured as wave packets of positive-energy Dirac plane waves. It is mathematically the simplest because, correspondingly, the operators $L_{+}^{U}(i)$ reduce to the Casimir projection operators

$$\Lambda_{+}(i) = (E(\vec{p}_i^{\,op}) + h_D(i))/2E(\vec{p}_i^{\,op}) \; . \qquad (2.12)$$

In $\vec{p}$-space $\Lambda_{+}(i)$ is just an algebraic matrix function of $\vec{p}$ and in $\vec{r}$-space its kernel can be given explicitly in terms of known functions.[9] The operator H_{+}^{U} simplifies to

$$h_{+} = \Sigma\, h_D(i) + \Lambda_{+}(V_{ext}^{tot} + \Sigma\alpha/r_{ij})\Lambda_{+} \qquad (2.13)$$

and the constraint $\Lambda_{+}(i)\phi = \phi$ can be used to reduce the equation

$$h_{+}\phi = E\phi \qquad (2.14)$$

to SP form exactly, without any expansion in $\vec{p}/m$. The resulting equation is, in $\vec{p}$-space, essentially no more complicated than the nonrelativistic Schroedinger equation.[5] Its major drawback is that it is not exact to all orders in αZ , because it neglects the virtual-pair effects coming from V_{ext}, which become important for large Z. For Z less than, say, 45, so that $(\alpha Z)^2 \lesssim 0.1$, these effects can be calculated perturbatively but beyond that this approach is likely to become less and less useful, especially for questions regarding inner electrons.

(ii) $\underline{U = V_{ext}}$. This choice corresponds to use of the Furry picture not only for one-electron atoms but also for atoms containing any number of electrons. Its main advantage is that no approximation is made, at least initially, with regard to V_{ext} so that it is exact to all orders in αZ . Its main disadvantage is that the projection operators are complicated. Thus reduction to SP form is possible only in a formal sense and will in practice require some sort of expansion or iteration technique (see, however, below).[10]

(iii) $\underline{U = V_{int}}$. One may choose U to be some potential intermediate between zero and V_{ext}. In particular, one may choose it to behave like (Z-N)/r for large r and like Z/r for small r, which is appealing on physical grounds. However, this choice has both the disadavantage of (i) -- not exact in αZ -- and the disadvantage of (ii) -- complicated projection operators.

At this stage use of either (ii) or (iii) appears to involve formidable complications but some of these disappear when one considers practical methods based on variational principles, as we will do in the next section. For the moment let us note that only the choice (ii) satisfies criterion (d). None of them satisfies criterion (e) but this is easily remedied by making the replacement $\alpha/r_{ij} \to \alpha/r_{ij} + B(i,j)$ where $B(i,j)$ is the Breit operator, in the definition (2.5) of H_+^U. The reduction to nonrelativistic form then contains all of the leading fine structure operators as well as some higher-order corrections. To avoid burdening the notation I will not make this modification in the following, but in practice it should be done: the old arguments against including $B(i,j)$ in H_{DC} are irrelevant here. Because of the CD problem they were in fact never cogent to begin with.[11]

D. DHF type equations

DHF-type equations can be obtained from the Hamiltonian H_+^U in a manner similar to that used for H_{DC}. The main difference is the need to take into account the constraint (2.4b). There is one approach in which this condition is automatically satisfied: Expand the wave function in terms of Slater determinants ϕ_Γ formed from a fixed basis set, as in the so-called relativistic configuration (RCI) method, and choose the Dirac orbitals $\phi_a(i)$ to be positive-energy eigenfunctions of $h_{D;U}(i)$. Then the trial wavefunction

$$\tilde{\phi} = \Sigma\, c_\Gamma \phi_\Gamma \tag{2.15}$$

satisfies

$$L_+^U(i)\tilde{\phi} = \tilde{\phi} \tag{2.16}$$

and in view of (2.16) and (2.10) we have

$$\langle\tilde{\phi}|H_+^U|\tilde{\phi}\rangle = \langle\tilde{\phi}|H_{DC}|\tilde{\phi}\rangle \; . \tag{2.17}$$

Thus the variation of the coefficients c_Γ yields the same equations as would be obtained from H_{DC}.

The major difference is that the wave function $\tilde{\phi}$ has now been given a clearcut field-theoretic meaning: it is an approximation to the function ϕ defined by eq. (2.3). One can therefore calculate corrections to the energy levels E by returning to field theory and taking into account previously neglected effects, e.g., the pair effects arising from the difference $V_{ext}-U$, transverse photon effects, etc., without fear of double counting.

An important application of these considerations comes on choosing U to coincide with the DHF potential V_{DHF}, obtained from H_{DC} with a single-determinant ansatz. It follows from the above that any RCI calculation which uses a basis generated by this V_{DHF} and applies the variational principle to H_{DC} can be reinterpreted as a calculation based on the corresponding H_+^U and thus can be given a straightforward field-theoretic interpretation. This choice of U has been advocated by M. Mittleman as being in some sense optimal.[6]

The field-theoretic interpretation of another method which has been much applied to H_{DC}, the so-called multiconfiguration Dirac-Fock (MCDF) method, is less clear-cut. In this approach the radial wave functions entering the Dirac orbitals P(r) and Q(r), associated with the upper and lower components respectively, are both left free to vary. As a consequence, it seems unlikely that there is a fixed reference potential U in terms of which the MCDF equations can be reinterpreted. Further study of this question is needed. Nevertheless the usual MCDF equations are not very different in their structure from those obtained for the present case by an obvious extension of the MCDF method. The constraint (2.4a) is satisfied whenever the individual orbitals $\psi_a(i)$ used in the construction of a trial wave function satisfy it. But for a rotationally invariant U the condition

$$L_+^U(i)\psi_a(i) = \psi_a(i) \tag{2.18}$$

simply serves to fix the relation between the upper and lower components of ψ_a and hence determines $Q_a(r_i)$ in terms of $P_a(r_i)$:

$$Q_a = X_{op} P_a \tag{2.19}$$

with X_{op} a fixed operator, dependent only on the choice of U. The modified equations are then obtained from the standard functional $\langle\tilde{\psi}|H_{DC}|\tilde{\psi}\rangle$ by varying both the coefficients c_Γ and $P_a(r)$, with the variation in Q_a determined by that of P_a via

$$\delta Q_a = X_{op}\delta P_a \ . \tag{2.20}$$

In the course of the usual MCDF calculations the ratio of Q_a to P_a is fixed at both large and small r in a way which incorporates physically reasonable boundary conditions.[12] Although this is a weaker condition than that required here it suggests that the numerical results of calculations based on the proposed modification of the MCDF method may not differ much from those obtained in the standard calculations. However this conjecture remains to be verified by actual computation and comparison.

III. APPLICATIONS

Healthy hamiltonians such as (2.11) and (2.13) have been successfully applied in both the distant and recent past to a variety of concrete physical problems in atomic and particle physics. I will describe these briefly, by way of showing that the approach being advocated has more than just formal value.

In atomic physics the applications have been primarily to the physics of two-electron atoms or ions, including:

(a) α^3 Ryd corrections to low-lying levels.

The formal starting point of this calculation was a four dimensional Bethe-Salpeter type equation for two electrons moving in an external field.[13] Its reduction to equal times yields, in the "Coulomb ladder approximation," an equation which, after a term corresponding to virtual-pair creation is dropped, serves as a benchmark for organizing the calculation of higher-order effects. This is just the "no-pair Coulomb ladder equation"

$$H_+\psi = E\psi \qquad (3.1)$$

where H_+ is given by (2.11) with N = 2.

(b) α^4 Ryd corrections to the fine structure of low-lying levels.

This heroic calculation of the spin-dependent terms was an extension of (a) which also made extensive use of (3.1).[14] The spin-independent terms of this order have apparently never been calculated.

(c) Decay rates for the transition $2\,^3S_1 \rightarrow 1\,^1S_0 + \gamma$.

The calculation of the rates for these relativistic or "hindered" M1 transitions is most conveniently based on the counterpart of (3.1) which involves free projection operators,

$$h_+\psi = E\psi\ , \qquad (3.2)$$

where h_+ is given by (2.13) with N = 2.[15] The theoretical values are in agreement with experiment from Z = 2 to Z = 36, which involves a range of lifetimes spanning fourteen orders of magnitude.[16]

(d) Level splitting of Rydberg states

The splitting of the n = 10 levels of He has been measured recently with great accuracy by Palfrey and Lundeen.[17] This affords an opportunity to test some aspects of the long-range forces which arise from two-photon exchange between charged and/or neutral particles.[18] A general theory of these effects has been developed, based on a combination of the dispersion-theoretic approach to long-range forces and the use of eq. (3.1), within the context of QED. Its application to the n = 10 levels yields results which are consistent with experiment.[19]

(e) Parity violation in heavy atoms.

The many-electron no-pair equation (2.11) has also been used as the basis for the development of a formalism for the calculation of parity-violating-effects in heavy atoms which starts from first principles.[20]

(f) Two-body bound states; strong coupling.

In contrast to the case of two (or more) electrons in an external field, for the pure two-body problem the projection operators are not essential, from the mathematical point of view, to the existence of bound states. Nevertheless, they arise naturally in any treatment based on field-theory, as should by now be amply clear. In particular, the dropping of virtual-pair effects in an equation of the Salpeter-type yields, in the c.m. system, a simplified form of (2.13) viz.

$$(\vec{\alpha}_1 \cdot \vec{p}_1 + \beta_1 m_1 + \vec{\alpha}_2 \cdot \vec{p}_2 + \beta_2 m_2 + \Lambda_+(1)\Lambda_+(2)V\Lambda_+(1)\Lambda_+(2))\phi = E\phi \quad (3.3)$$

with $\vec{p}_1 = -\vec{p}_2 = \vec{p}$. Eq. (3.3) has recently been used by G. Hardekopf and myself as a "laboratory" for testing numerical techniques in $\vec{p}$-space as well as for a comparison of virtual-pair effects in hydrogen and positronium.[21] As an extension of this work we have studied the limit of very strong coupling γ , with V either a pure-Coulomb or a Coulomb-plus-Breit interaction.[22] For $m_2 = \infty$, eq. (3.3) reduces to the Dirac equation, for which the maximal value of γ is unity. However if m_2 is finite, as it of course is in any physical situation, γ_{max} turns out to both differ from unity and to be independent of the finite value of m_2. This result may be of interest for the question of anomalous pair production in heavy-ion collisions. It needs to be extended to the case of a distributed charge for the heavy particle, to see if it is of physical relevance in this context.

For particle physics, no-pair equations of the type (3.3) have been applied to the description of the narrow resonances in the Gev region as (quark-, anti-quark) bound states. The first application was to the calculation of the amplitudes for M1 transitions.[23] More recently, applications have been made to the calculation of relativistic corrections to E1 transitions.[24] These corrections turn out to be sufficiently large so as to warrant a non-perturbative treatment of eq. (3.3).

IV. Concluding Remarks

I hope I have been able to convince you of the basic soundness and utility of the healthy-hamiltonian approach to the calculation of relativistic and QED effects in atomic physics. This is not to say that other approaches, some of which bypass the construction of configuration-space hamiltonians altogether, cannot be useful also.[25] One of the virtues of the hamiltonian approach is that one is not committed to some kind of central-field method from the outset: any method that has been found to be useful in the nonrelativistic hamiltonian theory will have a relativistic counterpart. One may imagine, for example, a relativistic version of the Hylleras method, for a two- or even a three-electron atom or a relativistic version of many-body perturbation theory, starting with H_+ or h_+ or their equivalent relativistic SP forms. The evaluation of radiative corrections for two-electron atoms was long ago carried out within the H_+-framework to an accuracy of order α^3 Ryd[13] and there do not appear to be any problems of principle in an extension to more than two electrons. However, the development of a convincing yet practical approach to the calculation of Lamb shifts in heavy atoms remains as a challenge to theorists.

With regard to the standard relativistic calculations for many-electron atoms, we have seen that those which use the RCI method can be reinterpreted as based on QED in such a way that hitherto neglected pair effects, arising from the difference $V_{ext}-$

U, can be calculated in an unambiguous way. While some crude numerical estimates of these have been made,[8] accurate calculations have yet to be carried out. The precise connection of the MCDF method with field theory also requires further elucidation.

As mentioned before, the constraint $L_+^U(i)\phi = \phi$ can be used to reduce the eigenvalue equation for ϕ to SP form and for the choice U = 0 the resulting equation is as simple as the non-relativistic Schroedinger equation in $\vec{p}$-space. This would seem to provide a good starting point for the study of the simultaneous effects of correlation and relativity in atoms with moderately large values of Z within the framework proposed.[26,27]

In conclusion I offer you a (taped) musical quote from the masters, lyrics made famous in the 1940's by the singer Johnny Mercer, composed jointly by him and Harold Arlen, which I propose as an official theme song for this workshop:

"You got to ac-cent-tchu-ate the positive,
Eliminate the negative,
Latch on to the affirmative,
Don't mess with Mr. In-between!"

Acknowledgment

This work was supported in part by the National Science Foundation.

References

1. See, e.g. C. Itzykson and J.-B. Zuber, Quantum Field Theory (McGraw-Hill, New York, 1980), Ch. 10.
2. J. Sucher, Phys. Rev. 107, 1448 (1957)
3. See P. Mohr, these Proceedings.
4. G.E. Brown and D.G Ravenhall, Proc. R. Soc. London, Sec. A 208, 552 (1951).
5. J. Sucher, Phys. Rev. A22, 348 (1980).
6. M. Mittleman, Phys. Rev. A24, 1167 (1981).
7. J. Sucher, in Proceedings of the Argonne Workshop on the Relativistic Theory of Atomic Structure, H.G. Berry, K.T. Cheng, W.K. Johnson and Y.-K. Kim, Eds. ANL-80-116 (Argonne National Laboratory, Argonne, IL, 1980).
8. For a more leisurely review, see (a) J. Sucher, in Proceedings of the NATO Advanced Study Institute on Relativistic Effects in Atoms, Molecules and Solids, G. Malli, Ed. (Plenum, New York, (1982) and (b) Int. Jour. Quant. Chem. XXV, 3 (1984).
9. See Ref. 7, pp. 9-14.
10. See e.g. Ref. 8b, p. 19.
11. See Ref. 7, eqs. (5.2) and (5.3).
12. I thank J.-P. Desclaux and I. Grant for an explanation of this point.
13. J. Sucher, Columbia University Ph.D. Dissertation (1957, unpublished) and Phys. Rev. 109, 1010 (1958); H. Araki, Prog.

Theoret. Phys. 17,1 (1957).
14. M. Douglas and N.M. Kroll, Ann. Phys. N.Y. 82, (1974).
15. G. Feinberg and J. Sucher, Phys. Rev. Lett. 26, 681 (1971); G.W.F. Drake, Phys. Rev. A5, 1979 (1972).
16. For a review with extensive references, see J. Sucher, Rep. Prog. Phys. 41, 1781 (1978).
17. S.L. Palfrey and S.R. Lundeen, Phys. Rev. Lett. 53, 1141 (1984).
18. E.J. Kelsey and L. Spruch, Phys. Rev. A18, 1511 (1978).
19. C.-K. Au, G. Feinberg and J. Sucher, Phys. Rev. Lett. 53, 1145 (1984) and paper in preparation.
20. J. Hiller, J. Sucher, G. Feinberg and B. Lynn, Ann. Phys. 127, 149 (1980).
21. G. Hardekopf and J. Sucher, Phys. Rev. A30, 703 (1984).
22. G. Hardekopf and J. Sucher, Phys. Rev. 31, 2020 (1985).
23. G. Feinberg and J. Sucher, Phys. Rev. Lett. 35, 1740 (1975).
24. G. Hardekopf and J. Sucher, Phys. Rev. D25, 2938 (1982); R. McClary and N. Byers, idid. 28, 1692 (1983).
25. See K. Dietz, these Proceedings.
26. Subsequent to this workshop I learned that such calculations have recently been carried out by B. Hess (Wuppertal preprint; to appear in the Physical Review), with encouraging results.
27. In response to questions raised at this workshop, I have constructed a solvable model which provides an explicit illustration of the problem of continuum dissolution (J. Sucher, UM preprint #86-24, to appear in Phys. Rev. Letters, September, 1985).

COMMENTS ON SUCHER'S TALK

I. P. Grant
Department of Theoretical Chemistry, Oxford University
1 South Parks Road, Oxford OX1 3TG, England

It is possible to attempt to clarify the present computational methods of relativistic atomic structure theory in the following way. Provisionally, we treat the nuclei as classical fixed sources of potential--a sort of Born-Oppenheimer approximation to be removed later--to make the calculation tractable.

We start from the Furry picture, assuming that the unperturbed electrons move in some classical potential--Z/r + u(r) in some given frame of reference; the representation in other frames of reference and gauges can be obtained by the usual transformations. Total charge is a constant of the motion but not the individual numbers in 'electron' and 'positron' states. We introduce the radiation field in the usual way, and the minimal coupling in the standard Lagrangian treatment, including, of course -u(r) as a counter-term in the perturbation. We now develop QED perturbation theory in the usual way; to treat bound states, we seek those terms which contribute to the energy of a system of N electrons, no positrons and no free photons, whose energy we can measure in the chosen frame of reference. These terms in lowest order yield an effective Hamiltonian

$$H = H^o + V$$

where

$$H^o = \sum_p :a_p^\dagger a_p: E_p$$

is the unperturbed quantized and normally ordered Hamiltonian with respect to the spectrum of the Dirac operator

$$h = c\vec{\alpha}\cdot\vec{p} + (\beta - 1)c^2 - Z_N(r)/r + u(r)$$

where $Z_N(r)$ represents some finite nuclear charge distribution, and V is a perturbation or "fluctuation" potential

$$V = \frac{1}{2} \sum_{p,q,r,s} :a_p^\dagger a_q^\dagger a_s a_r: \langle pq|g|rs\rangle - \sum_{p,q} :a_p^\dagger a_q: \langle p|u|q\rangle$$

g being the covariant Coulomb interaction (Coulomb potential plus retarded transverse interaction in the Coulomb gauge). The important point is that u, which defines the orbitals, and which appears here as a counter-term, is a fixed potential, so that there is no ambiguity about its meaning. However, we are at liberty to choose it in any convenient way; for example, we may require the lowest order correction to the energy for some closed shell system given by the effective potential V

0094-243X/85/1360017-3$3.00

$$\langle i|V|j\rangle = \sum_a [\langle ia|g|ja\rangle - \langle ai|g|ja\rangle] - \langle i|u|j\rangle$$

to vanish. One might choose the g-Hartree potential recently introduced by Dietz and Weymans as an alternative. In either case we can develop perturbation theory within this approximation, which will bring in an infinite number of terms that would anyway have been present in the original QED perturbation expansion, or continue in a non-perturbative fashion--without contradictions. Of course, the omitted terms--renormalizable self-energy and vertex corrections and terms which do not conserve particle numbers--may still be included as perturbations. This gives a framework for 'conventional' Dirac-Fock theory which seems likely to be as rigorous as can be attained at the present time. It is probably sufficient to add the nuclear motion terms as in, for example, the paper of Grotch and Yennie [Rev. Mod. Phys. 41, 350 (1969)], though this can hardly be regarded as an elegant solution. The problem of finding a workable method of treating self-energy in many-electron systems must also be solved. Indeed it is my view that this is where the real effort ought to be directed.

Sucher's question about the meaning of MCDF (or MCHF) as originally conceived remains to be answered. This is not such a problem provided one only uses the MCDF procedure in the above manner to define the effective potential for computing one-electron spinors, and to do the complete calculation in this basis. In Grant et al.'s first paper on MCDF [J. Phys. B 9, 2777 (1976)], it was pointed out that doing separate self-consistent MCDF calculations on different atomic energy levels is neither economical nor particularly consistent. For many purposes, it makes more sense simply to perform an SCF calculation using a functional based on some mean energy expression, which will be a compromise for all the states of interest. The level structure and atomic states are then obtained by doing a CI calculation within the MCDF orbital basis. If we include all the various symmetries of a configuration with the apprropriate statistical weighting this corresponds to what we now call an EAL calculation. It is simple, cost-effective, and often surprisingly accurate. One can also treat it as the first stage of a perturbation calculation which will include both correlation and radiative and other correction terms in a very straightforward fashion. Of course, working with the one-photon interaction kernel in the effective Hamiltonian defined above is just an application of the Coulomb ladder approximation--if one neglects the transverse part of the interaction. In fact, the transverse part can be included reasonably economically. Crossed diagrams cannot be handled at the moment, and electron self-energy is fudged, even in lowest order; no attempt has been made to deal with higher order corrections. And, nuclear motion remains a problem.

It is important to realize that any practical MCDF code imposes bound-state boundary conditions thereby excluding both positive and negative continuua from the wave function. Projection operators used in the manner suggested by Sucher are not necessary. There are no circumstances in which one would wish to embed bound states in the lower continuum as postulated in these models--except, of course, for super-heavy atoms with $Z \gtrsim 170$! In this case one has indeed a state of autoionizing type, but this is not a bound state as it is normally understood. What form does the projection operator take in these circumstances?

IMPLICIT PROJECTION OPERATORS IN BASIS-SET EXPANSIONS OF THE MOLECULAR DIRAC-FOCK-SLATER PROBLEM

W.-D. Sepp and B. Fricke
Department of Physics, University of Kassel
3500 Kassel/Germany

INTRODUCTION

In a classic paper Brown and Ravenhall[1] showed that the naive many-electron configuration space Dirac-equation

$$H_{CS}\ \Psi(1\ldots N) = E\ \Psi(1\ldots N) \tag{1}$$

with the Hamiltonian

$$H_{CS} = \sum_{j=1}^{N} h_j + \sum_{i>j=1}^{N} V_{ij} \tag{2}$$

has no normalizable solutions for $N \geq 2$. In (2) h_j is the single-particle Dirac Hamiltonian

$$h_j = c\ \vec{\alpha}_j \vec{p}_j + \beta_j mc^2 + V^{ext}(j), \tag{3}$$

V_{ij} the two-particle electron-electron interaction which in its simplest approximation is given by the Coulomb potential

$$V_{ij} = e^2/r_{ij} \tag{4}$$

and N is the number of electrons.
Despite this "Brown-Ravenhall desease" or "continuum dissolution" the Hamiltonian (2) was used in the past to obtain the relativistic Dirac-Fock (DF) equations

$$h_j^{DF}\ \psi_n(j) = \varepsilon_n\ \psi_n(j) \tag{5}$$

with

$$h_j^{DF} = c\ \vec{\alpha}_j \vec{p}_j + \beta_j mc^2 + V^{ext}(j) + V^{DF}(j) \tag{6}$$

where the Dirac-Fock single-particle potential is given by

$$V^{DF} = V^{Coul} + V^{Ex} \tag{7}$$

with the Coulomb potential

$$V^{Coul}(1) = \sum_{k \in occ} \langle \psi_k(2)\ \left| \frac{e^2}{|\vec{r}_1 - \vec{r}_2|} \right|\ \psi_k(2) \rangle \tag{8}$$

0094-243X/85/1360020-6$3.00

and the exchange potential

$$\langle\psi_m|V^{Ex}|\psi_n\rangle = \sum_{k\in occ} \langle\psi_m(1)\ |\langle\psi_k(2)|\ \frac{e^2}{|\vec{r}_1-\vec{r}_2|}\ |\psi_n(2)\rangle|\psi_k(1)\rangle. \quad (9)$$

Here the sums $k\in occ$ run over all occupied states.

The set of equations (5) to (9) (and its generalizations to the Multiconfiguration -Dirac-Fock method) proved to be very successful in calculations of atomic properties[2-5], where the DF-equations were solved by numerical integration techniques. On the other hand, direct application of the DF-equations (5) to (9) to molecular problems using basis-set expansion techniques yielded rather dissatisfactory results[6-13]. Schwarz and Wechsel-Trakowski[14] pointed out that this failure is due to the fact that the single-particle Dirac-Hamiltonian (6) has not only bound (and positive unbound) eigenstates but also negative unbound eigenstates (the Dirac-Hamiltonian (6) is unbound from below). To distinguish from the Brown-Ravenhall disease, which is caused by the two-body interaction V_{ij} in eq.(2), this failure was named "finite basis-set disease" or "variational collapse". The various attempts to avoid this variational collapse in basis-set expansions of the DF-equations have been reviewed by Kutzelnigg[15].

The purpose of this paper is twofold: First, we want to point out that both obstacles are basically connected to the same property of any single-particle Dirac-Hamiltonian h_j, namely the existence of the negative energy continuum of its spectrum. Both deseases are, therefore, cured by the same procedure.

Secondly, we want to show that the variational collapse can be avoided in molecular calculations if the basis-set is physically adapted to the problem.

PROJECTION OPERATORS AND QED

Already in their original paper Brown and Ravenhall[1] showed that the BR-desease can be removed by deriving a configuration space many-particle Dirac-Hamiltonian from quantum electrodynamics. In recent years their derivation was clarified and generalized in a series of papers by Mittleman[16-18], Sucher[19-22], and Grant[23]. The basic idea of this derivation is that the many-electron relativistic configuration space Hamiltonian should describe a system with a fixed number of "electrons". As the eigenstates of the QED-Hamiltonian are only states with a definite charge number, one has to project this state onto the subspace of a fixed number of "electrons" and no "positrons", where the definition of an "electron" and a "positron" is given by the positive and negative energy eigenstates of a reference single-particle Dirac-Hamiltonian h_j^r. The resulting Hamiltonian in this subspace in lowest approximation is then given by

$$H_{CS}^{np} = \sum_{j=1}^{N} \Lambda_+^r(j)\ h_j\ \Lambda_+^r(j) + \sum_{i>j=1}^{N} \Lambda_+^r(i)\ \Lambda_+^r(j)\ V_{ij}\ \Lambda_+^r(i)\ \Lambda_+^r(j), \quad (10)$$

where $\Lambda_+^r(j)$ is the projection operator onto the "electron" states of the reference Hamiltonian h_j^r and np means no pair. Note that both the one-particle operators h_j as well as the two-particle operators V_{ij} are sandwiched by projection operators. This is in contrast to Sucher's notation[21,22] of H_{CS}^{np} where projection operators are attached only to the two-particle operators V_{ij}. The corresponding many-electron Dirac-equation is then given by

$$H_{CS}^{np}\ \Psi^{np}(1,\ldots N) = E\ \Psi^{np}(1,\ldots N) \tag{11}$$

with the constraint

$$\prod_{j=1}^{N} \Lambda_+^r(j)\ \Psi^{np}(1,\ldots N) = \Psi^{np}(1,\ldots N). \tag{12}$$

The constraint (12) is the remainder of the fact that the configuration space state vector Ψ^{np} represents a many-electron state without any electron-positron pairs. In our opinion this constraint is the essential point in the derivation of the Hamiltonian H_{CS}^{np} from QED. This means that the relativistic many-electron configuration space Hamiltonian H_{CS}^{np}, and the configuration state $\Psi^{np}(1,\ldots N)$ are defined only in a certain subspace $\mathcal{H}_N^{np}$ of the total configuration Hilbert space $\mathcal{H}_N$ with

$$\mathcal{H}_N^{np} = \prod_{j=1}^{N} \Lambda_+^r(j)\,\mathcal{H}_N. \tag{13}$$

As the projection operators $\Lambda_+^r(j)$ are unity operators in this subspace we can omit all projection operators in the Hamiltonian H_{CS}^{np} of eq.(10), and recover the old expression H_{CS} of eq.(2).

Application of the usual Hartree-Fock variational method in this subspace then gives us back the set of DF-equations (5) to (9) with the additional constraint

$$\Lambda_+^r(j)\ \psi(j) = \psi(j) \tag{14}$$

for all single-particle states invloved[22].

In conclusion we may say that the original configuration space Hamiltonian H_{CS}, and the corresponding Dirac-Fock Hamiltonian h^{DF} are still given by the "naive" expression (1) and (6-9) but that their domain has to be restricted to the "electronic" subspace (13) and (14) of the configuration space. This reduction of the domain not only cures the BR-desease of the Hamiltonian H_{CS} but also the variational collapse of the h^{DF}, as h^{DF} in its restricted domain is now bound from below. Of course, the difficulty now is the construction of the appropriate projection operator Λ_+^r which seems to be very complex for arbitrary reference Hamiltonians h^r.

Various suggestions for the choice of the projection operator Λ_+^r have been made[18,22] ("free electrons", "external (nuclear field) electrons", and "SCF-electrons"). We adopt here the proposal of Mittleman[18] who showed by a variational method that the optimal choice for Λ_+^r in the context of a Dirac-Fock calculation is the use of projection operators onto the non-negative Dirac-Fock states

$$\Lambda_+^r = \Lambda_+^{DF}. \tag{15}$$

With this choice of the projection operator the projected DF-equations (5-9) plus eq.(14) are equivalent to the unprojected DF-equations (5-9) if the latter are solved only for non-negative energy eigenvalues using boundary conditions[18,22,23]. The great success of numerical Dirac-Fock calculations of atoms using this method may be regarded as a support of the suggestion of Mittelman.

PROJECTION OPERATORS IN MOLECULAR DIRAC-FOCK-SLATER CALCULATIONS

We solved the Dirac-Fock equation in Dirac-Fock-Slater approximation for various diatomic molecules using a MO-LCAO treatment. Details of the calculations may be found elsewhere[24-25]. The essential point in the present context is the following: The molecular orbital wave functions are expanded in positive atomic DFS-states

$$\psi_n = \sum_A \sum_{\lambda_+} b^A_{n\lambda_+} \chi^A_{\lambda_+} \tag{16}$$

where the atomic DFS-states are calculated by a numerical integration method. In eq.(16) the summation over A denotes the summation over the different sets of atomic basis states $\chi^A_{\lambda_+}$ centered at the constituent nuclei of the molecule and λ_+ denotes the summation over the positive energy states of the basis set A. Of course, the multicenter expansion (16) is over-complete. This difficulty can be removed by applying ortho-normalization procedures[25].

Each atomic basis state satisfies an atomic DFS-equation

$$h^A \chi^A_\lambda = \varepsilon^A_\lambda \chi^A_\lambda. \tag{17}$$

The restriction of the summation to positive energy states in eq.(16) can be removed by the introduction of atomic projection operators

$$\Lambda_+^A = \sum_{\lambda_+} \chi^{A\dagger}_{\lambda_+} \chi^A_{\lambda_+}. \tag{18}$$

Eq.(16) then reads

$$\psi_n = \sum_A \Lambda_+^A \cdot \sum_\lambda b^A_{n\lambda} \chi^A_\lambda. \tag{19}$$

Eq.(19) can be further simplified by introducing an effective molecular projection operator

$$\psi_n = \Lambda_+^M \sum_A \sum_\lambda b_{n\lambda}^A \chi_\lambda^A \tag{20}$$

which is defined by the equivalence of eq.(19) and (20).

Following the suggestion of Mittleman the correct ansatz for the molecular wave function should read

$$\tilde{\psi}_n = \tilde{\Lambda}_+^M \cdot \sum_A \sum_\lambda \tilde{b}_{n\lambda}^A \chi_\lambda^A \tag{21}$$

where $\tilde{\Lambda}_+^M$ is the molecular DF-projection operator.

With these two possibilities for the projection operator we get two different projected DFS Hamiltonians h^M and $\tilde{h}^M$ with corresponding solutions. These solutions are then given by eq.(19) and (21) where the coefficients $b_{n\lambda}^A$ and $\tilde{b}_{n\lambda}^A$ are now fixed numbers.

If the difference of the two molecular DFS Hamiltonians h^M and $\tilde{h}^M$ is assumed to be small, then we can calculate the differences of the two corresponding projection operators by perturbation theory[18]

$$\tilde{\Lambda}_+^M - \Lambda_+^M = \sum_{\substack{n+\\m-}} \psi_n^\dagger \frac{\langle\psi_n|\tilde{h}^M - h^M|\psi_m\rangle}{|\varepsilon_n - \varepsilon_m|} \psi_m + \sum_{\substack{n-\\m+}} \psi_n^\dagger \frac{\langle\psi_n|\tilde{h}^M - h^M|\psi_m\rangle}{|\varepsilon_n - \varepsilon_m|} \psi_m . \tag{22}$$

This equation shows that the effect of the different projection operators can be expected to be small for essentially non-relativisitic systems as the denominator is then close to $2mc^2$ or even larger for all states.

For highly relativistic states (inner shell states) the difference of the projection operators can be expected to be small only if the molecular orbital matrix elements

$$\Delta h_{nm} = \langle\psi_{n_\pm}|\tilde{h}^M - h^M|\psi_{m_\mp}\rangle, \tag{23}$$

which couple positive energy states to negative energy states, are small.

As inner shell orbitals are non-zero only in a small region of space, this means that the difference of the "true" and the approximate DFS-potential has to be zero or constant in that region in order to have small matrix elements (23) for highly relativistic systems, too. We expect this being the case if the basis set (16) is well adapted to the molecular problem.

This discussion shows two things: First, that, indeed, we really use a projected basis set, and second, that it is crucial to use a basis set which is (after pre-ortho-normalization) well suited to represent the inner shell molecular orbitals in order to avoid spurious contributions from the negative energy continuum.

REFERENCES

1. G.E. Brown and D.E. Ravenhall, Proc.Roy.Soc.(London) A 208, 552 (1951).
2. Proceedings of the Workshop on Foundation of the Relativistic Theory of Atomic Structures, Ed.: H.G. Berry, K.T. Cheng, W.R. Johnson, and Y.K. Kim, Argonne National Laboratory Report ANL-80-126, 1981.
3. Relativistic Effects in Atoms, Molecules, and Solids, Ed.: G.L. Malli, NATO ASI Series B, Vol. 87, Plenum Press, New York and London, 1983.
4. I.P. Grant, Adv.Physics 19, 747 (1970); I.P. Grant,in ref.3, p. 55.
5. J.P. Desclaux, in ref. 3, p. 15.
6. F. Mark, H. Lischka, and F. Rosicky, Chem.Phys.Lett. 71, 507 (1980).
7. F. Mark and F. Rosicky, Chem.Phys.Lett. 74 562 (1980).
8. H. Wallmeier and W. Kutzelnigg, Chem.Phys.Lett. 78, 341 (1981).
9. G. Malli and J. Oreg, Chem.Phys.Lett. 69, 313 (1979).
10. O. Matsuoka, N. Suzuki, T. Aoyana, and G. Malli, J.Chem.Phys. 73, 1320 (1980).
11. T. Aoyana, H. Yamakawa, and O. Marsuoka, J.Chem.Phys. 73, 1329 (1980).
12. S.N. Datta and C.S. Ewig, Chem.Phys.Lett. 85, 443 (1982).
13. Y.S. Lee and A.D. McLean, J.Chem.Phys. 76, 735 (1982).
14. W.H.E. Schwarz and E. Wechsel-Trakoski, Chem.Phys.Lett. 85, 94 (1982).
15. W. Kutzelnigg, Int.J.Quant.Chem. 25, 107 (1984).
16. M.H. Mittleman, Phys.Rev. A4, 893 (1971).
17. M.H. Mittleman, Phys.Rev. A15, 2395 (1972).
18. M.H. Mittleman, in ref.2, p. 27.
19. G. Feinberg and J. Sucher, Phys.Rev.Lett. 26, 681 (1971).
20. J. Sucher, Rep.Prog.Phys. 41, 1781 (1978).
21. J. Sucher, in ref.2, p. 1.
22. J. Sucher, in ref.3, p. 1.
23. I.P. Grant, in ref. 3, p. 73.
24. A. Rosén and D.E. Ellis, J.Chem.Phys. 62, 3039 (1975).
25. W.-D. Sepp, D. Kolb, W. Sengler, H. Hartung, and B. Fricke, to be published.

DISCUSSION

Marvin H. Mittleman
The City College of New York, Convent Avenue & 138th Street
New York, NY 10031

The positive energy projection operators, just described by Prof. Sucher, convert the "sick" Hamiltonian, H_{DC}, into a more robust one which can support bound states. They are however still a subject of some controversy.

Prof. Grant pointed out that existing computer codes produce remarkable accuracy in numerical calculations which start from H_{DC} (with no projection operators) and so he questioned whether these operators were indeed necessary. In response, it was pointed out by several people in the audience that the codes implicitely limit the Dirac-Hartee-Fock wave functions to a normalizable sub-space and that this operation can be described as a projection operator which has the effect of eliminating the "negative energy" states which are not normalizable. This operation is however, not any of the three projection operators described by Sucher and so the question arises as to the sensitivity of the results (for the energy and wave functions) to the particular projection operators which are used. This appears to be an open question.

It is clear from the numerical results to which Grant refers that the Dirac-Hartee-Fock treatment of H_{DC} is close to the correct result. But one need only go back into the history of the subject which was outlined by Sucher to see that such a calculation can run into trouble. If we return to the Helium calculations with a two body potential which includes the Breit interaction, V_B, we see an example of the trouble. The treatment of V_B by perturbation theory is certainly justified here. The codes would give accurate wave functions and energies and the first order inclusion of V_B is known to improve the results. However, its inclusion in second order without the positive energy projection operators, is known to give wrong results. Admittedly, the people who do such calculations are too sophisticated to make that error now but it is their artistry not the stability of H_{DC} which precludes this error.

Prof. Drake pointed out the strong similarity between H_{DC}, and the auto-ionization problem of conventional Atomic Physics. Mathematically, both are describable as a discrete state coupled to a continuum degenerate with it. In a time dependent picture, the discrete state will "leak" into the continuum and so its amplitude will decay with time. In the time independent picture the eigen value of the Hamiltonian describing either one of these problems will be complex with an imaginary part which is close to the width of the discrete state.

Another problem with the configuration space Hamiltonian which is used in these calculations is its lack of uniqueness.

For example, the procedure usually used for the extraction of the electron-electron potential interaction from quantum electrodynamics yields a potential V_{e-e} which depends upon the gauge in which the calculation is performed. The underlying theory, Q.E.D., is gauge invarient but the approximations which are usually made in extracting the configuration space Hamiltonian are not. For example, one can compare the results in the Coulomb (transverse) and in the Lorentz gauge. For non-relativistic situations there is general agreement that the Coulomb gauge is preferable since the dominant interaction, e^2/r_{12}, is treated exactly in that case. Prof. Fulton, pointed out the case of positronium as an example of that situation. For Ps. radiative corrections in the Coulomb gauge are well known to converge more rapidly than for the Lorentz gauge. However, it is difficult to generalize this to the high Z (relativistic) case.

The numerical dependence of the energy of a many-electron-atom upon the form of V_{ee} (in different gauges) has not been investigated. (However, see the presentation of Prof. Desclaux in these Proceedings).

In 1939 Primakoff and Holstein discussed the interaction of classical particles which is mediated by a field which propagates at finite velocity. They showed that the elimination of the field resulted in "action-at-a-distance" potentials but that three-body potentials appeared in addition to the usual two-body potentials. This was later extended to quantum mechanics with analogous results. The importance of these potentials in atoms has been crudely estimated at about a Rydberg for the very high Z case but a numerical evaluation has only been done recently. Dr. Zygelman will tell us about this and related subjects now.

QED THREE-BODY POTENTIALS IN HEAVY ATOMS

Bernard Zygelman
Harvard College Observatory
Cambridge Ma. 02138 USA

Marvin H. Mittleman
The City College of the City University of New York
New York, N.Y. 10031 USA

ABSTRACT

The conversion of quantum electrodynamics to a configuration space Hamiltonian formalism introduces three-electron potentials which are of relativistic origin. For high-Z Atoms we find that the contribution of these potentials to the inner shell binding energy is no more than 0.21 electron volts.

INTRODUCTION

An accurate calculation of the binding energy of a heavy atom must include relativistic effects which must be obtained in a systematic way from quantum electrodynamics. For historical reasons all such calculations have been done in a configuration space Hamiltonian (single-time) framework. This is not the only, or perhaps even the best calculational procedure, but it is the one that is used, so a necessary intermediate step is the extraction of the configuration space Hamiltonian from the Fock space formulation of QED. It has long been known[1] that a field theory will yield three-body potentials when the field theory is converted to a single time framework.

State of the art relativistic calculations that include the Breit interaction, approximate Lamb shift and nuclear size effects, for the binding energy of inner-shell electrons are not always in satisfactory agreement with experiment[2]. It has been suggested[3] by simple order of magnitude estimates of QED three-body potentials, that they might provide a resolution for these disagreements. The results of this paper suggest that this is not the case.

We will first give a short review of the classical derivation for the electrodynamic three-body potentials. The QED formulation for these potentials are presented and we will sketch out a non-relativistic limit where the Primakoff-Holstein potentials are recovered. We present the details for the evaluation of the contribution of the QED three-body potentials to the binding energy of the ground state of lithium-like ions with values of the nuclear charge in the range $80 < Z < 137$. Finally, we make some comments concerning ambiguities that arise when a Hamiltonian formulation for configuration space potentials is attempted.

0094-243X/85/1360028-7$3.00

CLASSICAL DERIVATION

Consider a triplet of electrons with coordinates r_i, r_j, r_k. The i'th electron probes at it's position $r_i(t)$, a vector potential produced by the motion of the other electrons. Due to the finite speed of light, this vector potential depends on the coordinates of electrons j, and k, at some other time t'. If the velocities of the particles is such that $v/c << 1$, we can expand the vector potential at $r_i(t)$ in terms of a power series in $1/c$. The lowest order term is the instantaneous approximation $t' = t$, and this will result in the pair-wise Coulomb interaction between the particles. The next order depends on the instantaneous velocities of the electrons and will be responsible for the pair-wise magnetic (Breit) interaction. Beyond this order, the vector potential at r_i will depend on the acceleration of the sources. In turn, because of Newton's second law, the acceleration of a source particle j, will depend on the instantaneous position of a third particle k. Primakoff and Holstein pointed out that this acceleration dependant term gives rise to an interaction that depends on the instantaneous coordinates of all three particles i,j,k. For electrons, this classical three-body action-at-a-distance potential was given by these authors by the following expression,

$$H_{P-H} = \frac{e^4}{8m^3c^4} \sum_{i,j,k} \frac{1}{|r_{ij}||r_{jk}|} \times$$
$$\{\vec{p_i} \cdot \vec{p_k} + (\vec{p_i} \cdot \hat{r}_{ij})(\vec{p_k} \cdot \hat{r}_{kj}) + (\vec{p_i} \cdot \hat{r}_{kj})(\vec{p_k} \cdot \hat{r}_{ij}) + (\vec{p_i} \cdot \hat{r}_{ik})(\hat{r}_{ik} \cdot \hat{r}_{kj})(\hat{r}_{kj} \cdot \vec{p_j})\}. \tag{1}$$

Here, $\vec{p_i}$ is the canonical momentum of the i'th electron. One can also derive (1) using quantum mechanical arguments. In this picture, the i'th particle can emit two simultaneous photons via the A^2 term in the nonrelativistic field-particle Hamiltonian. One photon can be absorbed by particle j via the $\vec{p_j} \cdot \vec{A}$ term, and the other photon can be reabsorbed by particle k via the same mechanism. Primakoff and Holstein showed that this quantum process reproduces the classical potential (1).

We consider the specialized case of a heavy atom with nuclear charge Z. The inner-shell electrons then move at the fraction $Z\alpha$ of the speed of light. Simple dimensional scaling of the expression (1) suggest that the contribution of the three-electron potential to the atom binding energy goes as $(Z\alpha)^4 Rydbergs$. For $Z > 80$ this could be a nonnegligible number, however, (1) is essentially a nonrelativistic expression, and for heavy atoms a generalization of (1) that also incorporates relativity must be derived.

QED FORMULATION

A many electron Hamiltonian derived from first principles i.e. QED has been given by Mittleman[3]. This Hamiltonian includes three-electron potentials and is given by,

$$H = \sum_{i=1}^{N} h_0(i)\Lambda(i) + \sum_{i<j} \Lambda(i)\Lambda(j)V(i,j)\Lambda(j)\Lambda(i) + \sum_{i<j<k} \Lambda(i)\Lambda(j)\Lambda(k)U(i,j,k)\Lambda(k)\Lambda(j)\Lambda(i). \quad (2)$$

Here, $h_0(i)$ is a single particle Dirac, central-field, Hamiltonian. For our purposes (see the discussion below) it will be convenient to take the central field to be the Coulomb potential of the nucleus with charge Z. $\Lambda(i)$ are projection operators that project onto the positive energy states of $h_0(i)$, and are necessary to prevent the Brown-Ravenhall disease[4]. $V(i,j)$ is a non-local two-body operator that arises because of the exchange of Coulomb and transverse photons among the electrons. Matrix elements of $V(i,j)$ with eigenstates of $h_0(i)$ are given by,

$$< nl|V|n'l' >= C_{nl,n'l'} + \eta_{nl,n'l'} + \eta_{ln,l'n'} \quad (3a)$$

$$C_{nl,n'l'} =< nl|\frac{e^2}{|r_{12}|}|n'l' > \quad (3b)$$

$$\eta_{nl,n'l'} = \frac{-e^2}{4} < nl|\alpha_1^a\alpha_2^b\{(\nabla^2\delta_{ab} + \nabla_1^a\nabla_2^b) \times (\frac{2hc}{W_{nn'}})^2 \frac{sin^2}{|r_{12}|}(\frac{W_{n,n'}|r_{12}|}{2hc})\}|n'l' > \quad (3c)$$

$$W_{n,n'} = E_n - E_{n'}. \quad (3d)$$

The $C_{nl,n'l'}$ are Coulomb interaction matrix elements and the η terms are matrix elements of the generalized Breit operator. In the limit where the expectation value of $\frac{W_{n,n'}|r_{12}|}{2hc}$ is much less than unity, $\eta_{nl,n'l'}$ reduces to $1/2 < nl|B(1,2)|n'l' >$ where $B(1,2)$ is the local Breit operator,

$$B(1,2) = \frac{-e^2}{2}(\frac{\vec{\alpha}_1 \cdot \vec{\alpha}_2 + \vec{\alpha}_1 \cdot \hat{r}_{12}\vec{\alpha}_2 \cdot \hat{r}_{12}}{|r_{12}|}).$$

The three-body operator $U(i,j,k)$ is given by,

$$U = U^{C-T} + U^{T-T} + U^{Pair}$$

with matrix elements,

$$U^{C-T}_{nml,n'm'l'} = P.V. \quad \sum_s \frac{(\eta_{nm,n's} - \eta_{mn,sn'})C_{sl,m'l'}}{W_{nn'} + W_{ms}} + H.c.$$

$$U^{T-T}_{nml,n'm'l'} = P.V. \quad \frac{1}{2}\sum_s \frac{(\eta_{nm,n's} - \eta_{mn,sn'})(3\eta_{ls,l'm'} + \eta_{sl,m'l'})}{W_{nn'} + W_{ms}} + H.c.$$

$$U^{Pair}_{nml.n'm'l'} = \sum_s^{(-)} \frac{(V_{12})_{nm,n's}(V_{34})_{ls.l'm'}}{W_{nn'} + W_m + |W_s|}. \quad (4)$$

U^{C-T} is the three-body potential resulting from the action of the two-body transverse potential and the Coulomb operator each acting once when one electron is in a virtual state. U^{T-T} results from the transverse operator acting twice. U^{Pair} arises from the double action of $V(i,j)$ with a negative energy intermediate state. All sums are over a complete set of Coulomb - Dirac states except for the U^{Pair} terms which is summed only over the negative energy states.

We will now briefly discuss how one can recover expression (1) as a nonrelativistic limit of the QED three-body potentials given above. If we add up all the terms in (4) that contain the transverse-transverse interaction, limit terms that contain only negative energy intermediate states, and impose the on-shell condition $W_{nn'} + W_{ll'} + W_{mm'} = 0$ one gets,

$$H_{NR} = \sum_{s}^{(-)} 4\frac{\eta_{nl,n's}\eta_{ms,m'l'}}{W_{nn'} + W_{ls}}.$$

An important observation for the above expression is the apparent cancellation of terms in the generalized Breit operator that depend on the frequency of the intermediate states. We can now proceed to carry out the nonrelativistic limit by replacing the $\eta_{ab,a'b'}$ matrix elements by the matrix elements of the local Breit operator, and replacing the energy denominator by $2mc^2$ we thus get

$$H_{NR} = \sum_{s}^{(-)} \frac{< nl|B(1,2)|n's >< sm|B(3,4)|l'm' >}{2mc^2}. \qquad (5)$$

We now impose the closure property of the Dirac-Coulomb states in the sum (5), $\sum_{s}^{(-)} |s >< s| = 1+O(Z\alpha)$. Neglecting terms to $O(Z\alpha)$ and using the anticommutation relations $\{\alpha^a, \alpha^b\}_+ = 2\delta_{ab}$ for the Dirac matrices we get,

$$H_{NR} =< nlm|H_{P-H}|n'l'm' >$$

where we have used the heuristic relation $\vec{\alpha}_i \rightarrow \frac{\vec{p}_i}{mc}$.

We will now obtain the lowest order contribution of the three-body potentials (4), to the binding energy of a lithium-like ion of configuration $(1s^2 2s)S_{1/2}$. For high Z ions, screening effects would be relatively unimportant. Therefore, the use of the Dirac-Coulomb Hamiltonian $h_0(i)$ in (2) is justified. The energy shift due to the three-body potentials is given by,

$$\Delta E =< \Psi|H_3|\Psi > \qquad (6)$$

where H_3 is the third term in the right hand side of (2). $|\Psi >$ is a suitably antisymmetrized product of Dirac-Coulomb spinors, $|(n, j, m, \kappa) >$. In this notation[6], n is the principal quantum number, j is the total angular momentum, m is the azimuthal quantum number, and κ is the Dirac quantum number. Expanding (6) results in a set of thirty-six distinct terms of the type given in (4). We can simplify this expression somewhat by replacing the $\eta_{nl,n'l'}$ matrix elements by,

$$\eta_{nl,n'l'} = \frac{-e^2}{2} < nl|\frac{\vec{\alpha}_1 \cdot \vec{\alpha}_2}{|r_{12}|}cos(\frac{W_{nn'}|r_{12}|}{hc}) - \frac{W_{ll'}}{W_{nn'}}\frac{1}{|r_{12}|}(1 - cos(\frac{W_{nn'}|r_{12}|}{hc}))|n'l' >$$

This result was obtained by integrating eqn. (3c) by parts and then using the Dirac equation. If we define the two generic integrals

$$g_{nl,n'l'} =< nl|\frac{1}{|r_{12}|}cos(\frac{W_{nn'}|r_{12}|}{hc})|n'l' >$$

$$h_{nl,n'l'} = < nl|\frac{\vec{\alpha}_1 \cdot \vec{\alpha}_2}{|r_{12}|} cos(\frac{W_{nn'}|r_{12}|}{hc})|n'l' >$$

the shift ΔE can then be completely expressed in terms of them. The shift can be further reduced to a three-dimensional quadrature problem and the rest of the calculation must be done numerically[5]. The results for this energy shift are tabulated in Table I. The values of the nuclear charge range from the value $Z \approx 137(Z\alpha = 1)$ down to $Z = 80$. The total contribution starts positive at high Z and for decreasing Z decreases, going negative around $Z = 100$ and reaches a minimum near $Z = 90$. It approaches zero as Z becomes smaller. As we have seen the entire effect is relativistic in origin because of because of the transverse coupling in $U^{C-T}$and U^{T-T} or the negative energy intermediate states in U^{Pair}. It therefore vanishes as $Z\alpha \to 0$. The major contribution is known to come from negative energy intermediate states and the principal configuration space region of contribution of these states will be in the neighborhood of the nucleus, where the electrons move most rapidly. However, as $Z\alpha$ grows the negative energy positron states are repelled from the region of the (positively charged) nucleus thereby depressing their contribution. These are therefore competing effects, the first growing with Z and the second decreasing for increasing Z. This is probably the reason for the structure in the results (Table 1) and their smallness.

Table I. - Contribution of the three-body potential to the binding energy of a lithium-like ion.

Z	Energy (ev.)
137	2.1×10^{-1}
130	3.4×10^{-2}
118	5.7×10^{-3}
110	9.6×10^{-4}
100	-9.9×10^{-4}
90	-1.3×10^{-3}
80	-1.1×10^{-3}

The configuration space Hamiltonian (2) was obtained from a straigthforward extension of a method originally proposed by Schwinger[7]. This method decouples, via a unitary transformation, virtual photons in the radiation field from the matter field. The virtual photons then manifest themselves in the appearance of fermion-fermion couplings in the transformed Fock space Hamiltonian. In turn, these terms give rise to electron-electron potentials in configuration space. We would like to point out some ambiguities in this method. Perhaps, the most obvious ambiguity is the gauge ambiguity. It is well known that a different choice of gauge for the radiation field in the QED Hamiltonian will result in different two-body potentials in configuration space. Persuasive arguments exist[8] for the use of the Coulomb gauge since it is the only one in which the Coulomb interaction is treated exactly. We have used the Coulomb gauge in our formulation and we will not dwell on this aspect of the ambiguity.

When one proceeds with the Schwinger transformation mentioned above, the exchange of virtual photons is governed by a propagator $\epsilon(\tau)D_{ab}(r,\tau)$. Here, $\epsilon(\tau)$ is a discontinous step function,

$$\epsilon = \theta(\tau) - \theta(-\tau)$$

where, $\theta(\tau) = 1 \quad \tau > 0, \quad \theta(\tau) = 0 \quad \tau < 0$ and, $D_{ab}(r,\tau)$ is the transverse photon commutator,

$$D_{ab}(r,\tau) = [A_a(r_1,\tau), A_b(r_2,0)] = 4\pi ihc \int \frac{d^3k}{(2\pi)^3}(\delta_{ab} - \hat{k}_a\hat{k}_b)e^{i\vec{k}\cdot\vec{r}_{12}}\frac{sin(ck\tau)}{k}.$$

One can show[5] that the Schwinger decoupling transformation is not completely general. Applying a more general transformation that depends on an arbitrary parameter β , the step function $\epsilon(\tau)$ is now replaced by $\beta\epsilon(\tau) - 2(1-\beta)\theta(-\tau)$. The propagator now becomes,

$$G^{\beta}_{ab}(r,\tau) = \beta\epsilon(\tau)D_{ab}(r,\tau) - 2(1-\beta)\theta(-\tau)D_{ab}(r,\tau).$$

For $\beta = 2, \quad G^{\beta=2}_{ab}(r,\tau) = 2\theta(\tau)D_{ab}(r,\tau)$, or G becomes a retarded propagator (it vanishes for $\tau < 0$). Likewise, for $\beta = 0$ we get an advanced propagator . For $\beta = 1$ we recover the Schwinger propagator. This arbitrariness is thus a consequence in the freedom of choice in choosing the boundary condition for the transverse photon propagator, and will be reflected in a more general two-body potential than that given in (3). The matrix element of this new potential is given by a replacement of the $\eta_{nl,n'l'}$ matrix elements in (3) by $\tilde{\eta}_{nl,n'l'}$, where

$$\tilde{\eta}_{nl,n'l'} = \eta_{nl,n'l'} + (1-\beta)\omega_{nl,n'l'}$$

$$\omega_{nl,n'l'} = \frac{-ie^2}{2} < nl|(\vec{\alpha}_1 \cdot \vec{\alpha}_2 + 1)\frac{1}{|r_{12}|}sin\frac{W_{nn'}|r_{12}|}{hc}|n'm' > .$$

One can show that the on-shell matrix elements $\omega_{nl,n'l'} + \omega_{ln,l'n'}$ vanish. Therefore, there is no ambiguity in the two-body potential at the lowest order.

It is not clear at this point what the value of β should be. We can make some suggestions by noting these observations[5]; (1) the configuration space two-body potential will be time-reversal invariant only for the choice $\beta = 1$, (2) values of $\beta \neq 1$ introduce singularities in higher order (e^4) potentials. This seems to suggest that the Schwinger choice ($\beta = 1$) is the best one.

REFERENCES

[1] Primakoff H. and Holstein T. 1939 Phys. Rev.55 1218

[2] Desclaux J.P. 1980 in Proc. of the Workshop on Foundations of the Relativistic Theory of Atomic Structure, Argonne Nat. Lab.

[3] Mittleman M.H. 1971 Phys. Rev. A 4 893, 1980 in Proc. of the Workshop on Foundations of the Relativistic Theory of Atomic Structure, Argonne Nat. Lab.

[4] Brown G.E. and Ravenhall D.G. 1951 Proc. Roy. Soc. A208 552

[5] Zygelman B. and Mittleman M.H. 1985 Journal of Phys. B submitted for publication

[6] Rose M.E. 1961 Relativistic Electron Theory J. Wiley N.Y.

[7] Schwinger J. 1948 Phys. Rev.74 1439

[8] Sucher J. 1980 in Proc. of the Workshop on Foundations of the Relativistic Theory of Atomic Structure, Argonne Nat. Lab.

ACKNOWLEDGEMENT

One of us (B.Z) would like to thank Dr. K.T. Cheng for one of his computer codes that has greatly facilitated the progress of this work.

COMMENTS ON ZYGELMAN'S TALK

Marvin H. Mittleman
The City College of New York
Convent Avenue & 138th Street, New York, NY 10031

The surprisingly small contribution of the three-body potentials to the energy has been attributed to the competition of two opposing effects. First, the coupling of the transverse photons to the electrons is a relativistic effect which grows as $Z\alpha \sim v/c$ gets larger. But as Dr. Zygelman has pointed out, the dominant contribution comes from diagrams in which negative energy states occur. These are equivalent to positively charged particles which are repelled from the region of the nucleus by the highly (positively) charged nucleus. This keeps the overlap of the negative energy states with the positive energy states small, decreasing with higher Z. These two competing effects are probably the reason for the small numbers for the energy and the structure in the curve of energy vs. Z.

Prof. Pratt pointed out that the calculation was done for a three electron, high Z, ion. In a neutral atom the screening due to outer electrons might be important in reducing the nuclear avoidance by the virtual positrons in the calculation. He pointed out that his calculations of threshold pair production is much enhanced for neutral atoms in comparison with the pure Coulomb predictions and that the enhancement grows rapidly with Z.

The effect that Pratt describes will be largest in the vicinity of threshold since the long range Coulomb potential has its greatest effect on the wave function at the origin near zero energy. In the calculation described by Zygelman these negative energy states enter virtually and a sum over their energy must be performed. The enhancement of the wave function at the origin due to screening is degraded by this sum and it is not clear that it will be large enough to make the three body potential contribution significant. This also appears to be an open question.

ON THE RELATIVISTIC THEORY OF INHOMOGENEOUS MANY-ELECTRON SYSTEMS

K. Dietz
Physikalisches Institut der Universität Bonn
Nussallee 12, 5300 Bonn 1, Federal Republic of Germany

ABSTRACT

Ansätze for relativistic approximation schemes for inhomogeneous many-electron systems, starting from the full QED Lagrangian, are discussed in detail. A new relativistic mean-field theory - the g-Hartree theory - is introduced. It is shown that mean-fields can be optimally chosen within this framework. Successful tests in atomic physics - relevant problems were solved for the first time - demonstrate the predictive power of this theory.

1. INTRODUCTION

Qualitatively speaking, the nature of fundamental questions pertaining to the many-body problem is twofold: many-body observables are used to disentangle the interaction laws of the constituents or, in the case where the latter are known, the equation of motion for the constituents is employed to establish dynamical characteristics of the many-body problem under consideration.

An electrically neutral, inhomogeneous many-electron system is of this kind: the fundamental interactions are the ones best known in all of physics, the physics of it solely refers to questions of many-body dynamics. The framework for treating this system is relativistic quantum electrodynamics (QED), a theory which is tested to the highest degree of accuracy. The problem is to construct an approximation scheme which allows to calculate physical observables of this many-body system in a systematic manner. By systematic approximations we simply understand approximations successively ordered according to a unique ordering principle, e.g. by powers in the perturbing interaction or by the number of loops with fully dressed propagators.

Two essentially different schemes can be anticipated; the starting point is the QED-action in both instances.

I) Relativistic N-electron Dirac-Schrödinger equations are derived [1,2,3], remaining interactions are treated by perturbing around their solutions.

II) Mean-field methods

We shall point out (chapter 2) that the first method, although useful for particular problems, is not viable since no renormalisation algorithm is known which would allow to remove occuring infinities in an unambiguous, physically motivated way. This difficulty and other problems to be discussed below seem to stem from the very idea of the construction.

The essential idea underlying the mean-field approach (chapter 3) is to invent a single particle basis which subsumes, more or less effectively, mutual interactions of the constituent electrons; remaining correlations have to be treated perturbatively. Several single particle basis are used in the literature: Hartree-Fock (HF), Dirac-Hartree-Fock (DHF) in the relativistic case, Xα etc. The g-Hartree mean-field [4,5] can be optimally adapted to the physical quantity considered: g can be chosen so that correlations (defined by deviations from the independent particle (zeroth order) picture) vanish. For instance, in computing total energies, $g=g_0$ can be chosen such that the total correlation energy in the g-Hartree basis vanishes; for the calculation of transition energies, such that "Koopmans' theorem" [6] is an exact formula.

This idea of optimisation has been employed in a twofold way (chapter 5): we determined g_0 from experiment and predicted [7,8,9], semiphenomenologically, other atomic observables; secondly, g_o can be determined ab initio by demanding that correlations vanish up to a certain chosen order of perturbation theory [10,11].

Applications of these two methods show rather striking success.

A closer analysis of the relativistic case revealed an interesting contribution to transition energies which we interpreted [12] as a Casimir effect in atoms induced by a deformation of the Dirac sea (chapter 4).

The g-optimisation procedure, furthermore, led us to the construction of a Landau-type basis [13] for the description of the equilibrium thermodynamics of the inhomogeneous (electrically neutral) electron gas - the g-Hartree basis for $g=g_o$ is a single particle basis which yields the exact thermodynamical potential.

Needless to say, quantum fluctuations calculated in a mean-field basis can be renormalised by standard methods.

Finally, we should like to stress that our g-Hartree computations proceed by a systematic hierachy of approximations. We avoid, for example, partial summations of selected diagrams leading e.g. to the replacement of DHF eigenvalues by ΔSCF energies; we do not have to ask whether a V(N) or V(N-1) basis should be used, etc. Of course, we are fully aware of the importance of semiphenomenological methods as the ones just mentioned for a variety of questions in atomic physics. There are, however, important issues which can be tackled only within the framework of a systematic theory, the quest for relativistic effects is an example.

2. THE "HAMILTONIAN METHOD"

A theory of bound states has to involve non-perturbative steps which yield the essential features of their energy spectrum. Whatever theory is considered, the decisive first step is to find a zeroth order ansatz which is amenable to non-perturbative methods of solution, which gives an approximate bound state spectrum. A systematic approximation scheme for the energy levels and other properties of the bound system then consists in the calculation of successive orders of perturbations around such zeroth order solutions.

Atoms and molecules are weakly bound, it makes good sense to conceive them as composed of N-electrons and positively charged nuclei acting as external fields; due to the weakness of the interaction pair production processes are sufficiently suppressed. A relativistic theory neglecting pair production in zeroth order seems a viable concept.

The so-called hamiltonian approach [1,2,3] translates this idea rather directly into the language of relativistic QED. Essentially it comprises two steps.

(i) A non-perturbative approach in a particle number conserving approximate theory

(ii) Particle number non-conserving parts of the interaction are treated perturbatively.

In the following we shall describe this method in appropriate detail in order to substantiate our critical remarks formulated in the introduction.

We work in the Schrödinger picture and start with the Coulomb gauge QED Hamiltonian (for notation see ref. (14))

$$H = \int d^3x \Big\{ :\psi^{+}(\vec{x}) \left(-i\vec{\alpha}\cdot\vec{D} + \beta m + V_{ext}(\vec{x}) \right) \psi(\vec{x}): + \frac{1}{2} :(\vec{E}_t^{\,2} + \vec{B}^{\,2}): + \frac{e^2}{8\pi} \int d^3x' \frac{1}{|\vec{x} - \vec{x}'|} :\psi^{+}\psi:(\vec{x}) :\psi^{+}\psi:(\vec{x}') \Big\}$$

$$\vec{D} = \vec{\partial} - ie\vec{A}\,, \tag{2.1}$$

where $\psi(\vec{x})$ is the electron spinor field, $\vec{E}_t$ the transverse electric and $\vec{B} = \text{curl}\,\vec{A}$ the magnetic field, double dots indicate normal ordering; $V_{ext}(\vec{x})$ stands for the nuclear Coulomb potential.

A realisation of the specifications (i) and (ii) is now constructed by splitting this Hamiltonian

$$H = H_0 + H_{int} \tag{2.2}$$

such that H_o conserves the particle-(electron-) number and H_{int} contains the remaining interactions, in particular pair-production and -annihilation processes.

The diagonalisation of H_o in a space of N-particle states is achieved if solutions of the following N-particle Dirac-Schrödinger equation can be found

$$\sum_{k=1}^{N} (-i\vec{\alpha}\cdot\vec{\partial} + \beta m + V_{ext})_{\gamma_k \alpha_k} P^{(+)}_{\alpha_k \beta_k} \eta_{\dots\beta_k\dots}(\dots\vec{x}_k\dots)$$

$$+ \frac{e^2}{4\pi} \sum_{i<j}^{N} \frac{1}{|\vec{x}_i - \vec{x}_j|} P^{(+)}_{\gamma_i \alpha_i}(\vec{x}_i) P^{(+)}_{\gamma_j \alpha_j}(\vec{x}_j) \eta_{\dots\alpha_i\dots\alpha_j\dots}(\dots\vec{x}_i\dots\vec{x}_j\dots)$$

$$= E \, \eta_{\gamma_1\dots\gamma_N}(\vec{x}_1 \dots \vec{x}_N) \; ; \tag{2.3}$$

the projection operators

$$P^{(+)}(\vec{x}) = \frac{\sqrt{-\Delta + m^2} + (i\vec{\alpha}\cdot\vec{\partial} + \beta m)}{2\sqrt{-\Delta + m^2}} \tag{2.4}$$

result from the restriction to free particle states, i.e. the positive energy part of the spectrum.

Even after the separation [2] of the spin-dependence by means of a Foldy-Wouthuysen transformation

$$\eta_{\gamma_1\dots\gamma_N}(\vec{x}_1 \dots \vec{x}_N) = \prod_{i=1}^{N} A_{\gamma_i \alpha_i}(\vec{x}_i) \, \varphi_{\alpha_1\dots\alpha_N}(\vec{x}_1 \dots \vec{x}_N) \tag{2.5}$$

with

$$A(\vec{x}) = \left(\frac{2\sqrt{-\Delta + m^2}}{\sqrt{-\Delta + m^2} + m} \right)^{1/2} P^{(+)}(\vec{x}) \, \frac{1+\beta}{2} \tag{2.6}$$

which leads to the much simpler equation

$$\Big(\sum_{i=1}^{N}\Big(\sqrt{-\Delta_i+m^2}+V_{ext}(\vec{x}_i)+\frac{e^2}{4\pi}\sum_{i<j}^{N}\frac{1}{|\vec{x}_i-\vec{x}_j|}\Big)\Big)\varphi(\vec{x}_1\ldots\vec{x}_N)$$

$$=E\,\varphi(\vec{x}_1\ldots\vec{x}_N), \qquad (2.7)$$

the determination of the N-particle wave function $\varphi(\vec{x}_1\ldots\vec{x}_N)$ or $\eta_{\gamma_1\ldots\gamma_N}(\vec{x}_1\ldots\vec{x}_N)$ is an absolutely formidable task.

The use of approximate solutions obtained from a Dirac-Hartree-Fock version [3] of (2.3) or (2.7) introduces theoretical errors which, in principle, could be controlled only by a Brückner-Hartree-Fock type expansion. This would imply that already the first step (i) of this kind of approach involves a perturbation expansion.

Even more fundamental difficulties appear if we formulate the perturbation series induced by H_{int}. We shall point out that, for the splitting (2.2) of H, available renormalisation algorithms are not applicable; there is no known procedure by which the infinities of this perturbation theory can be removed in a systematic and physically motivated manner. The problems involved do not depend on the special degrees of freedom of Spinor-QED and their couplings, rather general structural features lead to this ill-behaviour. We shall try to emphasise the generality of the problem and avoid superfluous, cumbersome notation by simply considering a self-interacting spinless fermion field. To elucidate our point we juxtapose the non-relativistic and the "relativistic" case. For our purpose the essential feature of the latter is that the energy spectrum is unbound from below, spin degrees of freedom not pertinent to our problem are simply decoupled (of course, Lorentz invariance is destroyed in this manner).

We take the unperturbed model Hamiltonian

$$H_0=\int d^3x\;\psi^{+}(\vec{x})\,D(\vec{x})\,\psi(\vec{x})$$

$$+\frac{1}{2}\int d^3x\int d^3x'\;\psi^{+}\psi(\vec{x})\,V(\vec{x},\vec{x}')\,\psi^{+}\psi(\vec{x}') \qquad (2.8)$$

where $\psi(\vec{x})$ is the field operator in the Schrödinger picture and

$$\mathcal{D}(\vec{x}) := \mathcal{D}_{kin}(\vec{x}) + V_{ext}(\vec{x}) \tag{2.9}$$

describes free particles with dispersion $D_{kin}(\vec{x})$ in an external potential $V_{ext}(\vec{x})$; $V(\vec{x},\vec{x}')$ stands for the self-interaction.

Green's functions of the full theory, i.e. vacuum expectation values of time-ordered products of the Heisenberg fields $\psi(\vec{x},t)$ are computed in the interaction picture. We have

$$H = H_0 + H_{int} \tag{2.10}$$

with

$$H_{int} = \int d^3x \, \mathcal{L}_{int}(\psi(\vec{x}), \psi^{+}(\vec{x}))$$

and $(x := (\vec{x}, t))$

$$\langle T(\psi(x_1) \cdots \psi(x_N)) \rangle$$

$$= N \langle T(\varphi(x_1) \cdots \varphi(x_N)\, e^{i\int d^4x \, \mathcal{L}_{int}(\varphi(x), \varphi^{+}(x))}) \rangle \tag{2.11}$$

where the interaction picture field $\varphi(x)$ obeys

$$-i\partial_t \varphi(x) = [H_0, \varphi(x)] \, . \tag{2.12}$$

Solving this equation with appropriate initial conditions at t=0, we find

$$\varphi(x) = \sum_{\alpha} \rho_{\alpha}(\vec{x})\, e^{-i\varepsilon_{\alpha}^{(1)} t} \int d^3x' \, \rho_{\alpha}^{*}(\vec{x}')\, \psi(\vec{x}') \tag{2.13}$$

$$+ \cdots ;$$

Let $\rho_\alpha(\vec{x}_1 \ldots \vec{x}_N)$ denote N-particle wave functions

$$\left(\sum_{i=1}^{N} D(\vec{x}_i) + \sum_{i<j}^{N} V(\vec{x}_i, \vec{x}_j)\right) \rho_\alpha(\vec{x}_1 \cdots \vec{x}_N) = \varepsilon_\alpha^{(N)} \rho_\alpha(\vec{x}_1 \cdots \vec{x}_N)$$

$N = 2, 3, \cdots$

$$D(\vec{x})\, \rho_\alpha(\vec{x}) = \varepsilon_\alpha^{(1)} \rho_\alpha(\vec{x}) \tag{2.14}$$

and $\varepsilon_\alpha^{(N)}$ the corresponding energy levels.
The free two-point function is now easily calculated.

$$\langle T(\varphi(x_1)\varphi^+(x_2))\rangle = \sum_\alpha \rho_\alpha(\vec{x}_1)\rho_\alpha^*(\vec{x}_2)\, e^{-i\varepsilon_\alpha^{(1)}(t_1 - t_2)}\, \Theta(t_1 - t_2). \tag{2.15}$$

It should be clear that, if particle states exhaust the entire spectrum, an N-particle eigenstate is given by

$$\int d^3x_1 \cdots \int d^3x_N\ \rho_\alpha(\vec{x}_1 \cdots \vec{x}_N)\, \psi^+(\vec{x}_1) \cdots \psi^+(\vec{x}_N)\, |0\rangle .$$

Matters get more intricate if we now extend our concept and introduce the notions of particles and holes (antiparticles). We distinguish occupied and unoccupied states by writing

$$\langle T(\varphi(x_1)\varphi(x_2))\rangle = \sum_\alpha \rho_\alpha(\vec{x}_1)\rho_\alpha^*(\vec{x}_2)\, e^{-i\varepsilon_\alpha^{(1)}(t_1 - t_2)} \left\{\Theta(t_1 - t_2) - n_\alpha\right\} \tag{2.16}$$

where n_α stand for the probability that the state $|\alpha\rangle$ is occupied. Hole states are constructed as follows: divide the set of indices $\{\alpha\}$ enumerating states $|\alpha\rangle$ into two disjoint sets $\{\alpha_p\}$ and $\{\alpha_h\}$ such that

$$\{\alpha\} = \{\alpha_p\} \cup \{\alpha_h\}$$

$$\{\alpha_p\} \cap \{\alpha_h\} = \emptyset \tag{2.17}$$

and define hole occupation numbers by

$$\bar{n}_\alpha := 1 - n_\alpha \quad \text{for} \quad \alpha \in \{\alpha_h\}. \tag{2.18}$$

Examples for this construction are well-known: for a system in thermodynamic equilibrium we assume a chemical potential μ, then

$$\alpha \in \{\alpha_p\} \quad \text{for} \quad \varepsilon_\alpha > \mu$$

$$\alpha \in \{\alpha_h\} \quad \text{for} \quad \varepsilon_\alpha \leq \mu \; ; \tag{2.19}$$

for relativistic electrons we have

$$\mu = -m_{\text{electron}}.$$

The propagator (2.16) then reads

$$\langle T(\varphi(x_1)\varphi^\dagger(x_2))\rangle =$$

$$\sum_{\alpha\in\{\alpha_p\}} \rho_\alpha(\vec{x}_1)\rho_\alpha^*(\vec{x}_2)\, e^{-i\varepsilon_\alpha^{(1)}(t_1-t_2)} \{\Theta(t_1-t_2) - n_\alpha\}$$

$$-\sum_{\alpha\in\{\alpha_h\}} \rho_\alpha(\vec{x}_1)\rho_\alpha^*(\vec{x}_2)\, e^{-i\varepsilon_\alpha^{(1)}(t_1-t_2)} \{\Theta(t_2-t_1) - \bar{n}_\alpha\}. \tag{2.20}$$

Empty particle states propagate forward, empty hole states backward in time. i.e. we recover Feynman boundary conditions.

Wick contractions in (2.11) can now be evaluated in two ways: either one performs t-integrations using (2.20) and obtains graphs with oriented lines - "old-fashioned" perturbation theory - or one uses the Fourier-transform

$$\Theta(t) = \frac{1}{2\pi i}\int dE \, \frac{e^{iEt}}{E - i\varepsilon} \tag{2.21}$$

to write the propagator (2.20) as

$$\langle T(\varphi(x_1)\,\varphi^{+}(x_2))\rangle =$$

$$-i\int \frac{dE}{2\pi}\, e^{iEt}\left\{\sum_{\alpha\in\{\alpha_p\}} \frac{\rho_\alpha(\vec{x}_1)\,\rho^{*}_\alpha(\vec{x}_2)}{E+\omega_\alpha - i\varepsilon} - \sum_{\alpha\in\{\alpha_h\}} \frac{\rho_\alpha(\vec{x}_1)\,\rho^{*}_\alpha(\vec{x}_2)}{E-\omega_\alpha + i\varepsilon}\right\} \tag{2.22}$$

$$+ \cdots$$

with

$$\omega_\alpha := |\varepsilon^{(1)}_\alpha - \mu| ,$$

i.e. in the Feynman representation; t-integrations are then performed and Feynman graphs result.

The latter have been employed for establishing a renormalisation algorithm for relativistic theories because of the well-known reason that in this case particle and hole propagation explicitly conspires to yield an asymptotic E^{-2}-behaviour for the E-integration in (2.22) whereas in "old-fashioned" perturbation theory this conspiracy is distributed over a set of graphs of the same order but with a different combinatorial arrangement of vertices. Renormalisation of "old-fashioned" perturbation series, hence, involves combining these different pieces to Feynman graphs and then employing the usual procedure.

We are now in the position to formulate the perturbation theory generated by (i) and (ii) and equation (2.2) and to discuss in a precise manner the impossibility of applying the well-established renormalisation schemes in this case. To this end we decompose the Schrödinger picture operator

$$\psi(\vec{x}) = \sum_{\alpha\in\{\alpha_p\}} \rho_\alpha(\vec{x}) \int d^3x' \, \rho^*_\alpha(\vec{x}') \, \psi(\vec{x}')$$

$$+ \sum_{\alpha\in\{\alpha_h\}} \rho_\alpha(\vec{x}) \int d^3x' \, \rho^*_\alpha(\vec{x}') \, \psi(\vec{x}')$$

$$=: \psi_p(\vec{x}) + \psi^\dagger_h(\vec{x}) \tag{2.23}$$

$\psi^\dagger_p(\vec{x})$ and $\psi^\dagger_h(\vec{x})$ act as particle and antiparticle, i.e. hole, creation operators. The unperturbed Hamiltonian H_o in (2.2) is simply

$$H_o = \int d^3x \, \psi^\dagger_p(\vec{x}) D(\vec{x}) \psi_p(\vec{x}) + \frac{1}{2}\int d^3x \int d^3x' \, \psi^\dagger_p \psi_p(\vec{x}) V(\vec{x}, \vec{x}') \psi^\dagger_p \psi_p(\vec{x}'). \tag{2.24}$$

The interaction picture field is given by (2.13) with the replacement

$$\psi(\vec{x}) \longrightarrow \psi_p(\vec{x})$$
$$\psi^\dagger(\vec{x}) \longrightarrow \psi^\dagger_p(\vec{x})$$
$$\rho_\alpha(\vec{x}_1 \cdots \vec{x}_N) \longrightarrow \rho^{(p)}_\alpha(\vec{x}_1 \cdots \vec{x}_N)$$

the wave functions $\rho^{(p)}_\alpha(\vec{x}_1 \ldots \vec{x}_N)$ are to be determined from projected equations like (2.3). In the propagators (2.20) or (2.22) the antiparticle contributions are not present after this replacement, only particle states are propagated forward in time.

Perturbing with H_{int} then necessarily leads to graphs with oriented lines. The characteristics of the perturbation theory defined by (2.2) and (2.24) are easily spelled out: only particle lines are leaving ρ_α or entering ρ^*_α, antiparticle lines only leave and enter vertices contained in H_{int}.

As we already stressed, to employ current renormalisation techniques we should combine particle and antiparticle lines such that graphs with propagators (2.22) - Feynman propagators - obtain. The biased emphasis on only N-particle non-perturbative contributions represented by solutions $\rho_\alpha^{(P)}(\vec{x}_1 \ldots \vec{x}_N)$ of projected equations like (2.3) makes such a combinatorial rearrangement into Feynman graphs, in general, prohibitively complicated. For the lowest order graphs this has been discussed in ref. (2).

Generally speaking the renormalisation of the pertubation series of the "Hamiltonian Method" requires the knowledge of the asymptotic behaviour of the Fouriertransform of the wavefunctions $\rho_\alpha(\vec{x}_1 \ldots \vec{x}_N)$ in all the momenta to an order increasing with the order of perturbation theory: a problem which cannot be handled by known methods.

Even if the renormalisation problem could be solved for low orders of perturbation, for the purpose of establishing a systematic approach in the sense described in the Introduction, the "Hamiltonian Method" is not a suitable framework. This is because, seen from a practical standpoint, it requires a Brückner-Hartree-Fock (BHF) perturbation expansion already in the non-perturbative sector defined by H_0 on top of which perturbations in H_{int} have to be performed. In a mean-field approach, on the other hand, a single-particle basis defines a BHF-expansion which contains all interactions, in particular pair-production and scattering terms, in a completely symmetric way.

A further problem is connected with the question of gauge-invariance. It is well-known that in order to define a Hamiltonian motion one has to define a cross-section in the relevant gauge-bundles, i.e. one has to fix a suitably chosen gauge to define a Hamiltonian. Hence, the discussion of consequences of gauge-invariance for physical observables is obviously impeded in this kind of approach.

3. THE g-HARTREE METHOD

The lesson to be learned from the previous chapter can be compactly formulated: the zeroth order non-perturbative step should involve a single particle basis comprising particle and antiparticle states. Hartree-like ideas supply the non-perturbative aspect. Such an approach has the chance to be accessible by available computer facilities, its particle-antiparticle symmetry allows for the application of well-established renormalisation procedures.

The g-Hartree method goes substantially beyond this idea: we show that under circumstances to be specified the single particle basis can be optimised such that physical observables chosen to be considered - total energies, transition energies, electron density etc. of atoms - are exactly determined by this single particle basis. Each physical quantity entails a different optimisation procedure.

To guarantee an effective book-keeping we shall use the functional integral formalism which also allows for conceptual clarity. As we already did in the previous chapter we shall expound our method by using a simplified model which disregards all spin and gauge degrees of freedom. After this exposition, it will be an easy matter to generalise to the case of relativistic QED.

The action functional from which the Hamiltonian (2.8) can be derived has the form

$$S[\psi,\psi^*] = \int_0^T dt \int_\Omega d^3x \; \psi^*(\vec{x},t)\left(i\partial_t - D + \mu\right)\psi(\vec{x},t)$$

$$-\frac{1}{2}\int_0^T dt \int_\Omega d^3x \int_\Omega d^3x' \; |\psi(\vec{x},t)|^2 \, V(\vec{x},\vec{x}') \, |\psi(\vec{x}',t)|^2 , \tag{3.1}$$

Ω denotes the volume of the system, $\psi(\vec{x},t)$ stands for the electron field, $V(\vec{x},\vec{x}')$ is the electron-electron potential

$$V(\vec{x},\vec{x}') = \frac{e^2}{4\pi}\,\frac{1}{|\vec{x}-\vec{x}'|} , \tag{3.2}$$

μ is the chemical potential and

$$D = D_{kin} + \mu_{ext}(x) \tag{3.3}$$

is the Schrödinger operator for electrons with kinetic energy D_{kin} in a potential $\mu_{ext}(\vec{x})$. The important distinction between the non-relativistic and the relativistic cases is for our purposes that the spectrum of D_{kin} is unbounded from below in the latter case whereas in the former it has a state of lowest energy; negative energy states giving rise to an antiparticle interpretation are responsible for the unboundness.

The method which we are going to use to extract energy levels from our theory is oriented by notions of statistical mechanics: the Laplace transform in the inverse temperature variable of the partition function of our system yields the energy levels as poles [15]. Laplace transforming

$$Z(\beta) = \mathrm{tr}\left(e^{-\beta H}\right) = \sum_{\text{states}} e^{-\beta E_n}$$

we arrive at

$$\hat{Z}(E) = \int_0^\infty d\beta\, e^{-\beta E} Z(\beta) = \sum_{\text{states}} \frac{1}{E - E_n} \,.$$

For the ground state energy, in particular, we have

$$E_0 = -\beta^{-1} \ln Z(\beta) \Big|_{\beta \to \infty} \,. \tag{3.4}$$

The important observation in this context is now that the partition function $Z(\beta)$ obtains as the analytic continuation $T \to i\beta$ (β is the (real) inverse temperature) of the functional integral

$$\tilde{Z}(T) = \underset{\text{antiperiodic}}{\int \mathcal{D}\psi \int \mathcal{D}\psi^*} \, e^{iS[\psi,\psi^*]} \tag{3.5}$$

i.e.

$$Z(\beta) = \tilde{Z}(i\beta) \,;$$

to take care of the Pauli Principle, the functional integration is performed over Grassmannian (anticommuting) fields ψ , ψ^* antiperiodic in t with period T.

This procedure can be looked upon in two ways: the use of the partition function is only a convenient trick or it describes the physical situation of an inhomogeneous (because of $\mu_{ext}(\vec{x})$) many-electron system in equilibrium with black-body radiation at temperature β^{-1}.

Using the Hubbard-Stratonowitch transformation $Z(\beta)$ is transformed [4,5] into

$$Z(\beta) = \int_{\text{periodic}} \mathcal{D}\phi \, e^{\frac{i}{2}\int \phi V^{-1}\phi \, + \, \mathrm{tr} \ln(i\partial_t - D + \mu - \phi)} \Big|_{T = i\beta} \tag{3.6}$$

$V^{-1}(\vec{x},\vec{x}')$ is the inverse of (3.2)

$$V^{-1}(\vec{x},\vec{x}') = -e^{-2}\,\delta(\vec{x}-\vec{x}')\,\Delta$$

We expand around a time-independent configuration $\phi_0(\vec{x})$

$$Z(\beta) = e^{\frac{i}{2}\int \phi_0 V^{-1}\phi_0 \;+\; \mathrm{tr}\ln(i\partial_t - \mathcal{D} + \mu - \phi_0(\vec{x}))}$$

$$\cdot \int \mathcal{D}\varphi\; e^{\int\{iV^{-1}\phi_0 + \frac{\delta}{\delta\varphi}\mathrm{tr}\ln(i\partial_t - \mathcal{D}+\mu-\varphi)|_{\phi_0}\}\varphi}$$

$$\cdot e^{\int\varphi\{iV^{-1} + \frac{1}{2!}\frac{\delta^2}{\delta\varphi\delta\varphi}\mathrm{tr}\ln(i\partial_t - \mathcal{D}+\mu-\varphi)|_{\phi_0}\}\varphi}$$

$$\cdot \;\ldots$$

$$=: Z_0 \cdot Z_{corr} \tag{3.7}$$

and demand that $\phi_0(\vec{x})$ be stationary, i.e. that the linear term in this expansion vanishes.

Introducing the eigenbasis

$$(\mathcal{D} + \phi_0)\,\psi_\alpha(\vec{x}) = \varepsilon_\alpha\,\psi_\alpha(\vec{x}) \tag{3.8}$$

$Z_0(\beta)$ can be calculated [16] $(\hat{\varepsilon}_\alpha := \varepsilon_\alpha - \mu)$

$$Z_0(\beta) = e^{-\sum_\alpha \ln(1+e^{-\beta\hat{\varepsilon}_\alpha}) - \frac{\beta}{2}\int\phi_0 V^{-1}\phi_0} \tag{3.9}$$

.

Since

$$\left.\frac{\delta \varepsilon_\alpha}{\delta \varphi(\vec{x})}\right|_{\phi_0} = |\psi_\alpha(\vec{x})|^2 , \tag{3.10}$$

the stationarity condition

$$\frac{\delta \ln Z_0}{\delta \phi_0(\vec{x})} = 0 \tag{3.11}$$

translates into

$$\phi_0(\vec{x}) = \int d^3x' \sum_\alpha n_\alpha |\psi_\alpha(\vec{x}')|^2 V(\vec{x},\vec{x}') \tag{3.12}$$

i.e. the stationary configuration is just the Hartree mean field induced by electrons thermally distributed over the energy spectrum $\{\varepsilon_\alpha\}$

$$n_\alpha = \frac{1}{e^{\beta \hat{\varepsilon}_\alpha} + 1} . \tag{3.13}$$

The energy spectrum has to be determined self-consistently from the Hartree equation (3.8), (3.12).

A more general class of mean-field configurations can be uncovered by following a more general line of argumentation. First we rearrange the interaction part of the action taking into account the anticommutativity of the Grassmann valued electron field $\psi(\vec{x},t)$. Abbreviating

$$x := (\vec{x},t) \quad , \quad \hat{V}(x,x') := \delta(t-t')\, V(\vec{x},\vec{x}')$$

we write

$$y := (x,x') \quad , \quad y' := (x'', x''')$$

$$\begin{aligned} S_{int} &= -\frac{1}{2} \int d^4x \int d^4x' \, |\psi(x)|^2 \, \hat{V}(x,x') \, |\psi(x')|^2 \\ &= -\frac{1}{2} \int d^8y \int d^8y' \, \eta(y) \, \tilde{V}(y,y') \, \eta(y') \end{aligned} \tag{3.14}$$

with

$$\eta(y) = \psi^*(x)\,\psi(x') \tag{3.15}$$

and

$$\tilde{V}(y,y') = g\,\hat{V}(x,x'')\,\delta(x-x')\,\delta(x''-x''')$$

$$-(1-g)\,\hat{V}(x,x')\,\delta(x-x''')\,\delta(x'-x''). \tag{3.16}$$

Following the same procedure as above [4], i.e. deriving a Bose field representation (3.6) in terms of the bilocal fields η(y) by an appropriate Hubbard-Stratonowitch transformation and determining the stationary configuration, we arrive at the g-Hartree equation (* denotes a convolution)

$$\left(\mathcal{D}(\vec{x}) + \eta_0^{(g)}(\vec{x})\right)\psi_\alpha(\vec{x}) = \varepsilon_\alpha\,\psi_\alpha(\vec{x}) \tag{3.17}$$

with

$$\eta_0^{(g)}(\vec{x}) = \int d^3x'\,V(\vec{x},\vec{x}')\sum_\alpha n_\alpha\left\{g\,|\psi_\alpha(\vec{x}')|^2 - (1-g)\,\psi_\alpha^*(\vec{x}')\,\psi_\alpha(\vec{x})\,*\right\}$$

Finally we have to evaluate the perturbative contributions to the thermodynamic grand canonical potential Z

$$\ln Z = \ln Z_0 + \ln Z_{corr} \tag{3.18}$$

which are derived from (3.7). We find

$$\ln Z_{corr} =: \sum \mathcal{U}_i$$

where the $\mathcal{U}_i$ denote perturbative terms ordered in powers of the electron-electron interaction and

$\mathcal{U}_1 = \frac{1}{2}(1-g)$  $- \frac{1}{2} g$ 

$\mathcal{U}_2 = -\frac{1}{2} g^2$  $+ g(1-g)$ 

$- \frac{1}{2}(1-g)^2$ 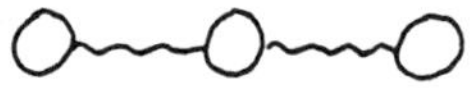

$- \frac{1}{4}$  $+ \frac{1}{4}$ 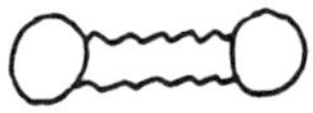 (3.19)

etc.,

wavy lines depict the electron-electron interaction, straight lines the electron propagators

$\vec{x}, t$ $\vec{x}', t'$

$$\hat{V}(x, x') = \delta(t-t') \, V(\vec{x}, \vec{x}')$$

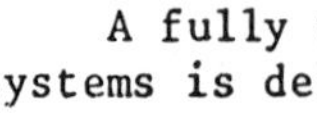

$\vec{x}, t$ $\vec{x}', t'$

$$G(\vec{x}, \vec{x}'; t-t') = -i \sum_\alpha \psi_\alpha(\vec{x}) \psi_\alpha^*(\vec{x}') \cdot e^{-i(\varepsilon_\alpha - \mu)(t-t')} \{ \Theta(t-t') - n_\alpha \}$$

A fully relativistic theory of inhomogeneous many-electron systems is defined by the classical action

$$S_{QED}[\psi, \psi^+, A_\mu] = \int d^4x \left\{ -\frac{1}{4} F_{\mu\nu} F^{\mu\nu} + \psi^+(x) (i \partial_t - D + \mu + e \gamma_0 \gamma_\mu A^\mu) \psi(x) \right\} \tag{3.20}$$

with

$$\mathcal{D} = i\gamma_0 \gamma_k \partial^k + \gamma_0 m + \mu_{ext}(\vec{x})$$

(notations are taken from ref. (14)). We shall now derive g-Hartree equations for this theory and start by fixing a covariant, ghost-free gauge

$$S'_{QED} = \int d^4x \Big\{ \psi^+(x) \left(i\partial_t - \mathcal{D} + \mu + e\gamma_0\gamma_\mu A^\mu \right) \psi(x)$$

$$- \frac{1}{2} \int d^4x' A^\mu(x) \mathcal{D}^{-1}_{\mu\nu}(x,x') A^\nu(x') \Big\}, \tag{3.20a}$$

$D_{\mu\nu}(x,x')$ is the photon propagator in this gauge; in Feynman gauge, for instance,

$$\mathcal{D}^{(F)}_{\mu\nu}(x,x') = - g_{\mu\nu} \frac{1}{\Box + i\varepsilon} \delta(x-x'). \tag{3.21}$$

The functional integral representation of the partition function of Spinor-QED

$$Z_{QED} = \underbrace{\int \mathcal{D}\psi \int \mathcal{D}\psi^+}_{\text{antiperiodic}} \underbrace{\int \mathcal{D}A_\mu}_{\text{periodic}} e^{iS'_{QED}} \tag{3.22}$$

can be put into a form which resembles (3.5) and (3.1) insofar as all electro-magnetic field degrees of freedom are eliminated: we integrate A_μ and find

$$Z_{QED} = \int \mathcal{D}\psi \int \mathcal{D}\psi^+ e^{iS_{eff}[\psi,\psi^+]} \tag{3.23}$$

with

$$S_{eff}[\psi,\psi^+] = \int d^4x\, \psi^+(x)\{i\partial_t - \mathcal{D} + \mu\}\psi(x)$$

$$+ \frac{e^2}{2}\int d^4x \int d^4x'\, j^\mu(x)\, D_{\mu\nu}(x,x')\, j^\nu(x')$$

(3.24)

and

$$j_\mu(x) = \psi^+(x)\gamma_0\gamma_\mu\psi(x).$$

The action $S_{eff}[\psi,\psi^+]$ is immediately seen to be a covariant generalisation of our toy model (3.1): instead of the pure Coulomb propagator

$$\Delta(\vec{x},\vec{x}') = -\frac{1}{4\pi}\frac{1}{|\vec{x}-\vec{x}'|}$$

we have the covariant extension $D_{\mu\nu}(x,x')$ which introduces the retardation required by relativity; at the same time, space components of the electro-magnetic current appear.

This analogy immediately leads to the observation that the stationary point of the A_μ-integration is a relativistic generalisation of the Hartree-field (3.12); bilocal electro-magnetic potentials lead to g-Hartree extensions. The latter turn out to be time-independent. It is important to note that the notion of time-independence is related to the choice of gauge-fixing - it has a dynamical meaning within a properly fixed gauge. The choice (3.21) leads to the relativistic g-Hartree equations

$$\left(\mathcal{D}(\vec{x}) - e\gamma_0\gamma^\mu\eta_\mu^{(g)}(\vec{x})\right)\psi_\alpha(\vec{x}) = \varepsilon_\alpha\psi_\alpha(\vec{x}) \qquad (3.25)$$

where

$$\eta_\mu^{(g)}(\vec{x}) = -e\int d^3x'\, D_{\mu\nu}^{(o)}(\vec{x},\vec{x}') \sum_\alpha n_\alpha \cdot$$

$$\cdot\left\{ g\,\psi_\alpha^\dagger(\vec{x}')\gamma_0\gamma^\nu\psi_\alpha(\vec{x}') - (1-g)\,\psi_\alpha^\dagger(\vec{x})\gamma_0\gamma^\nu\psi_\alpha(\vec{x})* \right\};$$

$\psi_\alpha(\vec{x})$ are Dirac spinors, $D_{\mu\nu}^{(o)}(\vec{x},\vec{x}') = D_{\mu\nu}^{(F)}(x,x')|_{t=t'}$, or

$$D_{\mu\nu}^{(o)}(\vec{x},\vec{x}') = g_{\mu\nu}\frac{1}{4\pi|\vec{x}-\vec{x}'|} \quad . \tag{3.26}$$

We see that retardation does not contribute to the stationary point. Since non-trivial solutions are generated by the inhomogeneous $\mu^{ext}(\vec{x})$ - the external potential - questions of covariance and non-retardation in a covariant gauge are out of place.

To bring the formal derivation of relativistic equations to an end we should only point out that the graphs representing the perturbative expansion of Z_{corr} are the ones displayed in (3.19) where Dirac spinors have to be inserted in the expression for the electron propagator. We have seen at the end of the last chapter that the necessary renormalisation proceeds through standard methods.

In the following paragraphs we shall describe the optimisation of mean-fields. Optimising simply means to suppress fluctuations contained in Z_{corr} (cf. (3.7)) such that the suppression mechanism is adapted to the physical problem under consideration. We distinguish three different situations.

(a) Equilibrium properties [16] of the inhomogeneous electron gas and, in particular, the total energy of the ground state

(b) Transition energies in bound states, e.g. atoms

(c) Time-dependent processes: relaxation properties of excited atoms, scattering processes etc.

As far as the equilibrium properties are concerned, it is well-known that integration over periodic or antiperiodic fields in the functional representation of the partition function Z leads to Bose-Einstein or Fermi-Dirac statistics, respectively. The use of Grassmannian fields in the latter case gives the correct sign of the thermodynamic potential K which is defined by

$$K = \beta^{-1}\ln Z = -P\cdot\Omega \tag{3.27}$$

for the grand canonical ensemble considered here; Ω denotes the volume and P the pressure.

We have shown [5,10,11] that the parameter g can be chosen such that

$$Z_{corr}\Big|_{g=g_0} = 1 \tag{3.28}$$

i.e. that the grand canonical potential is exactly given by a single particle basis, the g-Hartree basis for $g = g_0$. We then have

$$K = -\beta^{-1}\sum_{\alpha} \ln(1+e^{-\beta\hat{\varepsilon}_\alpha}) + K_{g\text{-}H} \tag{3.29}$$

where the spectrum $\{\varepsilon_g\}$ has to be determined by solving the g-Hartree equations (3.17) or (3.25) and K_{g-H} is the explicitly known energy of the g-Hartree mean field for $g = g_0$.

Equation (3.28) will have solutions

$$g_0 = g_0\,[\mu_{ext}(\vec{x});\, \beta,\, \Omega]$$

for a large class of chemical potentials $\mu_{ext}(\vec{x})$.

The effective single particle basis - a Landau-type basis [13]- representation [16] of the grand canonical partition function of an interacting, inhomogeneous electron gas allows us to derive an equation of state. We start from

$$P\Omega = -K\,[\mu_{ext}(\vec{x});\, \beta,\, \Omega] \tag{3.30}$$

to derive the particle density

$$n(\vec{x}) = \frac{\delta K}{\delta \mu_{ext}(\vec{x})} = \sum_{\alpha} n_\alpha |\psi_\alpha(\vec{x})|^2 + \frac{\beta B}{2}\,\frac{\delta g_0}{\delta \mu_{ext}(\vec{x})} \tag{3.31}$$

where

$$\mathcal{B} = \int d^3x \int d^3x' \, V(\vec{x},\vec{x}') \sum_{\alpha,\gamma} n_\alpha n_\gamma \left\{ |\psi_\alpha(\vec{x})|^2 |\psi_\gamma(\vec{x}')|^2 + \psi_\alpha^*(\vec{x}')\psi_\alpha(\vec{x})\psi_\gamma^*(\vec{x})\psi_\gamma(\vec{x}') \right\}$$

for the action (3.1). The number of particles

$$N = \int d^3x \; n(\vec{x}) = N[\mu_{ext}(\vec{x}); \beta, \Omega] \tag{3.32}$$

has to be considered as a functional of the inhomogeneity $\mu_{ext}(\vec{x})$. We define the Fermi energy ε_F by

$$\mu_{ext}(\vec{x}) - \mu = \mu_0(\vec{x}) + \varepsilon_F$$

$$\mu_0(\vec{x}) \longrightarrow 0 \quad \text{for } |\vec{x}| \longrightarrow \infty$$

The equation of state

$$F[\mu_0(\vec{x}); P, \Omega, \beta, N] = 0$$

is obtained by eliminating ε_F. One can even go farther and use the local structure of (3.31) to derive [16] the Hohenberg-Kohn-Mermin energy density functional in a Thomas-Fermi expansion.

The viability of the concept of optimisation of a g-Hartree basis by the realisation $Z_{corr}(g_0) = 1$ has been numerically checked for the ground state of atoms the total energy of which is given by

$$E = -K[V_{ext}(\vec{x}); \beta, \infty]\Big|_{\beta\to\infty} \tag{3.33}$$

and is, of course, g-independent; the parameter g only controls the relative magnitude of g-Hartree and perturbative contributions. Equation (3.28) is, of course, equivalent to

$$E_{corr}(g)\Big|_{g=g_0} = 0 \tag{3.34}$$

where

$$E_{corr} := E - E_{g\text{-}H} .$$

We have used this relation in two ways

a1) Determine g_0 from experiment by demanding

$$E_{exp} = E_{g\text{-}H}(g)\Big|_{g=g_0} \tag{3.35}$$

and compute other atomic observables.

a2) A scheme of ab initio calculations (only the fine-structure constant and the electron mass appear as parameters) emerges if we determine g_0 theoretically by solving

$$E^{(n)}_{corr}(g)\Big|_{g=g_0} = 0 \tag{3.36}$$

for g_0, n denotes that the correlation energy E_{corr} has to be calculated up to n-th order in the electron-electron interaction.

The second point (b) (see page) requires a slight generalisation: we go back to (3.7) where $\phi_0(\vec{x})$ is not necessarily a stationary point and denote the total energy expanded around the field ϕ_0(cf. (3.7)) for an atomic configuration characterised by a set of occupation numbers $\{n_\alpha\}$ by

$$-K\Big|_{\beta\to\infty} = E_{tot} =: E_{\phi_0}[n_\alpha]$$

the dependence on ϕ_0 is, of course, dummy and is of importance only if the expansion in (3.7) is cut at a finite order.

Let us now consider the atomic transition

$$\{n_\alpha^{(i)}\} \longrightarrow \{n_\alpha^{(f)}\} \;;$$

the transition energy is given by

$$\Delta E = E_{\phi_0^{(f)}}[n_\alpha^{(f)}] - E_{\phi_0^{(i)}}[n_\alpha^{(i)}] \tag{3.37}$$

where the reference fields ϕ_0, in general, can be chosen differently.

We choose

$$\phi_0^{(i)} = \phi_0^{(f)} = \eta_\mu^{(g)}[n_\alpha^{(i)}](\vec{x}) \tag{3.38}$$

i.e. the common mean field is given by the solutions of the g-Hartree equation (3.25) where $\{n_\alpha\}$ is replaced by $\{n_\alpha^{(i)}\}$. The transition energy is then

$$\Delta E = \sum_\alpha (n_\alpha^{(f)} - n_\alpha^{(i)})\,\varepsilon_\alpha + E_{corr}(g;\{n_\alpha^{(i)}\}\rightarrow\{n_\alpha^{(f)}\}) \tag{3.39}$$

where it is important to observe that the graphical expansion of E_{corr} contains additional graphs which result from the fact that the mean field (3.38) is <u>not</u> stationary with respect to the final configuration $\{n_\alpha^{(f)}\}$. For details we refer to ref. (11), we only note that an ab initio scheme for the calculation of transition energies in atoms results if we calculate g_0 such that

$$E_{corr}^{(n)}(g;\{n_\alpha^{(i)}\} \longrightarrow \{n_\alpha^{(f)}\})\Big|_{g=g_0} = 0 \tag{3.40}$$

where the correlation energy is assumed to be expanded up to n-th order in the electron-electron interaction.

As far as (c) is concerned, work is in progress, we shall report in a separate publication.

4. THE CASIMIR-EFFECT IN ATOMS

In this chapter we take a closer look at the relativistic case and unravel the influence of the structure of the physical vacuum on atoms. We shall point out that this influence results in measurable effects and, if measured, should give interesting insights.

The essential clue appears if we observe that in the relativistic case the spectral sums building the g-Hartree potentials (3.17) or (3.25) run over two branches:

$$\varepsilon_\alpha > -m \qquad \text{particle bound and scattering states}$$

$$\varepsilon_\alpha \leq -m \qquad \text{antiparticle scattering states.}$$

We employ the charge conjugation redefinitions (2.18) with (2.19) and use for atoms

$$\bar{n}_\alpha = 0 \tag{4.1}$$

i.e. we demand that the atoms considered do not contain positrons as constituents, and obtain for the g-Hartree potential (the n_α are, of course, the particle, i.e. electron, occupation numbers)

$$\eta_\mu^{(g)}(\vec{x}) = -e \int d^3x' \, D_{\mu\nu}^{(0)}(\vec{x}, \vec{x}') \cdot$$

$$\cdot \left\{ \sum_{\varepsilon_\alpha > -m} n_\alpha + \sum_{\varepsilon_\alpha \leq -m} \right\}$$

$$\cdot \left\{ g\, \psi_\alpha^\dagger(\vec{x}') \gamma_0 \gamma^\nu \psi_\alpha(\vec{x}') - (1-g)\, \psi_\alpha^\dagger(\vec{x}') \gamma_0 \gamma^\nu \psi_\alpha(\vec{x}) * \right\}. \tag{4.2}$$

For this to be a physically sensible expression we have to ensure, apart from the usual multiplicative renormalisations, that the sum over the Dirac sea does not induce a charge distribution.

To discuss the physical background of this condition let us look at the ground-state energy E at g = g₀ [(3.34)]

$$E = \sum_{\varepsilon_\alpha > -m} n_\alpha \varepsilon_\alpha + \sum_{\varepsilon_\alpha \le -m} \varepsilon_\alpha$$

$$+ \frac{1}{2} \int \eta_\mu^{(g_0)} D^{(0)\mu\nu} \eta_\nu^{(g_0)} . \tag{4.3}$$

The second term in this sum

$$E_c := \sum_{\varepsilon_\alpha \le -m} \varepsilon_\alpha \tag{4.4}$$

the sum over all negative energy modes, represents the Casimir effect [17] in atoms [12]. As is well known, the Casimir effect is an energy shift due to the deformation of the vacuum effected by conducting matter acting as boundary: the energy shift is determined by computing the sum over all modes corresponding to these boundary conditions; the sum over all modes corresponding to some reference boundary conditions has to be subtracted to obtain a properly renormalised energy shift. In our case the deformation is produced by the external nuclear Coulomb field plus an effective average potential due to the atomic electron configuration; the energy shift is measured by considering different atomic configurations corresponding to different deformations of the vacuum. In the simplest model, the latter is endowed with a non-trivial structure by the Dirac sea of positron states in the effective Coulomb potential. The sum (4.4) is precisely the analogue of the sum over all modes of the electromagnetic field originally considered by Casimir. As it stands it is clearly infinite, we shall now describe in some more detail the proper renormalisation procedure for the extraction of a measurable energy shift. Let us introduce a suitable regularisation prescription depending on a parameter Λ such that the regularised sum E_c^Λ is finite and $E_c^\Lambda \to E_c$ for $\Lambda \to \infty$. We then observe that E_c^Λ depends on the configurations $\{n_\alpha\}$ and so do the antiparticle-positron-scattering states. Interpreted in physical terms this means that the self-consistent potential induced by a certain atomic configuration deforms the Dirac sea, the energy shift produced by this deformation should be a measurable effect.In slightly more general terms, the shift of energy produced by the change of configurations in the transition

$$\{ n_\alpha^{(i)} \} \longrightarrow \{ n_\alpha^{(f)} \} \tag{4.5}$$

via the change of polarisation of the vacuum is expected to be a finite observable quantity. Indeed it can be shown [12] that the difference

$$\Delta E_c = \lim_{\Lambda \to \infty} \left(E_c^\Lambda [n_\alpha^{(f)}] - E_c^\Lambda [n_\alpha^{(i)}] \right) \tag{4.6}$$

is finite after the standard multiplicative renormalisations.

The effect should be most drastically seen if

$$n_\alpha^{(f)} = 0 \qquad \text{for all } \alpha,$$

i.e. if we consider the complete stripping of atoms. The difference (4.5) then measures the energy shift due to a Van der Waals-type potential built up by the change in vacuum deformation produced by the binding of electrons in the nuclear Coulomb field.

This procedure of extracting the finite, physically measurable energy shift ΔE_c is completely analogous to the calculation of the Casimir potential [18] between two conducting plates or the Van der Waals potential [19] in different geometries.

A measurement of ΔE_c would reveal interesting properties of the physical vacuum. To see this we must note that the expression (4.3) for the ground state energy has to be generalised

$$E = \sum_i \left\{ \sum_{\varepsilon_{\alpha,i} > -m_i} n_{\alpha,i} \varepsilon_{\alpha,i} + \sum_{\varepsilon_{\alpha,i} \leq -m_i} \varepsilon_{\alpha,i} \right\}$$

$$+ \frac{1}{2} \int \eta_\mu^{(g_0)} D^{(0)\mu\nu} \eta_\nu^{(g_0)} \tag{4.7}$$

where i enumerates all charged fields. In particular $n_{\alpha,i}$ is the occupation number giving the probability that the energy level $\varepsilon_{\alpha,i}$ is occupied by quanta of type (i). Now for atoms $n_{\alpha,i}$ is different from zero only if (i) corresponds to electrons; all fields(i), however, contribute to the total energy of the atom since their Dirac seas (we have tacitly assumed that the fundamental fields building up the physical vacuum are fermions!) are deformed by atomic self-consistent potentials. Since for $g = g_o$, (4.7) is the exact expression, ΔE_c can be unambiguously extracted from measured transition energies for the transition (4.6) using the g-Hartree formalism at $g = g_o$. The effect of a Dirac sea deformation should be the more pronounced, the larger the nuclear charge Z is taken and the more final and initial configurations differ in the transition (4.5).

To indicate the kind of questions that should be asked once experimental data of this kind are available we remind our readers that, for instance, unbroken supersymmetry predicts the vanishing of ΔE_c.

5. NUMERICAL RESULTS

This chapter is added to give a short review of the numerical tests which the g-Hartree method has already passed. In order not to blow up the frame of this report excessively, we shall give only short descriptions of the calculations done, for actual results and the details of how they are obtained we refer to our publications.

Let us first turn our attention to the first application (a1) explained in chapter 3. Using this semi-phenomenological approach we first considered [7] the 5d $\rightarrow$nf series of Ba^+, and computed g_o to match the quantum defects, spin-orbit splitting and dipole strengths were then predicted. The predictions reproduced the experimental values of the spin orbit splitting and the experimental behaviour of the dipole strength showing a typical shoulder. Then we turned to a g-Hartree mean-field analysis of inner-shell many-body effects for small and large Z-atoms [8]. We analysed the atomic many-body effects associated with single electron excitations: g-values are determined for inner- and outer-shell ionisation energies for the whole periodic table. Particulary for large atoms excellent agreement with experiment is obtained if average g-values are used to compute physical observables for several channels using the single particle g-Hartree basis. In a further publication [9], we showed that the optimised g-Hartree single particle potential for the excited states interpolates between purely V(N) and V(N-1) potentials and calculated oscillator strengths for the Principal Series of the alkali atoms Li, Na and K showed that the zero-order g-Hartree single particle prediction gives good agreement with experiment thus yielding a substantial improvement over HF and DHF calculations.

The scheme (a2) for ab initio calculations was tested [10] for "first row" atoms. We took the first non-trivial case, n= 2,to solve (3.36) together with the g-Hartree equations (3.25) to compute the total energies of Be - Ne, our second order g-Hartree values agree with experiments within the experimental errors. To get an even more severe test of our method we computed the non-relativistic "correlation" energies (correlation now refers to the HF-basis)

$$E_{"corr"} = E - E_{HF}$$

and compared our results with the CI-values, excellent agreement was obtained. For a proper assessment of these results, we emphasise that they were obtained in bona fide second order: no RPA-like partial summations to get "denominator shifts" etc. Second order means here second order in the electron-electron interaction.

Finally, we calculated [11] all ionisation energies for atoms with $Z \leq 18$ and a few levels for Xe using the second order iteration scheme defined by the ab initio determination of g_0 using (3.40) for $n = 2$ and the g-Hartree equations (3.25). Agreement with experiment was reached within the experimental and numerical errors for all the lines in question.

Since it has passed these tests with astounding success and since it fulfills the criteria formulated in the Introduction for a systematic approximation and, last not least, because of its numerical simplicity, the g-Hartree method has shown, we think, important advantages, sometimes even its superiority, over many of the current approaches for treating the inhomogeneous many-electron problem

REFERENCES

1. M. H. Mittleman, Phys. Rev. A4, 893 (1971), Phys. Rev. A24, 1167 (1981).
2. W. Buchmüller and K. Dietz, Z. Phys. C5, 54 (1980).
3. J. Sucher, Phys. Rev. A22, 348 (1980).
4. K. Dietz, O. Lechtenfeld and G. Weymans, J. Phys. B: At.Mol. Phys. 15 4301 (1982).
6. T. Koopmans, Physica 1, 104 (1934).
7. J. P. Connerade, K. Dietz, M. W. D. Mansfield and G. Weymans, J. Phys. B: At. Mol. Phys. 17, 1211 (1984).
8. J. P. Connerade, K. Dietz, M. Ohno and G. Weymans, Bonn preprint HE-85-04, to be published in J. Phys. B.
9. J. P. Connerade, K. Dietz and G. Weymans, Bonn preprint HE-85-03, to be published in J. Phys. B.
10. K. Dietz and G. Weymans, J. Phys. B: At. Mol. Phys. 17, 2987 (1984).
11. K. Dietz, M. Ohno and G. Weymans, in preparation.
12. K. Dietz and G. Weymans, J. Phys. B: At. Mol. Phys. 17, 4801 (1984).
13. L. D. Landau, Sov. Phys. JETP 3, 920 (1956).
14. J. D. Björken and S. D. Drell, "Relativistic Quantum Fields", New York, McGraw Hill 1965.
15. R. Dashen, R. Hasslacher and A. Neveu, Phys. Rev. D10, 4115 (1974).

16. K. Dietz and G. Weymans, to be published in Physica.
17. H. B. G. Casimir, Proc. Kon. Ned. Akad. Wetenshap 51, 793 (1948).
18. For a particularly transparent exposition of the actual computation of the Casimir shift see:
M. Fierz, Helv. Phys. Acta 33, 855 (1960).
19. F. London, Z. Phys. 63, 245 (1930).
H. B. G. Casimir and D. Polder, Phys. Rev. 73, 360 (1948),
I. E. Dzyaloshinskii, E. M. Lifshitz and L. P. Pitaevskii, Adv. Phys. 10, 165 (1961).

NEW EXPERIMENTS ON FEW-ELECTRON VERY HEAVY ATOMS

Harvey Gould
Materials and Molecular Research Division,
Lawrence Berkeley Laboratory,
University of California, Berkeley CA 94720

ABSTRACT

New experiments, to test quantum electrodynamics (QED) in strong Coulomb fields and to study atomic collisions at ultra relativistic energies, are proposed. A 0.1% measurement of the $2\,^2P_{1/2}$ - $2\,^2S_{1/2}$ splitting in lithium-like uranium (Z =92) and the $2\,^3P_0$ - $2\,^3S_1$ splitting in heliumlike uranium is proposed as a sub 1% test of the Lamb shift in a strong Coulomb field. Measurements of the hyperfine splitting of hydrogenlike thallium (Z=81) and the g_j factor of the ground state of hydrogenlike uranium are proposed as a test of the QED contribution to the magnetic moment of an electron bound in a strong Coulomb field. Measurements of capture cross sections for ultra-relativistic very heavy nuclei are proposed to look for the capture of electrons from pair production.

INTRODUCTION

The production[1] in 1983 of a beam of bare U^{92+} at the Lawrence Berkeley Laboratory's Bevalac[2] -the Bevatron and Super-HILAC operating in tandem-demonstrated the feasibility of experiments using few- electron uranium. X rays from n=2 → n=1 transitions in hydrogenlike uranium and heliumlike uranium were observed[3] in 1984 (Fig. 1) and a low-precision determination of the $2\,^3P_0$ - $2\,^3S_1$ splitting in heliumlike uranium was operating in 1985 using beam intensities of 10^5 ions per second[4]. Experience from these experiments and higher uranium intensities will make it possible to perform a sub 1% measurement of the Lamb shift at Z=92. In addition, new experiments are being developed to measure the QED contribution to the magnetic moment (anomalous magnetic moment) of an electron bound in a very strong Coulomb field. Beyond present accelerators, a proposed heavy ion facility at Brookhaven National Laboratory will use the Alternate Gradient Synchrotron to produce 15 GeV/nucleon very heavy ions[5], will extend atomic collisions of very heavy ions to ultra-relativistic energies. A new mechanisms for charge capture from pair production is expected. This mechanism may be very important to the physics of relativistic heavy ion colliders.

SELF-ENERGY AT VERY HIGH Z

At Z=92, the contributions to the Lamb shift ($2\,^2S_{1/2}$ - $2\,^2P_{1/2}$ splitting) (Fig 2) are the self-energy[6] of ≈ -57 eV, the vacuum polarization[7] of ≈ +14 eV and the finite nuclear size correction[7] of ≈ -33 eV. Vacuum polarization, but not self-energy, is well tested in muonic atom experiments. High-Z Lamb shift measurements primarily test the self-energy in a strong Coulomb field[8].

In addition to being the largest contribution to the Lamb shift, the self-energy makes up a large fraction of the splittings between states of $2p_{1/2}$ and $2s_{1/2}$ electrons in heliumlike and lithiumlike uranium. The $2\,^3P_0$ - $2\,^3S_1$ splitting in heliumlike uranium[9] (Fig. 3) is ≈ 254 eV and the $2\,^2P_{1/2}$ - $2\,^2S_{1/2}$ splitting in lithiumlike uranium (Fig. 4) is ≈ 283 eV.

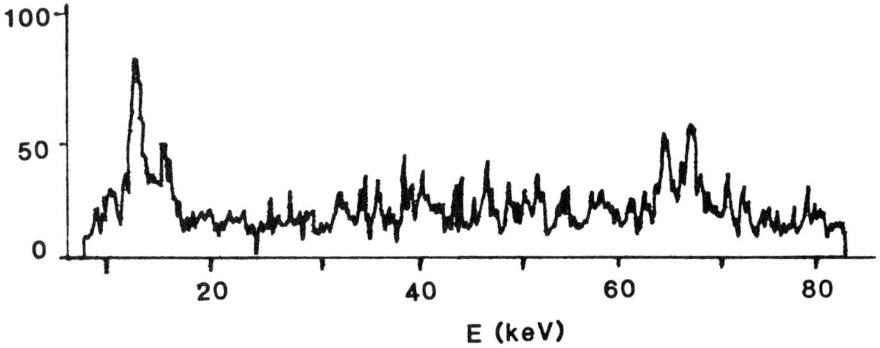

Fig. 1 - X rays from n=2 → n=1 transitions in hydrogenlike uranium (U 91+) and heliumlike uranium (U 90+) observed in a beam-foil experiment. The x-rays from 430 MeV/nucleon uranium were observed by a germanium x-ray detector looking perpendicular to the beam at a 15 mg/cm^2 mylar target. Because of relativistic Doppler shift, the x rays seen in the Laboratory are red shifted by a factor of 1.46. The peak at about 64 keV corresponds to transitions $2\,^2P_{1/2}$ and $2\,^2S_{1/2} \rightarrow 1\,^2S_{1/2}$ in hydrogenlike uranium plus $2\,^3S_1$ and $2\,^3P_1 \rightarrow 1\,^1S_0$ in heliumlike uranium. The peak near 67 keV corresponds to transitions $2\,^2P_{3/2} \rightarrow 1\,^2S_{1/2}$ in hydrogenlike uranium and the $2\,^3P_2$ and $2\,^1P_1 \rightarrow 1\,^1S_0$ in heliumlike uranium. The background is caused in part by bremsstrahlung photons produced when target electrons scatter off the uranium and by high energy electrons scattered out of the target.

Far more significant than the size of the self energy at Z=92 is that the self-energy contribution arises almost entirely from terms which are of very high order in $Z\alpha$ (where α is the fine structure constant). Because these terms are large only at very high Z (strong Coulomb field) they are not tested in present lower-Z Lamb shift and fine structure experiments.

The contribution of the higher order terms in the self-energy can be seen by comparing the series expansion, of the self energy with an evaluation of the $2\,^2S_{1/2}$ self-energy to all orders[6,10,11] in $Z\alpha$. If we write the self energy Σ_n in a power series in α and $Z\alpha$, we have:

$$\Sigma_n = n^{-3}(\alpha/\pi)\, m_o c^2 \Big[[A_{40} + A_{41} ln(Z\alpha)^{-2}](Z\alpha)^4 + A_{50}(Z\alpha)^5 + [A_{60} + A_{61} ln(Z\alpha)^{-2} + A_{62} \ln^2(Z\alpha)^{-2}](Z\alpha)^6 + A_{70}(Z\alpha)^7 + \textit{higher order terms} \Big] \qquad (1)$$

Where n is the principal quantum number and m_o is the electron mass. Values of the coefficients $A_{40} - A_{70}$ can be found in Ref. 11. Fig. 5 shows the ratio of the higher order terms in the self-energy to the total self energy. In neutral hydrogen the higher order terms in the self-energy contribute about 0.1 parts per million to the Lamb shift, nearly 100 times smaller than the uncertainty due to proton structure[12]. At Z=18 the contribution is only about 1% of the Lamb shift, equal to or smaller than the experimental uncertainty[13]. At Z=92 however, the higher order terms are essentially the entire self-energy contribution, and make up over half of the total Lamb shift.

Hydrogenlike Uranium

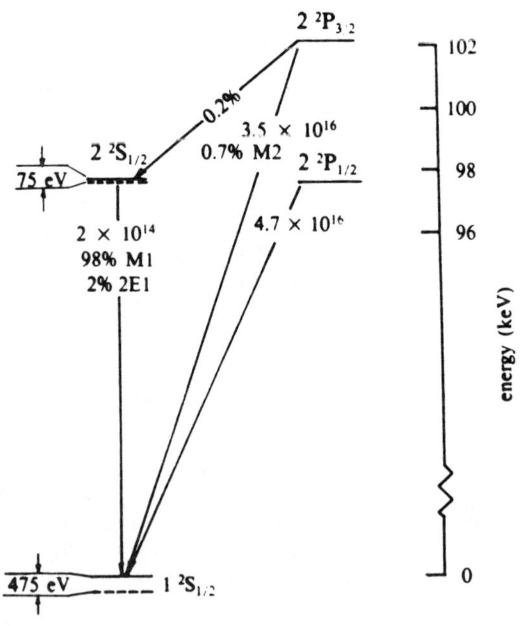

Fig 2 - Energy levels of n=1 and n=2 states of hydrogenlike uranium. The levels were tabulated using the Dirac energy[38] plus the self-energy[6], vacuum polarization[7] and finite nuclear size correction[7]. The radiative width of the $2\,^2P_{1/2}$ and $2\,^2P_{3/2}$ states is not indicated but is about the same as for the radiative width of the $2\,^3P_1$ and $2\,^1P_1$ states of heliumlike uranium (Fig. 3). E1 and magnetic quadrupole (M2) decay rates were calculated using matrix elements from Ref. 15. For the $2\,^2S_{1/2}$ state, the 2E1 rate is taken from Ref.'s 39,40 and the relativistic M1 rate from Ref. 40.

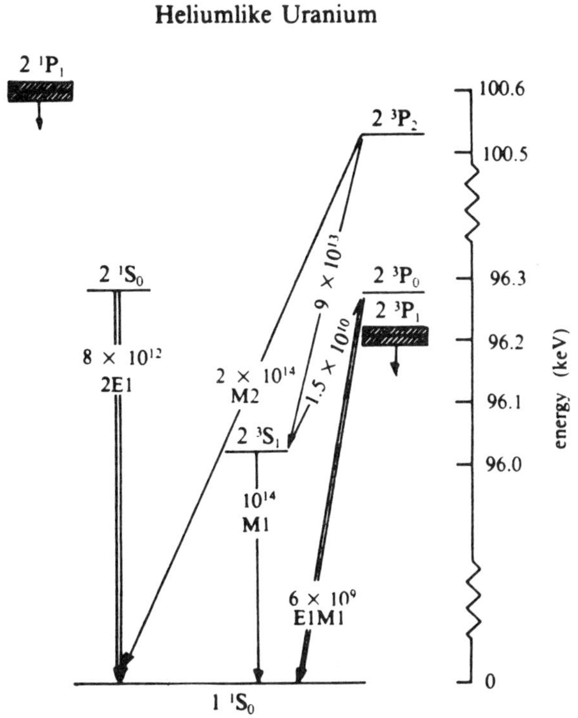

Fig. 3 - Energy levels of n=1 and n=2 states of heliumlike uranium. Decays without labels are E1 decays. The cross hatching on the $2\,^3P_1$ and $2\,^1P_1$ indicates the approximate radiative width. The non-QED contributions to the energy levels for heliumlike uranium were obtained from Ref. 16 to which were added the one-electron self energy and vacuum polarization[6,7]. The lowest order correction[8] to the QED terms for the presence of the second electron, of order 1/Z, is neglected. These values for heliumlike uranium are in general agreement with values calculated in Ref. 19 corrected for finite nuclear size effects. (The values used here are not intended to be complete. Terms of the order of 1 eV have been neglected and values have been rounded to the nearest eV.)

For the $2\,^3P_0$ state, the E1M1 decay rate is taken from Ref. 14 and the E1 rate to the $2\,^3S_1$ state is calculated using the E1 matrix element from Ref. 15. The E1 decays to the ground state have similar decay rates as the corresponding hydrogenlike transitions. The decay rate used for the $2\,^1S_0$ state is twice the decay rate for the 2E1 decay of the $2\,^2S_{1/2}$ state of hydrogenlike uranium. The $2\,^3S_1$ decay rate is taken from Ref. 16.

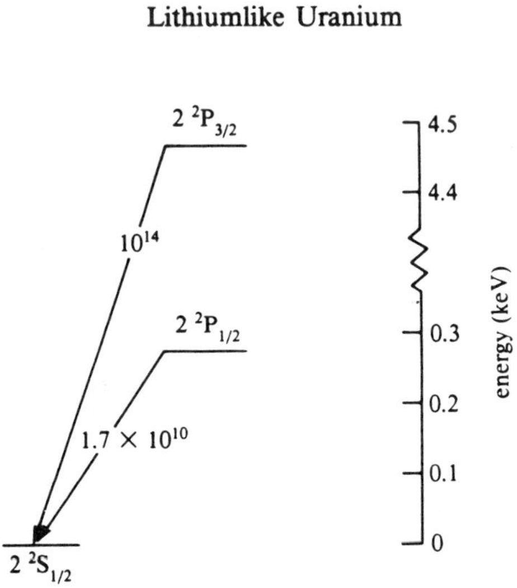

Fig. 4 - n=2 levels of lithiumlike uranium. Values of the energy levels were obtained from Ref. 17. Decay rates from Ref. 17 are in agreement with hydrogenic values taken from Ref. 14.

Previous Lamb shift experiments are compatible with both the series expansion and numerical calculations to all orders[6,10] in $Z\alpha$. Again, this is because the higher order terms in the self-energy are large only at very high Z and hence they are unmeasured in present QED experiments. A consequence is that a significant deviation from QED at high Z is unobservable in previous experiments and that the difference between the numerical calculation and the series expansion is not well tested.

2P - 2S SPLITTINGS

A low precision measurement of the $2\,^3P_0$ lifetime to determine the $2\,^3P_0$ - $2\,^3S_1$ splitting in heliumlike uranium is in progress[4]. The experiment takes advantage of several features of the heliumlike atom. The $2\,^3P_0$ state decays to the $2\,^3S_1$ state by an electric dipole (E1) decay and to the ground state by a two-photon electric-dipole, magnetic-dipole decay[14]. (Fig. 2). As the E1 decay to the $2\,^3S_1$ state is the dominant decay of the $2\,^3P_0$ state, the $2\,^3P_0$ lifetime is sensitive to the $2\,^3P_0$ - $2\,^3S_1$ splitting. Consequently, a measurement of the $2\,^3P_0$ lifetime plus the calculated E1 matrix element[15] and the calculated E1M1 decay rate[14] may be used to determine the $2\,^3P_0$ - $2\,^3S_1$ splitting. There are no significant QED corrections to the E1 matrix element or to the E1M1 decay rate.

A second feature of this system is that the $2\,^3S_1$ state decays to the $1\,^1S_0$ ground state with a rate of 10^{14} sec^{-1} -- much faster than the $2\,^3P_0$ decay. This allows the $2\,^3P_0$ lifetime to be measured by observing the 96 keV x ray from the $2\,^3P_0$ fed decay of the $2\,^3S_1$ state.

In addition to the Lamb shift terms, the interaction between the two electrons ($\approx$ +330 eV)[16] makes up the rest of the $\approx$ 254 eV $2\,^3P_0$ - $2\,^3S_1$ splitting in heliumlike uranium. The experiment is designed to determine the splitting to about 5 eV, or roughly 10% of the higher order self-energy.

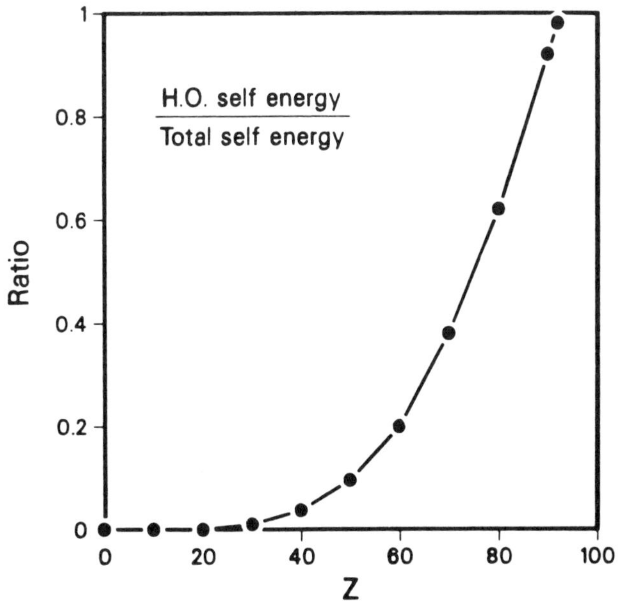

Fig. 5 - Ratio of the higher order terms in the self-energy to the total self-energy obtained by comparing the series expansion value through term A_{70} $(Z\ \alpha)^7$ with a numerical calculation to all orders[6,10] in $Z\alpha$. The series expansion changes sign near Z=60 allowing the ratio of the higher order self energy to the total self energy to exceed 1 for very high Z.

As more intense uranium beams become available, a Doppler-tuned UV spectrometer will be used to make a direct spectroscopic measurement of the ≈ 283 eV $2\,^2P_{1/2}$ - $2\,^2S_{1/2}$ transition in lithiumlike uranium[17] (or the $2\,^3P_0$ - $2\,^3S_1$ transition in heliumlike uranium). The goal is to achieve a precision of 0.2 eV (0.4% of the self-energy). Because the $2\,^2P_{1/2}$ - $2\,^2S_{1/2}$ transition occurs within the ground state manifold in lithiumlike uranium (Fig. 4), and contains fewer m_J states than n=2 heliumlike uranium, the lithiumlike $2\,^2P_{1/2}$ - $2\,^2S_{1/2}$ transition is expected to be at least an order of magnitude brighter than the heliumlike $2\,^3P_0$ - $2\,^3S_1$ transition. In addition, lithiumlike uranium can be produced at lower velocities than heliumlike uranium which will result in a relatively lower background from bremsstrahlung in the target. These are strong experimental incentives to choose lithiumlike ions in preference to heliumlike ions for a high-accuracy direct spectroscopic measurement.

In order to interpret a high-accuracy measurement of the $2\,^2P_{1/2}$ - $2\,^2S_{1/2}$ splitting of lithiumlike uranium in terms of the self-energy, theory for the $2\,^2P_{1/2}$ - $2\,^2S_{1/2}$ splitting in lithiumlike uranium must be improved. For lithiumlike uranium, Kim[18] has estimated that present calculations are accurate at the level of a few eV. For heliumlike atoms, agreement between theory[19] and experiment at low Z suggests that for heliumlike uranium present theory is reliable at about the 1 eV level. For hydrogenlike-, heliumlike- and lithiumlike- uranium, the uncertainty in the nuclear radius contributes an uncertainty of about 1 - 2 eV for the 2P - 2S transitions (and perhaps 5 - 10 eV for transitions involving the 1S ground state)[7,20]. Measurements on muonic uranium could greatly reduce the uncertainty in the nuclear size correction.

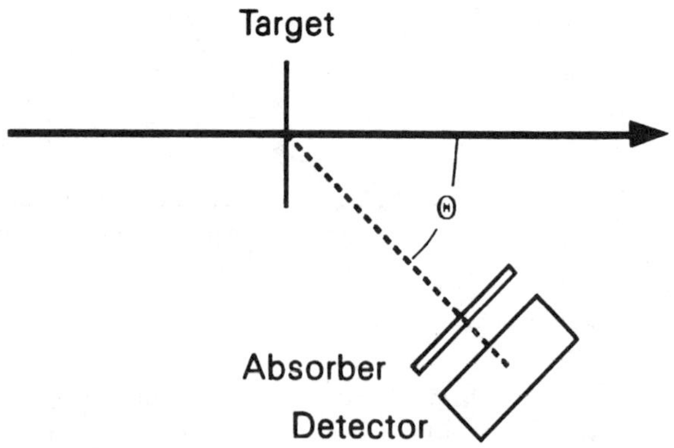

Fig. 6 - Schematic diagram of a Doppler tuned spectrometer.

One method of measuring the $2\,^2P_{1/2}$ - $2\,^2S_{1/2}$ splitting in lithiumlike uranium is with Doppler tuned UV spectrometer[21], shown schematically in Fig. 6. A detector views the UV photons from the decays in flight of the $2\,^2P_{1/2}$ state with a photon absorber imposed between the beam and a collimated detector. The detector-beam angle at which the photons are Doppler shifted across the K-edge of the absorber is found from the sharp change in the number of photons reaching the detector as a function of the beam - detector angle Θ. A photon with an energy of ω_0 in the rest frame of the uranium will be seen with an energy of ω_{LAB} where in the rest frame of the laboratory, where:

$$\omega_{LAB} = \omega_0 \frac{\sqrt{1-\beta^2}}{1-\beta\cos\Theta} \qquad (2)$$

where β = v/c and v is the beam velocity. If the absorber and beam velocity can be chosen to such that the photons are Doppler shifted across the K-edge at $\sin\Theta = \beta$, then the measurement is much less sensitive to uncertainties in the beam velocity.

At 400 MeV/nucleon ($\beta = 0.7$), and $\Theta = 45^o$ an angular displacement of 1 mR produces a Doppler shift of 1×10^{-3}. An angular divergence and horizontal beam width of 5 mR and 0.6 cm respectively, and a detector collimated to view 6 mR, located 1 meter from the beam yields an instrumental line width of about 5 eV. Splitting the line width to 1 part in 25 yields a 0.2 eV measurement.

QED CONTRIBUTIONS TO MAGNETIC MOMENTS OF BOUND ELECTRONS

In addition to the QED contribution to the mass of an electron in a Coulomb field (Lamb shift) there is also a QED contribution to the g-factor of the electron in a Coulomb field. This contribution is a bound state effect and is not tested by experiments which measure the g-factor of a free electron. The effect is observable in the hyperfine structure[22,23] and g-factor[24] of hydrogenlike atoms.

The QED contribution to the electron g-factor in a Coulomb field is tested in the hyperfine structure of hydrogen[23] and the hyperfine structure of muonium[23,25] and in the g-factor of the ground state of hydrogen[26]. (These experiments are discussed in the next section.) To the best of my knowledge it has not been observed for Z >1.

For the hyperfine splitting of hydrogenlike atoms the calculated terms are[22,23]:

$$E_F \frac{\alpha}{\pi} \Big[C_1(Z\alpha) + C_2(Z\alpha)^2 \ln^2(Z\alpha)^{-2} + C_3(Z\alpha)^2 \ln(Z\alpha)^{-2} + C_4(Z\alpha)^2 + \textit{higher order terms} \Big] \qquad (3)$$

The contribution to the total hyperfine splitting of the $Z\alpha$ and $(Z\alpha)^2$ terms at different Z computed from Eq. 3 is given in Table I. Terms of higher order than $(Z\alpha)^2$ have not yet been calculated.

Table I. Bound state QED contributions to hyperfine splitting

Z	$C_1(Z\alpha)$	$C_4(Z\alpha)^2$
1	1×10^{-4}	2×10^{-6}
19	2×10^{-3}	7×10^{-4}
81	8×10^{-3}	1×10^{-2}

The term of order $(Z\alpha)^2$ contributes about 1% of the hyperfine splitting at Z=81 (the anomalous magnetic moment of the free electron contributes roughly 0.1%). In addition, at Z=81, the $(Z\alpha)^2$ term is larger than the lower order $Z\alpha$ term. At very high Z terms of order $(Z\alpha)^3$ and higher could be larger than the lower order terms. The calculation of higher order terms presents a challenge because it is necessary to consider the energy of the electron bound by both strong Coulomb and magnetic fields[27].

The g_J factor of a bound electron also has QED contributions which are not present for a free electron and which become relatively large at high Z (Ref. 24). The leading term is $\alpha/\pi\,(Z\alpha)^2$ which contributes 3×10^{-8} in hydrogen and 3×10^{-4} in hydrogenlike uranium. The relative contribution to the g_J factor is smaller and of higher order than for the hyperfine splitting. This is because the g_J involves a uniform external magnetic field where as the hyperfine splitting arises in the magnetic field singularity at the nucleus.

HYPERFINE STRUCTURE AND G_J EXPERIMENTS

Tests of the QED contribution to the hyperfine splitting of an electron bound in a Coulomb field are limited in hydrogen at a few ppm due to the uncertainty in the proton polarizability and in muonium to a few tenths of a ppm due to uncertainties in the muon mass and fine structure constant[23]. These experiments test the term of order $(Z\alpha)^2$ to about 10% and are probably insensitive to higher order terms. By comparison the leading term in the Lamb shift in hydrogen is tested to about 10 parts-per-million.

Measurements of the g_J in the ground state of hydrogen[26] achieved a precision of 1×10^{-8} which tests the leading order term to about 30%. Experiments[28] in He^+ are not yet of sufficient sensitivity to see the contribution.

Measurements of the ground state hyperfine structure of hydrogenlike thallium using storage rings have been proposed by Bemis[29], and fixed target experiments have been proposed by Bemis and Gould[30]. The ground state hydrogenlike thallium (Fig. 7) (I = 1/2) F=1 - F=0 transition energy is calculated to be 3800 A° without QED corrections and the magnetic dipole decay (M1) rate for F=1 → F=0 is ≈ $10^3 s^{-1}$. Confinement of hydrogenlike thallium in a storage ring would then produce a spectra from the F=1 → F=0 allowed M1 decay and optical spectroscopy would be used to determine the ground state hyperfine interval. For a 15 meter radius ring capturing 10^9 ions per second, a highly collimated detector which intercepted 0.01% of the decay photons over 10 cm would detect 100 counts/sec. With 1 mR angular acceptance the Doppler width is of the order of 0.1% of the transition energy, which would immediately allow a measurement of the $(Z\alpha)^2$ term to 1% and almost certainly a sensitive test of higher order terms.

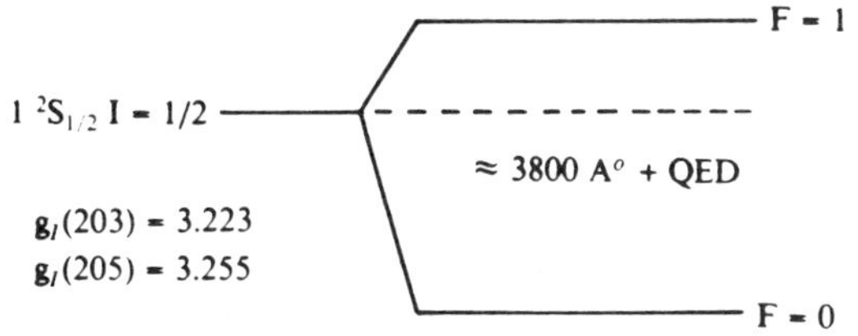

Fig. 7 - Hyperfine splitting of hydrogenlike thallium

The finite nuclear size correction to the hyperfine splitting is about 20% at Z=81. The ≈ 1% uncertainty in the nuclear radius presently limits a comparison between theory and experiment to a few- tenths of a percent. To utilize the precision which might be attained in a storage ring experiment a more precise value for the nuclear radius will be needed. Muonic atom experiments should be able to provide an improved value for the radius.

The g_J factor is much less sensitive to the finite nuclear size correction. It appears that in lowest order there is no nuclear size correction to the non-QED part of the g_J factor[31]. If the contribution of the nuclear size to the QED part is the same order as for the hyperfine splitting, then the nuclear size uncertainty is not a serious limitation to a g_J measurement.

An atomic beam resonance technique has been proposed for a Zeeman effect measurement of the g_J factor of hydrogenlike uranium[32] and an optical resonance measurement of the hyperfine splitting of hydrogenlike thallium[30]. The measurements require the production of a beam of polarized hydrogenlike atoms and analyzing the polarization to detect a resonance transition in an interaction region placed between the polarizer and analyzer. For g_J measurements the interaction region would consist of a uniform magnetic field and a pair of rf driven loops, while for hyperfine splitting measurements the interaction region would contain a high power LASER.

Experiments are being carried out at the Lawrence Berkeley Laboratory's Bevalac to produce polarized ground-state hydrogenlike uranium by capturing an electron onto bare uranium nuclei channeled through a magnetized single-crystal ferromagnetic foil. The analyzer would be a second foil. As the Pauli principle prohibits capture of a second electron in to the same ground state, the ratio of heliumlike to hydrogenlike uranium formed in the second target would depend upon whether the hydrogenlike uranium underwent a spin-flip transition between foils. At relativistic energies radiative electron capture (REC), the inverse of the photoelectric effect, becomes a significant capture mechanism.

Unlike nonradiative capture, which preferentially captures the K-shell electrons from the target, REC has a more equal probability of capturing the valence electrons. The valence electrons are aligned in a ferromagnetic target and channeling the ion enhances the probability of capturing them. The technique works at very low energies where K-shell capture is suppressed and has been used to produce polarized deuterium beams[33].

ULTRARELATIVISTIC PAIR PRODUCTION

The cross section for producing electron - positron pairs from the Coulomb field of two colliding (bare) nuclei is given for the limiting case of kinetic energies much larger than the electron (lepton) rest mass by[34]:

$$\sigma_{pair} = (28/27\pi)\, \alpha^2 Z_1^2 Z_2^2 r_0^2 \log^3\gamma \qquad (4)$$

where $\gamma = (1-\beta^2)^{-1/2}$ with $\beta = v/c$ and Z_1 and Z_2 are the nuclear charges, and r_0 is the classical electron radius.

Table II shows the cross sections computed from Eq. 1 for producing electron - positron pairs in uranium - uranium collisions at collider energies.

Table II. Pair production cross sections for U^{92+} on U^{92+}

energy/beam GeV/amu	equivalent fixed target energy GeV/amu	cross section kilo-barns
5	70	8
10	240	17
20	880	31
40	3400	54
100	20400	98

A transfer line between a tandem Van der Graff and the Brookhaven alternating gradient synchrotron (AGS) is presently under construction. With the addition of a proposed booster synchrotron[5], very heavy ions will be accelerated to energies of over 10 GeV/nucleon. Ultimately the AGS may serve as an injector for a 100 GeV/nucleon on 100 GeV/nucleon relativistic heavy ion collider[35]. For a 100 GeV/nucleon collider operating with a luminosity of $10^{27}\,cm^{-2}\,s^{-1}$ bare uranium nuclei, about 10^{6} electron - positron pairs per second are produced in the collision region. As number of muon pairs scales roughly as $(m_e/m_\mu)^2$, then in the limit of point nuclei, several thousand muon pairs would also be produced. (Additional discussion of atomic physics effects in relativistic heavy ion colliders may be found in Ref. 35,36.)

It is possible for the electron produced in pair production, to be captured into the K-shell of one of the uranium atoms which produced the pair. Classically, capture occurs if the electron is found within the potential well of the Coulomb field and has a kinetic energy less than the K-shell binding energy. In uranium this means within ≈ 580 fm and a kinetic energy of less than 130 keV. (If the two uranium are closer than 580 fm, then the binding energy of the combined system will be larger than 130 keV. In collisions at energies close to the Coulomb barrier, uranium nuclei which approach to within 35 fm have a combined binding energy in excess of 1 MeV.)

Experiments can be done to look for capture from pair production in fixed target experiments on ≈ 15 GeV/nucleon gold (Z=79). At this energy radiative and non-radiative capture cross sections for gold in a uranium target are both roughly equal and are a few parts in 10^4 of the K-shell ionization cross section[37]. The fraction of one-electron ions emerging from a uranium target would increase by 50% if 1% of the electrons from the pair production were captured by the gold projectile.

ACKNOWLEDGMENTS

I am happy to acknowledge the help given by, and to thank, among others: Dr. Curtis Bemis Jr., Professor Gordon Drake, Professor Peter Mohr, Mr. Charles Munger, and Professor Jonathan Sapirstein, This work was supported by the Director, Office of Energy Research; Office of High Energy and Nuclear Physics, Nuclear Science Division of the U.S. Department of Energy under Contract No. DE-AC-03-76SF00098.

REFERENCES

[1]H. Gould, D. Greiner, P. Lindstrom, T.J.M. Symons, and H. Crawford, Phys. Rev. Lett. 52, 180 (1984) (Errata- Phys. Rev. Lett. 52, 1654 [1984]).

[2]J.R. Alonso *et al*, Science 217, 1135 (1982).

[3]C. Munger and H. Gould, Bull. Am. Phys. Soc. 30, 860 (1985).

[4]H. Gould and C. Munger (private communication).

[5] "Proposal for a 15A-GeV Heavy Ion Facility at Brookhaven" Brookhaven National Laboratory Report No. BNL 32250 (1983).

[6]P.J. Mohr, Phys. Rev. A26, 2338 (1982).

[7]see for example, P.J. Mohr, Atomic Dat. and Nuclear Dat. Tables 29, 453 (1983).

[8]A wide range of strong field QED experiments are discussed in S.J. Brodsky and P.J. Mohr, "Quantum Electrodynamics in Strong and Supercritical

Fields" in *Topics in Current Physics: Quantum Electrodynamics in Strong and Supercritical Fields*, edited by I.A. Sellin (Springer, Berlin, 1978) Vol. 5, p.3.

[9]There is a near degeneracy between the $2\,^1S_0$ and the $2\,^3P_0$ states in helium-like uranium (Fig. 3). It is intriguing to consider a measurement of the $2\,^1S_0$ - $2\,^3P_0$ slitting since even a large fractional error yields a very sensitive test of the Lamb shift contribution. (The uncertainty in the energy levels is a few eV, too large to determine how close the levels really are.) Since the $2\,^3P_0$ state decays with a lifetime of ≈ 50 ps, any mechanism which produced a significant mixing of the $2\,^3P_0$ and $2\,^1S_0$ states would produce an experimentally observable signal. However no single photon transition can connect these j=0 states for atoms with no hyperfine splitting, while in atoms with large hyperfine effects, the $2\,^3P_0$ state is no longer long-lived.

[10]A.M. Desiderio and W.R. Johnson, Phys. Rev. A3, 1267 (1971); G.W. Erickson, Phys. Rev. Lett. 47, 780 (1971); K.T. Cheng and W.R. Johnson, Phys. Rev. A14, 1943 (1976); W.R. Johnson, and G. Soff, "The Lamb-Shift in Hydrogenlike Atoms $1 \leq Z \leq 110$", Submitted to Atomic Dat. and Nuclear Dat. Tables.

[11]For a discussion ot the series expansion and values for the coefficients see for example, P.J. Mohr, Ann. Phys. (N.Y.)88, 26 (1974); J. Sapirstein, Phys. Rev. Lett. 47, 1723 (1981).

[12] see for example, S.R. Lundeen and F.M. Pipkin, Phys. Rev. Lett. 46, 232 (1981).

[13]H. Gould and R. Marrus, Phys. Rev. A28, 2001 (1983); O.R. Wood II, C.K.N. Patel, D.E. Murnick, E.T. Nelson, M. Leventhal, H.W. Kugel, and Y. Niv, Phys. Rev. Lett. 48, 398 (1982); H.D. Sträter, L. von Gerdtell, A.P. Georgiadis, D. Müller, P. von Brentano, J.C. Sens, and A. Pape, Phys. Rev. A29, 1596 (1984). P. Pellegrin, Y. El Masri, L. Palffy, and R. Priells, Phys. Rev. Lett. 49, 1762 (1982);

[14]G.W.F. Drake, Bull. Am. Phys. Soc. 29, 781 (1984).

[15]M. Hillery and P.J. Mohr, Phys. Rev. A21, 24 (1980).

[16]C.D. Lin, W.R. Johnson and A. Dalgarno, Phys. Rev. A15, 154 (1977); W.R. Johnson and F. Parpia, private communication.

[17] K.T. Cheng, Y.-K. Kim, and J.P. Desclaux, At. Data Nucl. Data Tables, 24, 111 (1979); Y.-K. Kim and J.P. Desclaux, Phys. Rev. Lett. 36, 139 (1976); see also L. Armstrong, Jr., W.R. Fielder, and D.L. Lin, Phys. Rev. A14, 1114 (1976). (The one electron QED corrections are included in the 1979 paper but not the 1976 papers.)

[18] Yong-Ki Kim, (private communication).

[19]S.P. Goldman and G.W.F. Drake, J. Phys. B17, L197 (1984) (Lawrence Berkeley Laboratory report LBL-16466); G.W.F. Drake, (private communication); see also G.W.F. Drake, "Quantum Electrodynamic Effects in Few-Electron Atomic Systems" in *Advances in Atomic and Molecular Physics* edited by D. Bates and B. Bederson (Academic, N.Y., 1982) Vol. 18, p. 399.

[20] R. Hofstadter and H.R. Collard, in *Nuclear Radii*, Group I, Vol. 2 of the Landolt-Börnstein new series, H. Schopper, ed., (Springer-Verlag, Berlin, 1967), p. 21.

[21]R.W. Schmieder and R. Marrus, Nucl. Instrum. Methods, 110, 459 (1973); R.W. Schmieder, Rev. Sci. Instrum. 45, 687 (1974).

[22]S.J. Brodsky and G.W. Erickson, Phys. Rev. 148, 26 (1966)

[23]J.R. Sapirstein, Phys. Rev. Lett. 51, 985 (1983).

[24]H. Grotch, Phys. Rev. Lett. 24, 39 (1970); H. Grotch and R. Hegstrom, Phys. Rev. A4, 59 (1971).

[25]F.G. Marion ***et al.***, Phys. Rev. Lett. 49, 993 (1982).

[26]J.S. Tideman and H.G. Robinson, Phys. Rev. Lett. 39, 602 (1977).

[27]S.J. Brodsky, and J. Primack, "The Electromagnetic Interaction of Composite

Systems" Ann. Phys. (N.Y.) 52, 315 (1969).
[28]C.E. Johnson and H.G. Robinson, Phys. Rev. Lett. 45, 250 (1980).
[29]C.E. Bemis Jr. (private communication).
[30]H. Gould and C.E. Bemis Jr., Bevalac Proposal 719H *Polarized Hydrogenlike Uranium*, unpublished.
[31]A detailed treatment of the Zeeman effect in hydrogenlike atoms is found in H.A. Bethe and E.E. Salpeter, *Quantum Mechanics of One- and Two- Electron Atoms* (Springer-Verlag, Berlin, 1957), Chap III.
[32]H. Gould, C. Munger, and C.E. Bemis Jr., (private communication).
[33]M. Kaminsky, Phys. Rev. Lett. 23, 819 (1969); C. Rau, and R. Sizmann, Phys. Lett. 43A, 317 (1973); L.C. Feldman, D.W. Mingay and J.P.F. Sellschop, Radiat. Eff. 13, 145 (1972).
[34]E.J. Williams, Kgl. Dansk. Vid. Selsk. 13, No. 4 (1935); C.F. v. Weizsäcker, A. Phys. 88, 612 (1934); H.J. Bhabha, Proc. Roy. Soc. A152, 559 (1935); H.J. Bhabha, Proc. Camb. Phil. Soc. 31, 394 (1935); L. Landau and L. Lifshitz, Phys. Zs. Sov. U. 6, 244 (1934); Y. Nishina, S. Tomonaga and M. Kobayasi, Sci. Pap. Ins. Phys. Chem. Research, Japan 27, 137 (1935).
[35]"RHIC and Quark Matter: Proposal for a Relativistic Heavy Ion Collider at Brookhaven National Laboratory" Brookhaven National Laboratory Report No. BNL 51801 (UC-28) [Particle Accelerators and High Voltage Machines - TIC-4500], Aug. 1984.
[36]H. Gould, "Atomic Physics Aspects of a Relativistic Nuclear Collider" Lawrence Berkeley Laboratory Report No. LBL-18593 (UC-28), Nov. 1984.
[37]W.E. Meyerhof, R.Anholt, J. Eichler, H. Gould, Ch. Munger, J. Alonso, P. Thieberger, and H.E. Wegner, to be published in Phys. Rev. A32 (Dec 1985); H.A. Bethe, Ann. Phys. (Leipzig) 5, 325 (1930); C. Moller, Ann. Phys. (Leipzig) 14, 531 (1932).
[38] W. Gordon, Z. Phys. 48, 11 (1928); C.G. Darwin, Proc. Roy. Soc. London Ser A118, 654 (1928).
[39]S.P. Goldman and G.W.F. Drake, Phys. Rev. A24, 183 (1981).
[40]F.A. Parpia and W.R. Johnson, Phys. Rev. A26, 1142 (1982).

ACCURATE SPECTROSCOPY OF SINGLE-ELECTRON AND SINGLE-VACANCY IONS

Richard D. Deslattes
Quantum Metrology Group
National Bureau of Standards
Gaithersburg, MD 20899

ABSTRACT

This report focuses on one-electron and one-vacancy spectroscopy in the X-ray region where data are of adequate or nearly adequate quality to be of possible interest to this workshop, i.e. data where relativistic and QED effects are not merely noticeable but where measurements of significance may be found or at least hoped for. Several experimental difficulties are discussed including production of clean and interpretable spectra; securing appropriate wavelength normalization; and problems of spectator electrons and spectator vacancies. Available experimental results are summarized and an attempt made to combine information from single-electron and single vacancy spectra. Brief discussion of some directions in which future progress may be anticipated is also included.

KEY WORDS

X-ray spectra, hydrogen-like ions; wavelength standards; Lamb-shift.

INTRODUCTION AND OVERVIEW

This contribution to the workshop is mainly from the experimental side and focusses on information currently available or obtainable in the near term using existing technologies. On the other hand, it seems useful to try to communicate certain perspectives about the current situation as well as supportable speculations about future developments.

In the case of what is now known about high Z systems, there is available much more information about the more complex case of single vacancy spectra than about single electron spectra. Because of spectator-vacancy effects and availability of alternative (Auger) decay modes, vacancy spectra are not only more difficult to interpret satisfactorily but they are also more difficult to measure precisely. On the other hand, single-electron spectra (which except in certain cases are also troubled by spectators, in this case electrons) are much less widely available especially towards higher Z.

Because of the the above-noted complexities, spectroscopic limitations and the lack of experimental data for high Z 1-electron systems, I include in this report some discussion of scaling

0094-243X/85/1360080-14$3.00

rules which appear helpful in allowing one to combine one-electron and one-vacancy data.

Beside the need to summarize and organize the present situation this workshop is an opportunity and invitation to consider possible future developments. Most issues come from the theoretical side but there are also a few from the experimental aspect whose nature and significance need to be indicated. For example, it is clear that we have, in principle, access to spectral lines whose width can be chosen to be smaller than (almost) any preassigned value. Because the traditional production procedures for high Z, few electron systems have involved substantially unmodified heavy ion machines, limitations due to Doppler effects have predominated but this situation need not persist. There are opportunities, for instance using decelerated beams of naked ions, to bring Doppler and spectator problems under control at the same time. This possibility, which has already been demonstrated in the case of "once through" accelerators, will be considerably enhanced with the advent of heavy ion cooler rings and possibly by developments in the area of advanced technology ion sources.

When the above possibilities are realized there will be a need for improved spectroscopy. Aside from other problems of less obdurate character, the use of X-ray lines as transfer standards or markers surely stands in the way of measurement below the 1 ppm level. There are, however, certain new developments regarding both γ-ray spectroscopy and the fabrication of crystal filters that appear capable of removing these limitations as will be described below.

One then comes finally to a summary proposition namely that it appears possible to remove technical obstacles to optically based spectroscopy in the X-ray region to the level of a few parts in 10^8. Presumably at least some measurements will be carried out taking advantage of these possibilities. What is then being studied? Perhaps the most obvious and difficult response acknowledges that one's interest increases with increasing nuclear charge which **inter alia** entails that "finite nuclear size" effects are no longer small corrections but must be explicitly studied in themselves including not only detailed intra-nuclear charge distributions but also their polarization response. This is surely a difficult matter but is a least made less hopeless by the general availability of different isotopic forms for most nuclear charge values. Is it worth the corresponding large effort? I suspect that that is to be one of the central issues of this workshop.

REFERENCE THEORETICAL DATA BASES

Since the workshop focuses on theoretical issues it would be out of place for an experimentalist's comments to be other than modest and brief. Firstly, it is a distinct convenience if we can have ready access to effectively all-Z calculations both for inner shell vacancy energies and for one-electron levels. Although there is evidently some minimum requirement for the refinement of these exercises, their ultimate correctness is not an important question provided that they are widely available or readily calculable. Fortunately this happy situation obtains for both single-electron and single-vacancy spectra due to relatively recent work in both cases. (There is also a growing literature of 2-electron calculations of which brief mention will follow). This section aims to give a skeletal indication of the reference calculations and point out certain common features which emerge under appropriate scaling.

In the case of one-electron systems, the reference calculations are due to Walter Johnson and Gerhard Soff.[1] These are in the form of extensive tables that are presently in the publication process but have been widely and generously circulated by the authors. Salient features include the use of well-defined nuclear charge distribution including diffuse boundary with universal $A^{1/3}$ scaling. The Dirac levels were obtained by numerical integration and dominant QED corrections evaluated according to Mohr's procedures. Numerical accuracy is adequate for all experimental results in the region $Z > 10$. In most cases transition energies are more readily accessible to experiment than are level values. It is thus a modest additional convenience that principal transition energies are also tabulated. It is a minor misfortune that greatest interest will likely center on transitions having upper state principal quantum numbers outside the group calculated as will be described below.

In the case of inner vacancy energy levels and transition arrays, an all Z tabulation at the DHFS level was prepared almost 10 years ago by Huang, et. al.[2] Regions of this were subsequently revised to include various improvements but the whole has not been recast. On the other hand approximate multi-configuration self consistent field calculations have been made available as Fortran programs, especially by Ian Grant and collaborators.[3] My colleague, Ernest Kessler, who is also of primarily an experimental persuasion, was able (with occasional advice from Ken Dyall, Yong-Ki Kim and Andrew Weiss) to get this program operating on the recently departed NBS computer and generate such transition energies as corresponded with available accurate data, as well as those needed to examine certain possible trends. As Kessler and I have discussed at length elsewhere,[4] experimental transition energies (allowed emission line spectra) are far more accessible and reliable over broad ranges of Z than are procedures for the separate estimation of energy levels. Once again, the Grant, et. al.[3] programs permit convenient use of realistic nuclear models and include QED corrections in an approximate way not inconsistent with the Johnson and Soff approach.

To conclude this section, it is appropriate to give an overview of the main results and potential difficulties in both forms of spectra. The main energies and QED corrections for single electron levels and for single vacancy states are different in size at each Z mainly because of the effect of electron screening. Rather than attempt to estimate the effect of this by invoking screening rules, I would simply call attention to the empirical fact that, when expressed as a fraction of the transition energies, the main Q.E.D. effects are approximately equal for one-vacancy and one-electron spectra at the same Z. This is illustrated in Fig. 1 where the self-energy and vacuum polarization effects for $K\alpha_1$ transitions are shown separately and in combination. The estimates for these contributions to the

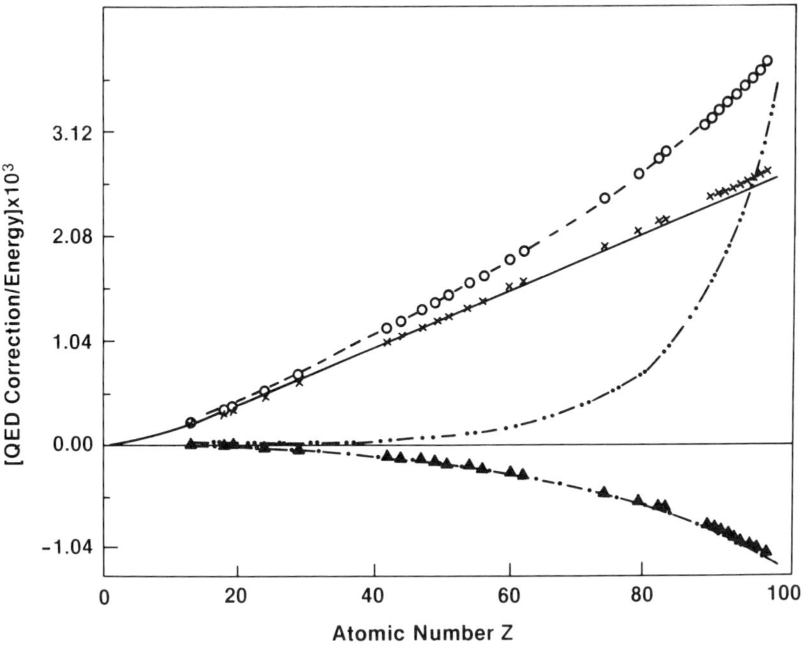

Fig 1 Variation with Z of QED corrections for both single-electron (continuous curves) and single vacancy (points) levels; self-energy is shown by - - -, 0; vacuum polarization by - • -, ▲; sum of previous two terms ——— , +. Also shown are representative data for nuclear finite size effect - •• -.

one-vacancy transitions were obtained using the Grant, et al. program. Also shown in Fig. 1 is the behavior of the main nuclear finite size effect. Clearly when one gets to the high Z end, this is a very big problem.

From a somewhat global, but still experimental point of view, there is one other question, namely to what extent are the QED and finite nuclear effects measurable? Aside from technical problems (which are non-trivial, of course), one ultimately confronts some kind of Q-value - in this case a ratio of shift,

S, to line-width, Υ. For inner-vacancy transitions, line-widths vary slowly from 0.3×10^{-3} in the range $10 < Z < 30$ to nearly 1×10^{-3} in the region around $70 < Z < 90$. The variation of Q value with Z is shown in Fig. 2 by the dashed line for Kα X-ray lines. In the case of Lyman transitions in one-electron ions

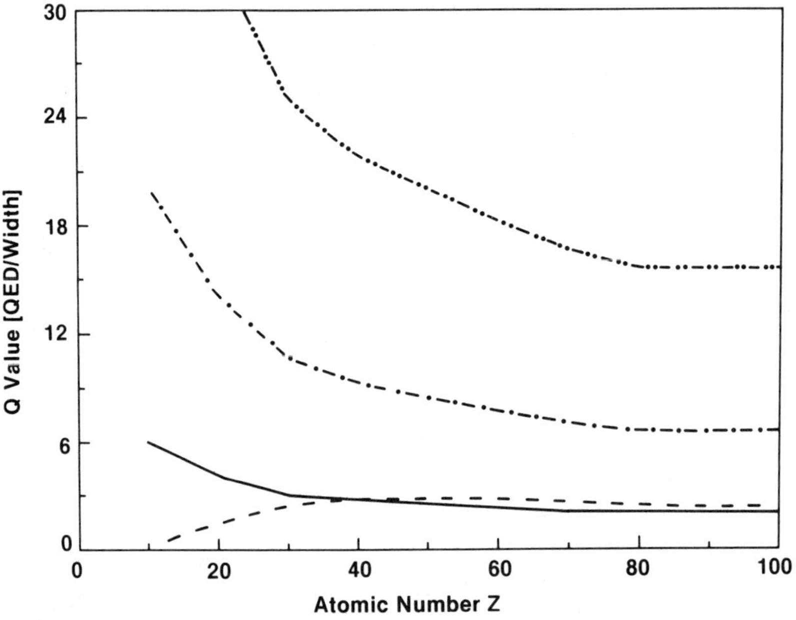

Fig. 2 Approximate Lamb shift Q values ($Q = S/\Upsilon$, see text) for Kα ---- lines and several members of the Lyman series, 2 p ——— , 3p - • - and 4p - •• -.

the situation is rather different even for the α lines which have smaller linewidths and correspondingly greater Q values in all cases. For the case of higher members of the Lyman series, β, Υ, δ etc. the situation improves consistently (in principle) as is also illustrated in Fig. 2. It is not illustrated but the spontaneous emission line 2s → 1s is of adequate intensity for experiment study from the region of $Z \approx 40$ upward. These lines are exceedingly narrow yielding Q values $\sim 10^4$ near $Z = 50$. It thus appears assured that when other matters are adequately cared for there is no conceptual limit to the accuracy with which one-electron transition energies and hence QED shifts can be measured.

SUMMARY OF EXPERIMENTS

In what follows the reader may note selective de-emphasis of several important groups of experiments. Aside from matters of space and time there are reasons which are roughly the following. Hydrogen, muonium, positronium are omitted since the workshop focuses on $Z \gg 1$. Experiments in systems $Z \gg 1$ structured in analogy to Lamb and Retherford's work on hydrogen but using lasers in place of RF to deplete metastable 2s populations are largely omitted since (1) they are already difficult below Z = 20 and become more so with increasing Z and (2) the Q values potentially possible in these experiments are always smaller than the worst case involving 1s levels (namely Lyman α) since they share a common line width, γ, while n^{-3} scaling of the dominant self energy contribution gives an eight fold increase for the 1s level.

As a second general comment, there is an important problem of wavelength (or energy) normalization (which is largely absent in 2s experiments) that is difficult in experiments involving either 1s single-vacancy levels or 1s single-electron levels. This problem is an essential one since all that one can possibly mean by a "measurement of the 1s Lamb shift" is the result of taking the difference between a calculated transition energy neglecting QED and observation. To obtain even modestly sensitive results regarding this difference requires both calculation and spectroscopy of considerable delicacy. This level of refinement calls specifically for a robust connection between experimental measurements in the X-ray region and the Rydberg constant since this constant serves as the customary and natural basis for calculation. The Rydberg constant in turn follows from numerical transformation of experimental data on one or more hyperfine components in hydrogen Balmer α. The wavelength (or frequency) reference in all recent experiments has been a 633 nm HeNe laser stabilized with respect to one or another Doppler-free saturated absorption feature in molecular iodine - a so called "iodine stabilized laser". Although totally devoid of fundamental significance, such I_2 stabilized lasers are by now well-connected with all fundamental standards of length and of frequency and are also involved directly or indirectly with several determinations of fundamental constants [5] as well as optical spectroscopy at its highest level.[6]

In establishing a "bootstrap" to X-rays and γ-rays we need as well to begin with such an I_2 stabilized laser. A diagram outlining our efforts in this regard is indicated in Fig. 3. In the first step, the wavelength from an I_2 stabilized laser is used to measure the interplanar spacing of a rather perfect sample of single crystal silicon. Even though there is a significant disagreement between this laboratory and the PTB in Braunschweig concerning interpretation of the numerical value obtained in this step, there is agreement that the controversy can be resolved and that the resulting value will readily achieve

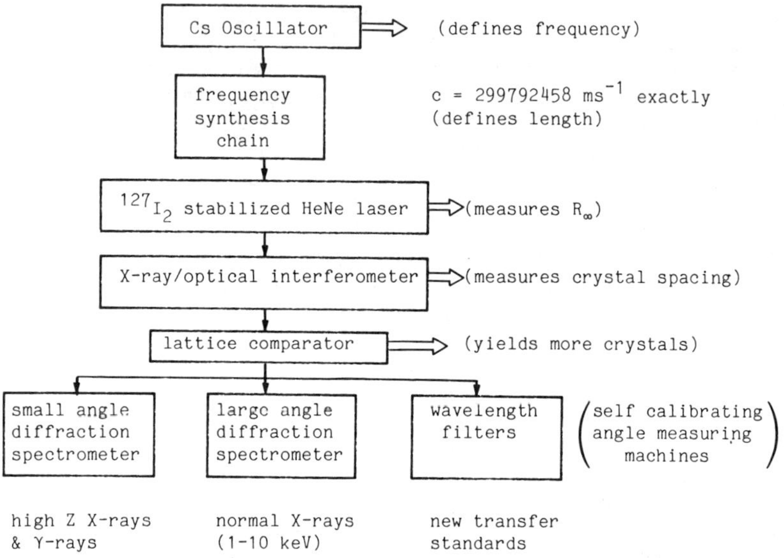

Fig. 3 Schematic diagram of the current bootstrap by which transitions measured in the X-ray region can be connected with the spectrum of atomic hydorgen and in this way with the Rydberg constant.

an accuracy level better than 5×10^{-8}.[7] In the second step we (and PTB colleagues as well) exploit some very fine-scale dynamical diffraction effects to achieve sensitive comparisons between the interferometrically calibrated samples and others whose sizes and shapes are better suited to subsequent use in X-ray and γ-ray measurements. Thereafter, in a third step we (and thus far only we) have made use of sensitive and well-calibrated angle measuring instruments to determine wavelengths in the X-ray and γ-ray region.

Inner shell emission lines of natural (often solid) targets are generally easy to produce with sufficient intensities that "direct" wavelength measurements in the fashion described above are possible. This makes such results readily available both to permit studies of Z systematics in relation to a reference theory or to permit them to serve as markers in experiments on more exotic systems such as one-electron ions. In such cases one encounters rather low intensities which generally require efficient imaging (focussing) spectrometers that in turn require use of rather close-lying reference lines. Unfortunately the reference lines will in these cases be less well-defined than the spectral lines of interest leading to limiting accuracies below the threshold of interest. Nuclear γ-ray lines are an attractive alternative since they are quite narrow even in comparison with lines from one-electron ions. Furthermore, at least near large research reactors, sources of sufficient strength to permit direct wavelength measurements are readily available. Unfortunately the number of such γ-lines is rather small and to some extent of diminishing utility toward lower energies because of strong self-absorption and internal conversion.

Because of the above described limitations of both X-ray and γ-ray secondary standards, considerable interest attaches to an alternative procedure labelled in Fig. 3 as synthetic λ-filters. These are relatively easy to construct to yield almost any energy in the range $1 < E < 100$ keV and are at least modestly convenient to calibrate especially if the passed λ is inside an X-ray line. Their output λ is also calculable in terms of crystal spacing and an angle measurement but is limited by index of refraction effects in the lower energy region.

Following this perhaps overly long digression on the standards problem I can turn to what we have in the way of experimental data bases first for X-ray emission lines and subsequently for Lyman α lines in one-electron ions. While X-ray emission line systematics surely is one of the oldest atomic physics pursuits (Mosely), it has not been until rather recent times that results from a wide range of Z became available on a consistent and accurate scale. It transpires that while the dominant improvements occurring over the past 20 years (ca 20ppm) are noticeable at the level where comparison with theory is sensible, the remaining controversy (about 1 ppm) is below the level at which meaningful interpretation can proceed. Data useful for systematic investigations comes at the present time from three sources as Kessler and I have described in more detail elsewhere.[4] Firstly, there are a number of Kα and Kβ lines which we have measured using direct reading instruments and interferometrically calibrated crystals. Since this group includes, **inter alia**, all reference lines used in traditional X-ray measurements, we can extend this first small group to a larger domain by use of selected collections of wavelength ratios especially among those data sets which ranked highly in Bearden's reevaluation.[8] Finally there is a group of X-ray spectra which for certain historical reasons were well connected to γ-ray transitions especially the 411 keV line emitted by the Hg daughter of ^{198}Au. The X-ray spectra involved are K series lines belonging to high Z atoms from thorium through most stable transuranic elements. Since we measured the 411 keV line directly the whole is knit together. There was no attempt in this work to get data on each and every element. Instead our objective was to obtain simply an adequately dense sample over a wide enough range in Z that trends of agreement/disagreement with theory could be established.The extent to which this effort has succeeded will be evident from the comparisons given in the next section.

In the case of one-electron spectra the universe of available data is rather small especially when discussion is restricted (as it is here) to transitions including a 1s final state. Most experiments have been of the "beam-foil" type wherein the nucleus of interest is (mostly) stripped and excited in one or more foils where initial energies are sufficiently high that an appreciable fraction of the initial beam winds up as one-electron ions. Work of this sort has included studies of chlorine,[9,10] argon,[11] iron,[12] krypton,[13] and uranium.[14] In all cases X-ray emission lines served as secondary standards but in no case were these the main limitation. Instead there were (and are) two other complications which interact with one another in an unpleasant way. There is, of course, the Doppler effect which is high on the list of everyone's favorite troubles. Even for observations made near 90° it is much larger than the QED shifts of interest. The second problem is that these experiments are almost incapable of yielding purely single electron ions. Instead, one is confronted with outer-orbit spectator electrons which perturb the emission lines generating noticeable asymmetries and contributing essentially unmodelable fine detail. The unpleasant interaction is that distortions produced by spectator electrons are reduced by use of higher projectile velocities thereby enhancing Doppler problems. At the same time going to lower energies to reduce Doppler problems entails noticeably enhanced asymmetries presumably reflecting greater spectator electron involvement.

One partial remedy for these problems involves use of a primary beam of highly charged heavy projectiles to strip and excite stationary target atoms by the recoil mechanism. The main advantage of this approach is that energy transfer to a target, e.g., argon incidental to a stripping encounter with U^{+66} at 6MeV/amu is small (ca 10 eV or less), thereby effectively eliminating Doppler corrections. The main disadvantage is that many electrons have the opportunity and more than a few have inclination to remain as spectators to the final emission process. Once again, substantial asymmetry and likely unmodelable fine detail conspire to limit interpretable accuracies in such experiments to values from which little of theoretical interest can emerge.

A constructive alternative to the above procedures has been demonstrated which for accidental reasons has thus far failed to deliver results commensurate with its potentiality. In this case, ions are raised to sufficiently high energies that foil or gas stripping yields a substantial fraction of naked ions which are separated from other charge states by magnetic analysis. The naked ion beam is then decelerated to a sufficiently low energy that Doppler corrections become manageable and resonant electron transfer can be used to populate excited levels of interest. In our initial trial, He gas served as an electron donor and observed profiles showed no asymmetries.[10]

DISCUSSION OF RESULTS

In this section I first offer separate comparisons between the reference theoretical estimates and experimental data. Thereafter, I invoke a certain scaling of all available data which appears to allow both single electron data and single-vacancy data to contribute to an overall assessment of what is known today. The issue of "scaling" is not as trivial as it might seem at first since one or another procedure may emphasize different regions and levels of agreement or disagreement. The objective is firstly to achieve a compact presentation that gives an accurate and more or less uniformly sensitive picture over all Z. To a first approximation, since both line widths and spectroscopic quality tend to be slowly varying over wide ranges of Z, the procedure of choice is evident, although different hypotheses about Z dependence of discrepancies might dictate a non-linear Z scaling which is not used here.

Figures 4a and b show the overall x-ray situation for the main K series α and β lines. The vertical axis is the dimensionless ratio between the deviation (theory-experiment) and the theoretical transition energy expressed in ppm. In assessing the

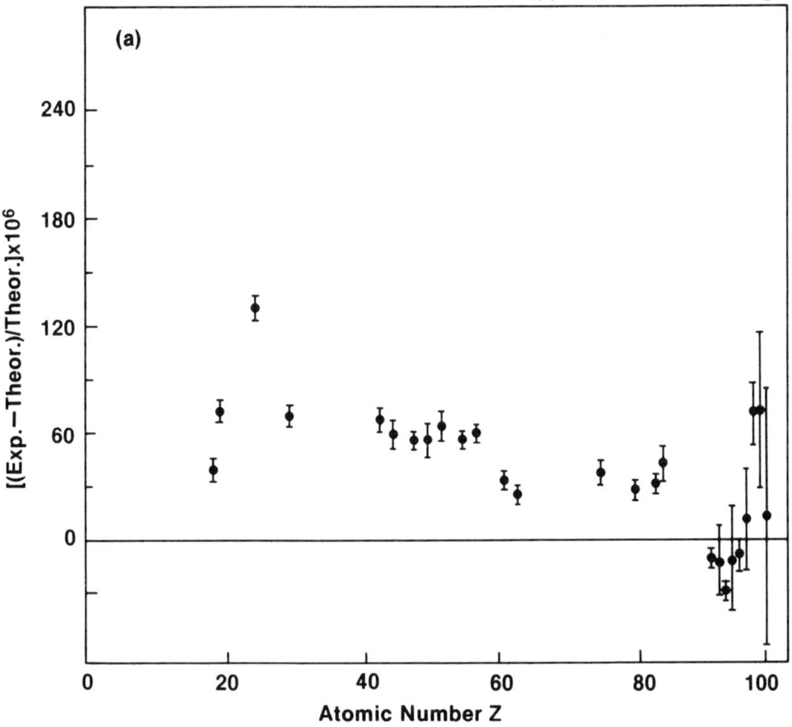

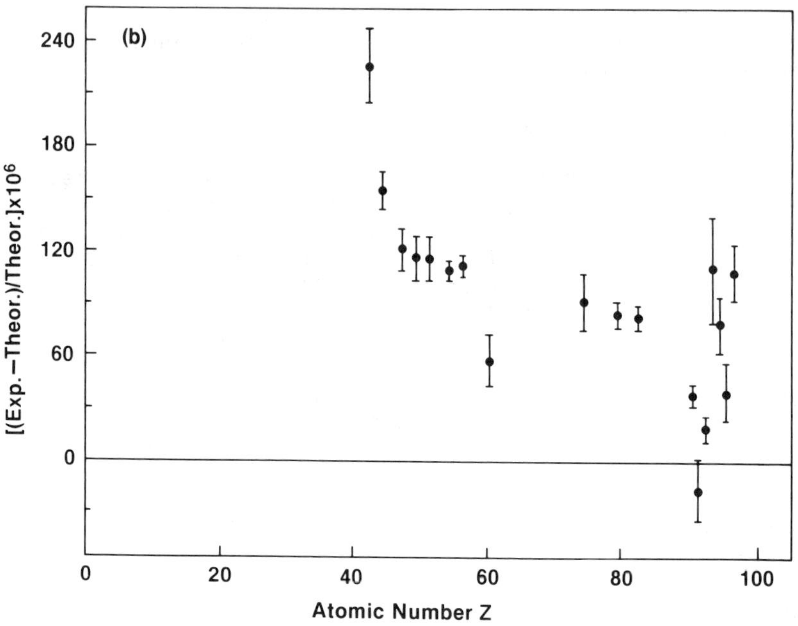

Fig. 4a & b Comparison of experimental results for Kα and Kβ x-ray lines with a reference theoretical data base (see text).

theoretical values for β lines, the dynamical shifts suggested by Chen et al.[15] and by Ohno[16] have not been included. Assuming this problem to have been adequately treated, shifts in the $1s^{-1}$ level would affect all the indicated lines equally whereas difficulties in outer vacancy levels can be inferred from differences between errors shown for α and β lines as can be confirmed by studies of L series spectra involving these same levels and a common outer state. Although further analysis may be informative, the main messages of the data in Fig. 4 are clear: Except for some needed work in the actinide region and possible study of 2s → 1s lines the experimental data base cannot be significantly improved. In most cases it is already at or near technical and interpretational limits. Unless trends become clearer through improved measurements near Z ~ 50 and above Z ~ 90 little will be learned of a fundamental character from continued study of x-ray emission lines. In addition, with all its limitations, the structure and level of approximation in present day relativistic SCF calculations approach those needed to treat x-ray data.

The relatively sparser set of one-electron data is partly illustrated in Fig. 5.

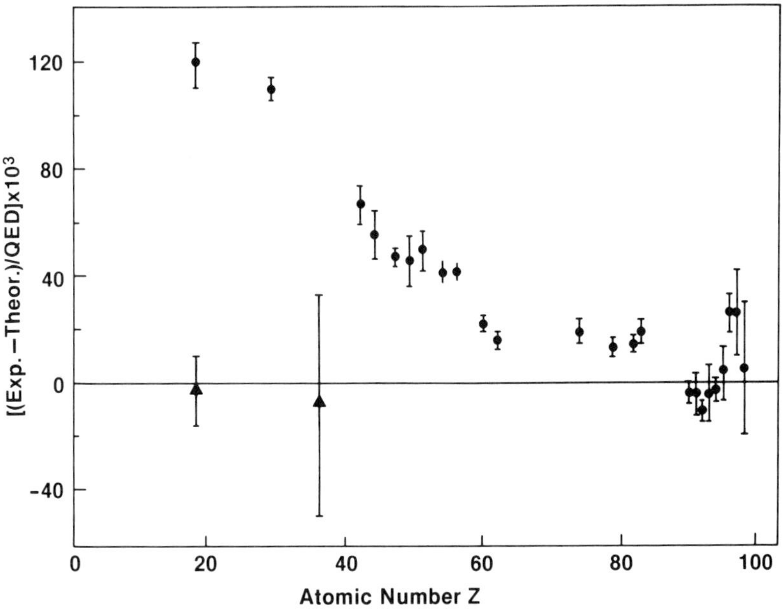

Fig. 5 Lamb shift measurements for one-electron ions (▲) and estimates based on a naive use of theory for the case of inner-shell transitions (●).

Other data on Fe and Cl have estimated errors too large to be of significance in the present discussion. What is displayed is the fractional difference between theory and experiment for the total radiative correction. The reference calculation is that due to Johnson and Soff.[1] The highest Z value is due to the recent work on krypton by Briand, et al.[13], while the most precise point indicated is the recoil experiment of Beyer, et. al.[11] There is no sense in which this data array represents a challenge to theory at the present time. In fact the most sensitive measurement has an imprecision only equal to the difference between the calculations of Erickson and Mohr. On the other hand, if one takes account of the fact that considerably improved accuracy lies close at hand and that newer accelerator facilities, especially heavy ion cooler rings, will permit access to experiments in much higher regions of Z, a challenging situation is likely ahead.

Pending these advances, is there anything which can be seen by an appropriate joining of X-ray data which is already available essentially over all Z < 94 with these results from one-electron experiments? In an attempt to speculate along these lines,

I have added in Fig. 5 the Kα x-ray results expressed in the same fashion as the one-electron results. One properly hesitates to do this since the inner vacancy data depends not only on electrodynamic corrections and Dirac theory but contain as well the machinery needed to account for the many electron system. Nontheless if this is done, the picture shown emerges.

While fairly complex analysis would be required for any attempt to sort out contributions to discord or excuses for agreement, one overall surmise seems justified by inspection. If one neglects at first weakness in the many-electron machinery, Fig. 5 may be taken to indicate levels of imprecision below which one-electron experiments must penetrate before they have the possibility to reveal weaknesses in our treatment of the fundamental interactions. To assert the converse would require wide-ranging cancellations among energy contributions having very different physical origins.

One can, in fact, proceed somewhat farther invoking only the general characteristics of many-electron calculations. For example, the problems (or absence of problems) associated with present-day many-electron calculations are **a priori** unlikely to be in error to a greater extent than the overall comparisons indicate. Accordingly, we should even now be looking for many-electron calculations formally functional at the 10 ppm level. At the same time one cannot fail to recommend that attention be directed to the question of why Kα and Kβ lines behave differently. This must represent something simple such as the effects considered by Chen et al.[15] and by Ohno.[16]

In a broad sense, we appear to be making progress although the difficulties are far from trivial. In fact further efforts toward carefully staged confrontations between theory and experiment appear to be the most likely agent for change.

ACKNOWLEDGEMENT

I wish to express special appreciation for the many contributions made to this work by Ernest G. Kessler, Jr. He participated in many of the studies and has contributed generously to the discussions and illustrations contained in this report.

REFERENCES

1. W.R. Johnson and Gerhard Soff, Atomic Data and Nuclear Data Tables 33 #2 (1985).

2. K.-N. Huang, M. Aoyagi, M.H. Chen, B. Crasemann and H. Mark, Atomic Data and Nuclear Data Tables 18, 243 (1976).

3. I.P. Grant, B.J. McKenzie, P.H. Norrington, D.F. Mayers and N.C. Pyper, Comput. Phys. Commun. 21, 207 (1980).
B.J. McKenzie, I.P. Grant and P.H. Norrington, ibid p233.

4. R.D. Deslattes and E.G. Kessler, Jr., Experimental Evaluation of Inner Vacancy Level Energies for Comparison with Theory in Atomic Inner Shell Physics, B. Crassmann, ed. (Plenum Publishing Co, in press) see also X-84 X-ray and Inner-shell Processes in Atoms, Molecular and Solids, Leipzig, DDR 1984, A. Meisel and J. Finster, eds. (Karl-Marx-Universität, Leipzig) p. 165.

5. Examples can be found in: Precision Measurement and Fundamental Constants-II, B.N. Taylor and W.D. Phillips, eds., N.B.S. Spec. Pub. 612, US GPO (1984).

6. C.J. Sansonetti and W.C. Martin, Phys. Rev. A 29, 159 (1984).

7. Subsequent to contributions which may be found in Ref. 5, additional reports have appeared including: P. Becker, P. Seyfried and H. Siegert , Zeits für Phys. B 48, 17 (1982); H. Siegert, P. Becker and P. Seyfried, ibid., 56, 273 (1984).

8. J.A. Bearden, Rev. Mod. Phys. 39, 78 (1967).

9. P. Richard, M. Stockli, R.D. Deslattes, P. Cowan, R.E. LaVilla, B. Johnson, K. Jones, M. Moran, R. Mann and K.-H. Schartner, Phys. Rev. A 29, 2939 (1984).

10. R.D. Deslattes, R. Schuch and E. Justiniano, Phys. Rev. A, in press.

11. H.F. Beyer, R.D. Deslattes, F. Folkmann and R.E. LaVilla, J. Phys. B 18, 207 (1985).

12. J.P. Briand, M. Tavernier, P. Indelicato, R. Marrus and H. Gould, Phys. Rev. Lett. 50, 832 (1983).

13. M. Tavernier, J.P. Briand, P. Indelicato, D. Liesen and P. Richard, J. Phys. B 18, (1985).

14. H. Gould, preceding contribution to this workshop.

15. M.H. Chen, B. Crasemann, N. Martensson and B. Johansson, Phys. Rev. A 31, 556 (1985).

16. M. Ohno, Phys. Rev. A 30, 1128 (1984).

RECENT WAVELENGTH MEASUREMENTS IN 2- AND 3- ELECTRON SYSTEMS - a brief report -

H. G. Berry

Physics Division, Argonne National Laboratory, Argonne, IL 60441

Abstract

We present 3 recent precision measurements of the $1s2s\ ^3S$ - $1s2p\ ^3P$ wavelengths in the helium-like, two-electron systems Li II, Ne IX and Ti XXI. We also comment on an old (1981) comparison between theory and experiment for the $1s^22s\ ^2S$ - $1s^22p\ ^2P$ transition wavelengths in the lithium-like, three electron isoelectronic sequence.

Introduction

Precision tests of QED and relativistic atomic structure continue to be made in measurements of transition wavelengths and fine structures in two- and three-electron systems, where only approximate hamiltonians of the systems have generally been evaluated. In this pap are presented some measurements using two different techniques for thre different ions of the helium isoelectronic sequence. The measurements presented are to be published elsewhere very shortly, but the results a assembled here for convenience for this workshop. I am grateful to my workers on the various measurements for allowing the publication of the results here.

The three-electron system is also becoming of more interest i making comparisons between experiment and the different relativistic ma body calculations now available. Edlén[1] has made a thorough semi-empirical analysis of this isoelectronic sequence, and has made some comparisons with ab initio theory. In this paper, we point out interesting differences between relativistic Hartree-Fock calculations a experiment for the resonance line transition energy, and also the fine structure of the 2p states.

Results

Riis et al.[2] have made a fast-beam laser resonance fluorescenc measurement in Li II. Using a collinear geometry, the resonance wavelength was measured for both parallel and antiparallel directions of the exciting laser. The two wavelengths were precisely calibrated in terms of specific hyperfine components of saturated absorption spectra o iodine. Measurements on three different hyperfine components led to results on the two fine structure separations and the absolute wavelengt separations. The results are consistent with previous experiments of

0094-243X/85/1360094-6$3.00 Copyright 1985 American Institute of Physics

Holt et al.[3] and Bayer et al.[4], but a factor of 5-10 more precise. Unfortunately, the theoretical precision of the absolute wavelength is limited by the calculation of the non-relativistic energy difference, (about a factor of 100 worse than experiment). The fine structures are also not calculated with sufficient precision. Table I shows some of the comparisons. Once the non-relativistic calculation is improved, the experiment is sufficiently accurate to test QED effects to a precision of about 70 ppm.

Table I Measurements of the 2s - 2p Transition in Li II

Transition		Result (cm^{-1})	Reference
1s2s $^3S_1$ - 1s2p ^3P_o		18231.30200 (15)	Riis et al.[2]
		.3030 (12)	Holt et al.[3]
		.3028 (8)	Bayer et al.[4]
	(theory)	.313	Drake[7]
^3P, J = 0 - 2		3.10265 (21)	Riis et al.[2]
	(theory)	3.10455	Accad et al.[12]
$^3P_1$, J = 2 - 1		2.08730 (21)	Riis et al.[2]
	(theory)	2.08988	Accad et al.[12]

Standard beam-foil measurements have been used to obtain the wavelengths of the same transitions in helium-like neon[5] and titanium[6]. These measurements test the QED corrections in these ions to about two percent, and show good agreement with the latest calculations of Drake et al.[7] and Hata and Grant[8]. They are also in general agreement with previous measurements. Table II shows the results, and a comparison of experiments with theory along the isoelectronic sequence is shown in Fig. 1.

Table II Measurements of the 1s2s $^3S_1$ - 1s2p $^3P_2$ transition in Ne IX and Ti XXI

Ion	Transition (Å)	Authors
Neon	1248.09 (5)	Berry and Hardis[5]
	1248.17 (3)	Klein et al.[13]
	1248.03 (9)	
	1248.12 (2)	Englehardt and Sommer[14]
(theory)	1248.10	Drake[7]
Titanium	389.49 (7)	Galvez et al.[6]
(theory)	389.57 (4)	Drake[7]

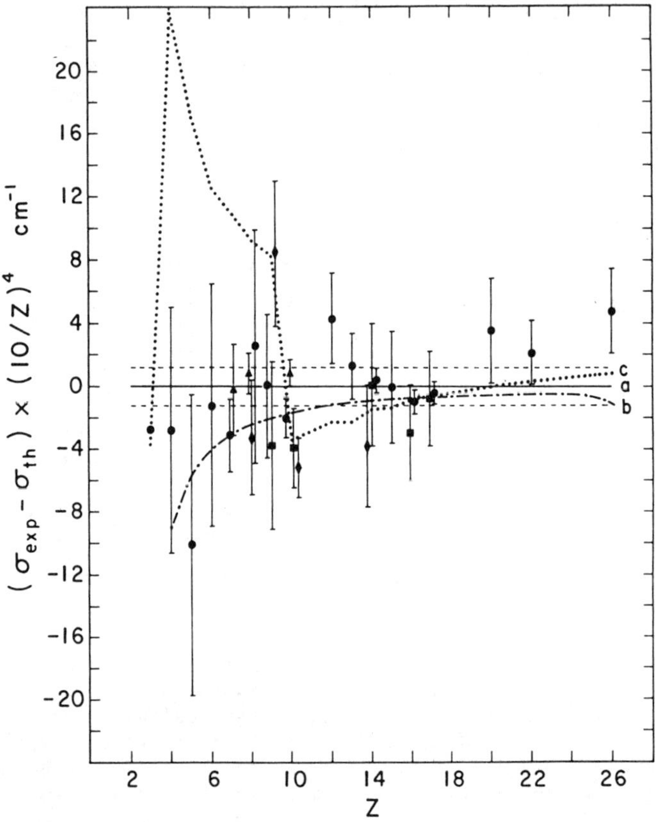

Figure 1 Comparison of measurements and calculations of the 1s2s $^3S_1$ - 1s2p $^3P_2$ transition energy for helium-like ions. The theories are from ref. 7 (full line), ref. 9 (dot-dashed line) and ref. 8 (dashed line).

The three electron lithium-like sequence is more of a challenge for theory, and most comparisons with experiment have been with semi-empirical calculations, as for example, the recent very complete analyses of Edlén[1]. However, we have previously[9] made a direct comparison of the experimental transition energies and fine structures of the 2s - 2p transitions in this sequence up to Z = 36, and our results are updated and presented again here.

One more recent experiment in this sequence has been completed for Krypton, Z = 36.[11] First in Fig. 2, we compare theory with experiment for the 2p fine structure. Most significant is the comparison with the relativistic Hartree-Fock calculations of Cheng et al.[10], represented by the crosses in the figure. The difference scales as Z^4, and the full curve which fits the data is the hydrogenic QED difference for the $2p_{1/2}$ and $2p_{3/2}$ states. This curve includes no screening by the 1s electrons; that is, it assumes the 2s wavefunctions are exactly hydrogenic, in particular, near the origin. This is in strong contrast to the helium-like system, where electron shielding has been shown to be significant, both experimentally and theoretically - see above.

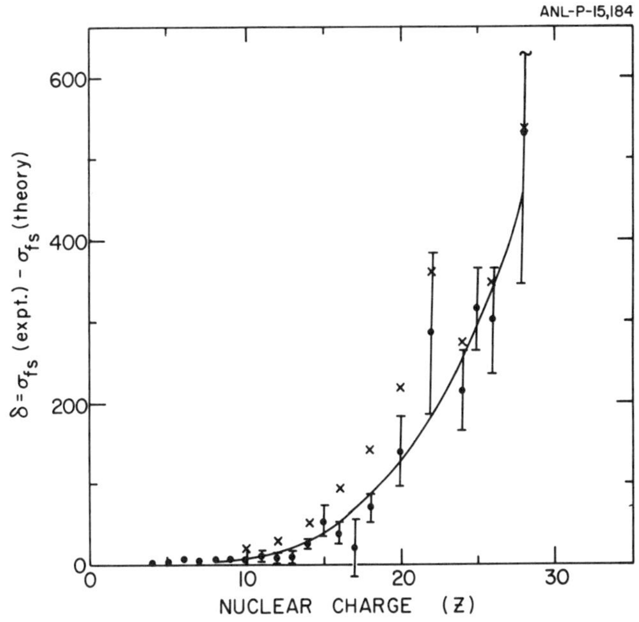

Figure 2 The fine structure of the $2p^2P$ state of the Li I isoelectronic sequence. The difference of experiment and the relativistic splitting (the crosses from Dirac-Fock calculations, the filled circles from Snyder screening, with error bars from experiments; the same error bars have been omitted from the crosses for clarity) is plotted as a function of nuclear charge Z. The curve represents the QED correction for hydrogenic $2p_{1/2}$ - $2p_{3/2}$ electrons.

The same comparison is made for the total transition energy in Figure 3. Note that the difference between Hartree-Fock values and experiment differs by the Z^4 term plus a Z-independent term of approximately 1200 cm^{-1}. This last part arises from correlation omitted in the HF approximation, and has been estimated by Safronova as 1190 cm^{-1}. The one electron QED terms shown by the solid curve then give an excellent fit to the data. No screening is included in the QED correction.

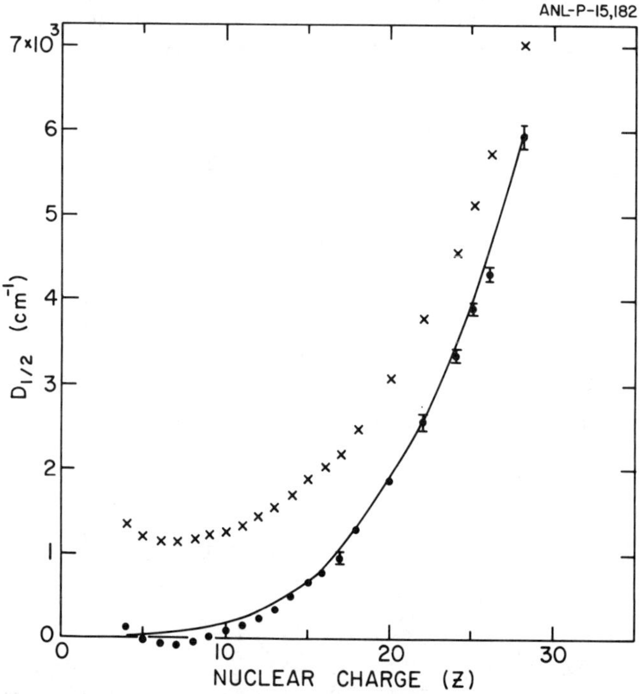

Figure 3 The wavenumber of the 2s $^2S_{1/2}$ - $2p^2P_{1/2}$ transition. The difference between experiment and calculations omitting all QED corrections is plotted against nuclear charge Z. The crosses represent Hartree-Fock 1/Z nonrelativistic plus the screened relativistic calculations. For the filled circles, the HF calculation is replaced by variational 1/Z nonrelativistic theory. The curve represents the QED correction (i.e., Lamb shift) for the hydrogenic $2s_{1/2}$-$2p_{1/2}$ electrons. The error bars are taken from the experiments.

This work was supported by the U.S. Department of Energy (Office of Basic Energy Sciences) under contract W-31-109-Eng-38.

References

1. B. Edlén, Physica Scripta 19, 255 (1979).
2. E. Riis, H. G. Berry, O. Poulsen, S. A. Lee and S. Y. Toung, submitted to Phys. Rev. A.
3. R. A. Holt, S. D. Rosner, T. D. Gaily and A. G. Adam, Phys. Rev. A22, 1563 (1980).
4. R. Bayer et al., Z. f. Phys. A192, 329 (1979).
5. H. G. Berry and J. E. Hardis, submitted to Phys. Rev. A.
6. E. J. Galvez, A. E. Livingston, A. J. Mazure, H. G. Berry, L. Engström, J. E. Hardis, L. P. Somerville and D. Zei, submitted to Phys. Rev. A.
7. G. W. F. Drake, private communication; S. P. Goldman and G. W. F. Drake, J. Phys. B17, L197 (1984).
8. J. Hata and I. P. Grant, J. Phys. B17, 931 (1984); J. Phys. B16, 523 (1983).
9. H. G. Berry, R. DeSerio and A. E. Livingston, Phys. Rev. A22, 998 (1980).
10. K. T. Cheng, Y. K. Kim and J. P. Desclaux, At. and Nucl. Data Tables 24, 111 (1979).
11. D. D. Dietrich, J. A. Leavitt, H. Gould and R. Marrus, Phys. Rev. A22, 1109 (1980).
12. Y. Accad, C. L. Pekeris and B. Schiff, Phys. Rev. A4, 516 (1971).
13. H. A. Klein et al., J. Phys. B15, 4507 (1982).
14. W. Englehardt and J. Sommer, Astrophys. J. 167, 201 (1971).

RECENT AND FUTURE PROGRESS IN QUANTUM ELECTRODYNAMICS

J. R. Sapirstein
Department of Physics
University of Notre Dame
Notre Dame, IN 46556

ABSTRACT

A review of the present status of Quantum Electrodynamics for lepton anomalies and bound states is presented. Emphasis is given to future limitations in progress due to strong interaction uncertainties and the possible extension of precision tests to many electron systems.

I. INTRODUCTION

I would like to discuss today two features of Quantum Electrodynamics (QED). The first is the present status of the agreement between experiment and theory in some of the classic tests of QED. Here I want to concentrate on where recent progress has been made, and where future progress is most necessary. The second feature is the possible development of QED in the direction of the physics being discussed at this conference, that is the possibility that the successes of QED in one electron atoms may be extended to many-electron atoms.

As regards the first feature, there is considerable progress to report in almost all the classic, and some newer tests of QED, both on the theoretical and experimental fronts. I will cover first two of the most sensitive tests, electron g-2 and muonium ground state hyperfine splitting (hfs), then turn to tests more limited by strong interaction effects, namely the Lamb shift, muon g-2, and hydrogen hfs, closing with the very clean, but also theoretically very demanding tests being opened up by new accurate experiments in the purest QED atom, positronium. I will try to indicate the potential for future progress in all these tests, and the point at which non-QED physics may enter and form a limit to the extent QED can be applied.

The second topic is more speculative in nature, but of more direct relevance to this conference. QED is well known to be the most successful fundamental field theory in nature, and serves as a model for any physical theory. Since precise atomic spectroscopy of one electron atoms plays such a fundamental role in the classic tests of QED, a natural question is to what extent this success can be extended to many electron atoms. The same experimental techniques that yield the many significant digit measurements in hydrogen, muonium, and positronium can and have been applied also to many electron atoms. Given the wealth of accurate data about these atoms, what is the likelihood that sensitive QED tests can be made?

The organization of the talk is as follows: in section II I present a standard QED review. The tables are presented in a somewhat unconventional manner, where successively smaller theoretical contributions are subtracted from the experimental value. This presentation serves to emphasize the step by step understanding of physical effects that are simultaneously smaller and more intricate provided by these QED tests. Electron g-2, muonium hfs, the Lamb shift, muon g-2, hydrogen hfs, and positronium hfs, fine structure, and decay rate will be discussed in turn. Precision tests in He will not be treated here, but are discussed in other contributions to this conference. In section III, a many-body perturbation approach to heavy atomic systems is presented as a possible way of beginning a systematic attack on the problem of providing atomic theory accurate enough to allow interesting tests of QED in this area.

II. CLASSIC QED TESTS

While the electron anomaly and the Lamb shift were both important in the original development of QED, the former has become by far the more accurate test. On the experimental front, the use of a Penning trap has allowed for measurements of the anomaly to a few ppb[1]:

$$a_{e^-} = 1\ 159\ 652\ 193(4) \cdot 10^{-12}$$

It is worth noting that at this great accuracy subtle theoretical corrections need to be applied: recent work by Fischbach et al.[2] and Brown et al.[3] has indicated that the presence of metallic walls in the trap alters the photon propagator in the one loop result enough to affect the measurement at the level of the quoted error. The present comparison between theory and experiment is presented in Table 2.1. This table uses for the fine structure constant $\alpha^{-1} = 137.035981(12)$[4]. This is a new value that differs from the standard Josephson junction value[5] $\alpha_J^{-1} = 137.035963(15)$. The reason for the change is the discovery that the standard ohm maintained at NBS has been decreasing by .06 ppm per year. When the older value is used the remainder at the bottom of the table is $-255(128)(43) \cdot 10^{-12}$ which, while not enough to claim a breakdown of QED, is large enough so that any decrease in error has the potential of uncovering a significant discrepancy. The present situation is entirely satisfactory.

I want to call attention here to the sheer size of the fourth order calculation[6]. 891 Feynman diagrams contribute, each of which involves up to a 10 dimensional integration of functions involving up to 15000 terms. While use of symmetry, and grouping together classes of vertex diagrams into self energy type graphs substantially reduce the number of graphs, this is an extremely large scale computation. The entire apparatus of renormalization theory must be applied correctly to render the graphs finite: up to 15 subtraction terms can be required. More than 40 days of CDC 7600 time were used to get the quoted result, and presently a vector supercomputer with greatly increased speed is being used to reduce the error[7]. The point I wish

Table 2.1 Electron g-2 in Units of 10^{-12}

a_e-(exp)	=	1 159 652 193 (4)
$.5(\frac{\alpha}{\pi})$		1 161 409 886 (102)
		-1 757 693
$-.328478966(\frac{\alpha}{\pi})^2$		-1 772 306
		14 613
$1.1765(13)(\frac{\alpha}{\pi})^3$		14 745 (16)
		-132
$-.8(1.4)(\frac{\alpha}{\pi})^4$		- 23(40)
		-109(102)(43)

to stress is that calculations of enormous size can now be attacked with the aid of the extremely powerful computers that are becoming available. This point will be returned to in section III, where many-body perturbation theory is discussed.

The largest source of error in the electron anomaly is presently the $102 \cdot 10^{-12}$ error due to the solid state determination of α^{-1}. This will provide a severe challenge to solid state experiment. It would also be of interest to see if at these refined levels of accuracy, any theoretical solid state corrections enter. The $43 \cdot 10^{-12}$ error from QED theory can be reduced with supercomputers, but since the integration programs have errors that are statistical in nature, this will be an expensive process: analytical progress would be very welcome.

One way to finesse the relatively large error arising from the uncertainty in α^{-1} is to assume the validity of QED, and use the electron anomaly to measure a 'QED value' of α^{-1}. This value is then $\alpha^{-1}_{g-2} = 137.035993(10)$. To make this a useful construct, it is necessary to find another system in which QED can be applied at a level of accuracy comparable to that of the electron anomaly. The best such system is muonium ground state hfs. The accurate measurement by Hughes et al.[8] is compared with different levels of theory in Table 2.2. The agreement with experiment is again very good. Here the major source of error arises from the lack of knowledge of the muon mass: the best determination comes from the same experiment that gives the hfs: a more accurate experiment would be of great value.

The calculation of bound state properties has a different character of difficulty than the anomaly calculations. Here one typically

Table 2.2 Muonium Ground State hfs in Units of kHz

$\Delta\nu$ (exp) =	4 463 302.88 (16)
E_F	4 459 033.44 (154)
	4 269.5
$a_e E_F$	5 170.9
	-901.4
$\alpha^2 E_F$	- 72.9
	-828.5
$\alpha \frac{m_e}{m_\mu} E_F$	-801.2
	- 27.3
$\alpha^3 E_F$	- 33.0
	5.7
$\alpha^2 \frac{m_e}{m_\mu} E_F$	6.0
	- .3 (.16)(1.54)

is concerned with one or two loop effects, so that the complications of lengthy numerators and intricate renormalization are not so severe. However, the new difficulty is in the treatment of the bound state. A careful treatment of recoil effects is needed, since numerically m_e/m_μ acts like a single power of α, and in addition care must be taken to avoid false expansions, in which Coulomb singularities cause graphs nominally of a high order in α to be of the order of interest. The framework in which these problems can be handled has been around since the early 50's, in the form of the Bethe-Salpeter equation[9]. However, it is only fairly recently that formalisms powerful enough to allow calculations at the ppm level have been developed. The problem is that, although in any formulation of the B-S equation, one can generally make approximations and see the non-relativistic Schrodinger picture result, it is much more difficult to cleanly isolate nonrelativistic terms and calculate higher order terms. Recent advances have been made in the framework of three dimensional formalisms[10-14]. These have the advantage of turning the bound state problem into a form where the lowest order wave function is exactly the Schrodinger or Dirac Coulomb wave function, with perhaps slightly modified mass and coupling parameters. Perturbation theory can be applied in a straightforward manner at this point. This allowed some time ago for the first reliable calculations of the logarithmic terms[10,11] $\alpha^2 m_e/m_\mu \ln\alpha E_F$ and more recently the calculation of the constant under the log[13,14]. This term, which I refer to as a 'pure recoil' term, has been calculated so far only in muonium and positronium hyperfine splitting: as will be pointed out later, these calculations must be extended to several other situations. Other corrections that have only recently been evaluated are the 'radiative recoil' terms, of the same order as pure recoil but involving one radiative photon, and 'two-loop non-recoil' terms, which involve two radiative photons and two Coulomb exchanges. The latter have not yet been evaluated, and form the largest theoretical uncertainty in muonium hfs, being perhaps as large as a kHz. Representative graphs for pure recoil, radiative recoil, and two loop non-recoil are shown in Fig. 1.

Turning now to the muon anomaly, the problem of the lack of an accurate calculational framework for strong interaction physics first enters. The situation is summarized in Table 2.3, where it is seen that pure QED corrections leave a large discrepancy. The relatively large mass of the muon allows for the probing of the strong interactions by the virtual photon. The uncertainties in the strong interaction contributions are seen to provide the dominant error in this QED test. While accurate e^+e^- experiments can reduce the error to a level where weak interaction effects may be seen, it is clear that at some point study of this system will be yielding more information about the electromagnetic interactions of hadrons than the QED of leptons. This is a sort of 'end' to QED: suppose that the hadron uncertainties cannot be pushed below, say, $20*10^{-10}$. Then it certainly pointless to contemplate a five loop calculation, since it would be impossible to distinguish the effect from the errors due to hadron physics. It is probably more accurate to say that progress in QED will simply slow to the level of progress in whatever other branch of physics that is 'contaminating' the QED test. In any case, there is certainly important work to do in the muon anomaly in the near future: first, the measurement should be improved by an order of magnitude, secondly, new experiments to measure the hadronic contributions are needed to reduce that error by a factor of 2 or 3, and finally the QED theory error should be reduced by the same factor. The reward is a test of the Weinberg-Salam model at the one-loop level, which has not yet been decisively made with other standard electroweak tests.

Turning now to the $2s_{1/2}$-$2p_{1/2}$ splitting in hydrogen, the Lamb shift, we see that it is no longer a particularly stringent QED test. The greatest experimental accuracy claimed[15] is only 2 ppm, to be compared with the .03 ppm level of muonim hfs or .003 ppm of the electron g-2. In forming Table 2.4 we quote the experimental result of Lundeen and Pipkin[16], with a more conservative 9 ppm error estimate. Here again, as in the muon g-2 case, uncertainty about hadronic electromagnetic interactions is limiting the interpretation of QED. Use of older proton size determinations[17] gives an 11 kHz discrepancy, while the newer Mainz determination[18] gives

Table 2.3 Muon g-2 in Units of 10^{-10}

$a_{\mu^+}(\text{exp})$	=	11 659 110 (110)
$\frac{1}{2}(\frac{\alpha}{\pi})$		11 614 099 (1)
		45 011
$+.76585810(\frac{\alpha}{\pi})^2$		41 322
		3 689
$+24.073(11)(\frac{\alpha}{\pi})^3$		3 017(2)
		672
$140(6)\ (\frac{\alpha}{\pi})^4$		41(2)
		631
Hadronic contribution		703(19)
		- 72(110)(20)
Weak contribution		20

Table 2.4 Lamb Shift in Units of MHz

S(exp)	1057.845(9)	
$\alpha^3\text{Ry}$	1050.560	
	7.285	
$\alpha^4\text{Ry}$	7.129	
	.156	
$\alpha^5\text{Ry}$	-.329	
	.475	
$\alpha^3\text{Ry}\ \frac{m_e}{m_p}$	.359	
	.116	
$\alpha^2\text{Ry}\ m_e^2\langle R^2\rangle$	.127(17),	.145(18)
	-.011 ,	-.029

a 29 kHz discrepancy. Unfortunately, the theoretical situation is not yet at the level reached in muonium hfs: pure recoil, radiative recoil, two loop non-recoil, and a three loop contribution to the Dirac slope are all uncalculated, and can all enter at the 10 kHz level (note that Bhatt and Grotch[19] have begun the radiative recoil calculation). Future progress in the Lamb shift requires that all these calculations be performed. It would then be desirable to have another 2-3 kHz Lamb shift measurement: at that point it may be possible for the Lamb shift to rule decisively on the proton size discrepancy. As in muon g-2, we are no longer learning about QED, but rather about hadronic electromagnetic interactions.

The situation is entirely parallel in hydrogen hfs, except that here experimental uncertainty is entirely negligible. A sizeable discrepancy is left over after QED terms are subtracted out in Table 2.5 that is mostly accounted for by 'static' proton structure, indicated by the graph in Fig. 2. It would be desirable to reduce the large error associated with uncertainty in the proton form factors. If this can be done, and the same terms mentioned in the Lamb shift calculated (Bodwin and Yennie have recently calculated the logarithmic part of the pure recoil term[20]), information about the 'dynamic polarizability' of the proton, which involves graphs like those in Fig. 3 can be obtained: one can already determine $\delta = 1.6\ (9)$ ppm. This parameter can be related to polarized γ-p scattering with sum rules: a contribution has been determined by Hughes[21]. It is of interest that spectroscopy of hydrogen, involving energies of 10^{-6} eV can provide information complementary to high energy (10^{10} eV) scattering of electrons off photons. However, there is again no precision test of QED being made here.

I close this section with a brief review of positronium. This purely QED atom is quite insensitive to hadronic physics, and a great deal of progress can be made in testing QED. The 1s-2s triplet splitting has recently been measured to 12 ppb[22]. This confirms an α^3Ry calculation by Fulton and Martin[23], which is basically the lowest order Lamb shift for the equal mass case. The difference between theory and experiment is 12 ± 15 MHz: once experimental error gets down to the 1 MHz level, it will be possible to test the as yet uncalculated α^4Ry contributions.

Another important recent development is the measurement of ground state hfs by Ritter et al.[24],

$$\Delta\nu = 203\ 389.10\ (74)\ \text{MHz}$$

A recent calculation by Caswell and Lepage[14] of the pure recoil term leaves 16 MHz to be explained by annihilation graphs like those in Fig. 4, which are the only remaining uncalculated $\alpha^2 E_F$ terms. Because the electron and positron have equal mass, a rigorous test of relativistic bound state formalisms is being provided by this system.

Table 2.5 Hydrogen hfs in Units of MHz

$\Delta\nu(\mathrm{exp})$	= 1 420 . 405 751 766 7(9)
E_F	1 418 . 840 832
	1 . 564 920
$a_e E_F$	1 . 645 359
	- .080 439
$\alpha^2 E_F$, $\alpha^3 E_F$	- .033 655
	- .046 784
'Static' Proto Structure	- .049 092(1277)
	.002 308

Finally I mention the decay rate of orthopositronium, which is 2.5 standard deviations above theory[25]. If this discrepancy remains, either a very large coefficient of the uncalculated $\alpha^2 \Gamma_o$ terms must be present, or perhaps new physics may finally be entering.

III. REMARKS ON THE MANY-ELECTRON PROBLEM

In this section I want to sketch out a possible approach to many-electron atoms in which, at least in principle, some of the progress discussed above for one electron atoms could be extended to these more complex systems. I leave out He, where in fact the progress has been extended to a great degree because the techniques used there cannot be straightforwardly extended to atoms with more than 2 electrons. Now, in the discussion of the Bethe-Salpeter equation in section II, it was mentioned that use of the Schrodinger equation can be rigorously justified in the framework of three-dimensional forms of that equation. One can use a Schrodinger wave function to get lowest order results, and a systematic perturbation theory exists that allows corrections of higher order in α to be evaluated. While I know of no explicit generalization of this to the many electron case, it seems very probable that one can justify using the many-electron Schrodinger equation,

$$H\Psi = \sum_i \left(\frac{\vec{p}_i^{\,2}}{2m} - \frac{Z\alpha}{r_i}\right)\Psi + \frac{1}{2}\sum_{ij} \frac{\alpha}{|\vec{r}_i - \vec{r}_j|}\Psi = E\Psi \qquad 3.1$$

in the same way. Here, however, one has no analytic solution for Ψ As a straightforward partial differential equation, 3.1 is clearly intractable. However, many-body perturbation theory, pioneered in atomic physics by Kelly, provides a systematic expansion of the wave function and physical observables based on

$$H = H_o + V_c$$

$$H_o = \sum_i \left(\frac{\vec{p}_i^{\,2}}{2m} + V(r_i)\right) \qquad 3.2$$

$$V_c = \frac{1}{2}\sum_{ij} \frac{\alpha}{|\vec{r}_i - \vec{r}_j|} - \sum_i \left(V(r_i) + \frac{Z\alpha}{r_i}\right)$$

Here, unlike the discussion of section II, we have no small expansion parameter: the convergence of an expansion in V_c depends entirely on how carefully H_o is chosen. H_o should of course be chosen to be separable and central, but beyond that there is a large degree of flexibility in its choice. The most common choice is of course the Hartree-Fock potential, but various model potentials can also be used. Unfortunately, rather little seems to be known about the rate of convergence of this perturbation series. As W. R. Johnson will explain in his contribution to this conference, we have encountered in the investigation of effects of weak interactions in atomic physics, the problem of making predictions for alkali atoms accurate at the few percent level[26]. This level of accuracy is necessary if forthcoming accurate experiments of parity violation in Cs and Tl[27] can be used to provide information about the Weinberg angle at this few percent level. However, the problem is general: can one make predictions of any many-electron atomic property reliably at the few percent level or under? We are advocating the use of several different models of H_o along with many body perturbation theory as a means of answering this question. Our hope is that the situation in Fig. 5 will be realized: that is, a relatively wide spread of results between different potentials in lowest order will systematically be narrowed to a unique value as successively higher orders of perturbation theory force the different models together to form the true picture of the atom. Unfortunately, as can be seen in Johnson's talk, working to only first order in V_c is clearly inadequate. A possible reason for hope that second order in V_c will be adequate is provided by the following results found by Dzuba et al.[28] for valence energies in Cs:

6S	.12737 (lowest order)	→ .14445 (second order)	.94% above exp.
7S	.05519 (lowest order)	→ .05856 (second order)	.15% below exp.

Note that going in a relativistic theory to second order involves some of the questions of principle raised by Sucher in this conference.

While this program is aimed at weak interaction properties, if it turns out that this many-body perturbation theory is rapidly convergent, the possibility of picking up QED effects in heavy atoms is raised. Such effects have been studied in high Z one electron atoms[29] and in inner shell binding energies[30]. However, in the first case the wave function is exactly known, and in the second correlation effects are greatly suppressed due to the deep binding of the inner shell. In general, it has not been possible to study QED effects in many-electron atoms due to the lack of knowledge of the wave function. However, if the approach discussed above succeeds in reducing wave function uncertainties to well under the 1% level, a wealth of experimental data on many electron atoms may be brought to bear on tests of QED. While α^2 corrections are probably too small to be distinguished from wave function uncertainties, $Z\alpha^2$ terms should be detectable for high Z atoms. A very fundamental understanding of the relativistic many-electron binding problem may be required to successfully predict these effects. For example, all present numerical work in the relativistic treatment of heavy atoms starts with generalizing 3.1 to a sum of Dirac Hamiltonians. As pointed out by Sucher at this conference, there is no fundamental basis for this assumption. The question then arises as to whether some scheme can be constructed in which the results obtained with the relativistic form of 3.1 can be justified as some lowest order result, with unambiguous predictions for the corrections due to a more proper treatment of the relativistic binding problem. If wave function uncertainties remain above the 1% level, this question must unfortunately remain moot: if, however, more accurate predictions can be made, a very fascinating mix of QED, correlation effects, and the many electron relativistic binding problem can be tested in the study of many-electron atoms.

REFERENCES

1. R.S. Van Dyck, Jr., P. B. Schwinberg, and H.G. Dehmelt, Proceedings of the Ninth International Conference on Atomic Physics, Seattle, Washington, July 23-27, 1984.

2. E. Fischbach and N. Nakagawa, Phys. Rev. D30, 2356 (1984).

3. L. Brown, G. Gabrielse, K. Helmerson, and J. Tan, University of Washington preprint 40048-05, 1985.

4. B. N. Taylor, Journal of Research of the National Bureau of Standards 90, #2, 91 (1985).

5. E. R. Williams and P. T. Olsen, Phys. Rev. Lett. 42, 1575 (1979).

6. T. Kinoshita and W. B. Lindquist, Phys. Rev. Lett. 47, 1679 (1981).

7. Significant increases in speed over CDC 7600 speed has been obtained on a HITAC machine (T. Kinoshita, private communication).

8. F. G. Mariam et al., Phys. Rev. Lett. 49, 993 (1982).

9. E. E. Salpeter and H. A. Bethe, Phys. Rev. 84, 1232 (1951); M. Gell-Mann and F. Low, Phys. Rev. 84, 350 (1951).

10. G. P. Lepage, Phys. Rev. A16, 1863 (1977); W. E. Caswell and G. P. Lepage, Phys. Rev. A20, 36 (1979).

11. G. T. Bodwin and D. R. Yennie, Phys. Rep. 43, #6 (1978).

12. R. Barbieri and E. Remiddi, Nucl. Phys. B141, 413 (1978).

13. G. T. Bodwin, D. R. Yennie, and M. Gregorio, Phys. Rev. D29, 2290 (1984).

14. W. E. Caswell and G. P. Lepage, Cornell U. preprint CLNS-85/641.

15. V. G. Palchikov, Yu. L. Sokolov, V. P. Uakovlev, Lett. Jour. Tech. Phys. 38, #7, 347 (1983).

16. S. R. Lundeen and F. M. Pipkin, Phys. Rev. Lett. 46, 232 (1981).

17. D. J. Drickey and L. N. Hand, Phys. Rev. Lett. 9, 521 (1962); L. N. Hand, D. J. Miller, and R. Wilson, Rev. Mod. Phys. 35, 335, (1963).

18. G. G. Simon, Ch. Schmitt, F. Borkowski, and V. H. Walther, Nucl. Phys. A333, 381 (1980).

19. B. Bhatt and H. Grotch, Pennsylvania State University preprint (1984).

20. Private communication from D. R. Yennie.

21. V. W. Hughes and J. Kuti, Ann. Rev. Nucl. Part. Sci. 33, 611 (1983).

22. S. Chu, A. P. Mills, Jr., and J. L. Hall, Phys. Rev. Lett. 52, 1689 (1984).

23. T. Fulton, Phys. Rev. A26, 1794 (1982); T. Fulton and P. C. Martin, Phys. Rev. 93, 903 (1954); 95, 811 (1954).

24. M. Ritter, P. O. Egan, V. W. Hughes, and K. A. Woodle, Phys. Rev. A30, 1331 (1984).

25. D. W. Gidley, A. Rich, E. Sweetman, and D. West, Phys. Rev. Lett. 49, 525 (1982).

26. W. R. Johnson, D. S. Guo, M. Idrees, and J. Sapirstein, to be published in Phys. Rev. A.

27. C. E. Wieman, S. Gilbert, R. Watts, and M. C. Noecker, JILA preprint, June 1985 and C. E. Wieman, E. D. Commins, private communications.

28. V. A. Dzuba, V. V. Flambaum, P. G. Silvestrov and O. P. Sushkov, J. Phys. B18, 597 (1985).

29. P. J. Mohr, Ann. Phys. (N.Y.) 88 26, 521 (1974).

30. W. R. Johnson and K. T. Cheng, Atomic Inner-Shell Physics, ed. Bernd Crasemann (Plenum, New York, 1984) pp. 1-47.

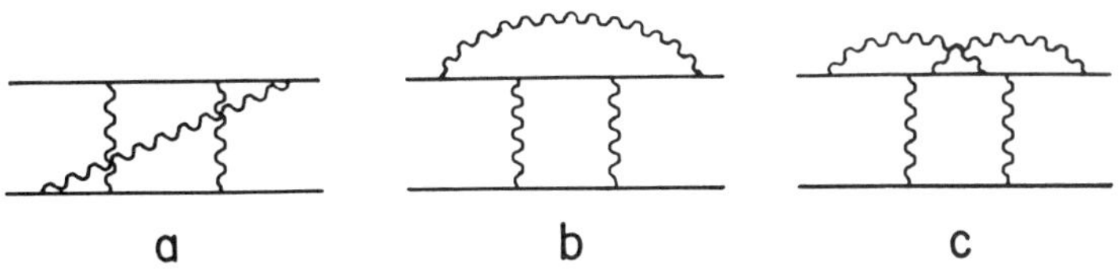

Fig. 1a Representative pure recoil graph.
1b Representative radiative recoil graph.
1c Representative two loop non-recoil graph.

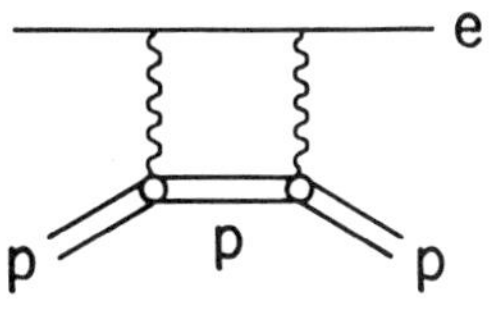

Fig. 2

Contribution to static proton structure.

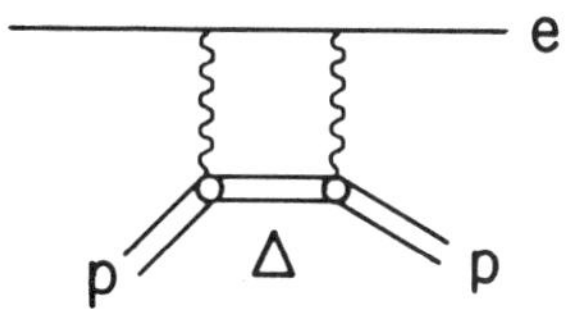

Fig. 3

Contribution to proton polarizability term.

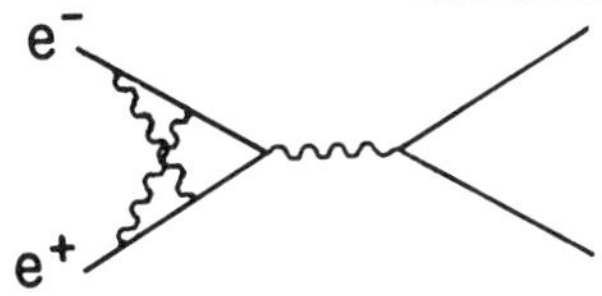

Fig. 4 $\alpha^2 E_F$ graph contributing to positronium hfs.

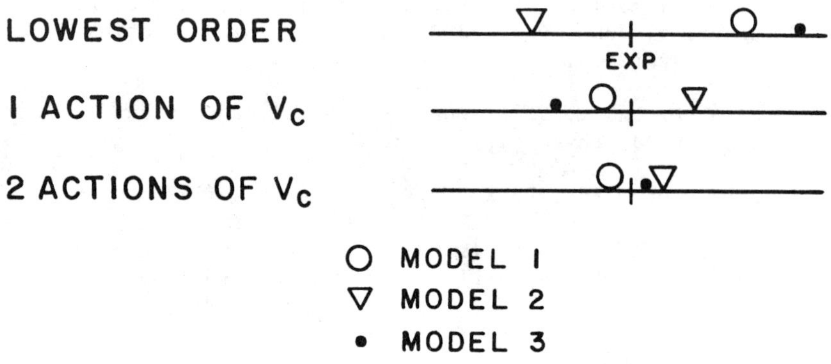

Fig. 5 Convergent many-body perturbation theory results.

QUANTUM ELECTRODYNAMICS OF HIGH-Z ONE- AND TWO-ELECTRON ATOMS

Peter J. Mohr
J. W. Gibbs Laboratory, Yale University, New Haven, CT 06511

ABSTRACT

A formulation of quantum electrodynamics suitable for high-Z few-electron atoms is described. Interaction of the bound electrons with the radiation field is treated as a perturbation that results in electron-electron interactions and radiative corrections. The leading corrections are written as Fock space operators that can be evaluated in states with any number of electrons. The extension of this approach to include higher-order corrections is discussed. The special cases of one- and two-electron atoms are examined in detail.

INTRODUCTION

This paper examines a formulation of quantum electrodynamics (QED) that is suitable for describing high-Z few-electron atoms. Perturbation theory is written here as an expansion about basis states of bound noninteracting relativistic Dirac electrons. Radiative corrections and electron-electron interactions are produced by interactions of the electrons with the quantized radiation field. For a few strongly bound electrons in a high-Z Coulomb field, these perturbations are small compared to the binding energy, and the perturbation expansion converges rapidly. In particular, the relevant expansion parameter is 1/Z, and this method provides a generalization of the 1/Z expansion methods of Layzer and Bahcall[1] and of Dalgarno and Stewart[2] to include higher-order relativistic effects and radiative corrections. In this approach, the energy level is developed as

$$E = [f_0(Z\alpha) + \frac{1}{Z} f_2(Z\alpha) + \frac{1}{Z^2} f_4(Z\alpha) + \cdots](Z\alpha)^2 mc^2 \qquad (1)$$

where

$$\lim_{Z\alpha \to 0} f_n(Z\alpha) = constant \qquad (2)$$

Each of the functions f_n is determined exactly, including radiative corrections, to all orders in $Z\alpha$ in this formulation, and the perturbation series in 1/Z converges rapidly for large Z.

This approach to high-Z two-electron atoms has been briefly discussed by Bethe and Salpeter[3] and by Sucher[4]; it is of renewed interest due to the rapid rise of experimental activity in this area. More recent studies of electron interaction corrections in this framework have been made, for example, by Ivanov, Ivanova, and Safronova.[5]

The present paper outlines the basic theory and indicates the derivation of the leading corrections. These corrections are written as Fock space operators valid for any number of electrons. Corrections in the special cases of one- and two-electron atoms are examined and the extension to higher orders is discussed.

BASIC FORMULATION

Energy levels of high-Z few-electron atoms are described here within the framework of the Furry bound interaction picture of quantum electrodynamics.[6] The zero-order states are the eigenfunctions of the Dirac equation (in units in which ħ, c, and m have unit magnitude)

$$[-i\vec{\alpha}\cdot\vec{\nabla} + V(x) + \beta - E_n]\phi_n(\vec{x}) = 0 \qquad (3)$$

for an external Coulomb potential with source charge Ze

$$V(x) = -\frac{Z\alpha}{x} \tag{4}$$

In the Furry picture, the electron-positron field operator $\psi(x)$ is expanded in terms of electron annihilation operators a_n and positron creation operators b_n^* as

$$\psi(x) = \sum_{E_n>0} a_n\phi_n(x) + \sum_{E_n<0} b_n^*\phi_n(x) \tag{5}$$

where

$$\phi_n(x) = \phi_n(\bar{x})e^{-iE_nt} \tag{6}$$

The creation and annihilation operators satisfy the Fermi anticommutation relations. The zero-order Hamiltonian is

$$H_0 = \sum_{E_n>0} a_n^*a_nE_n - \sum_{E_n<0} b_n^*b_nE_n \tag{7}$$

where the E_n are the Dirac eigenvalues.

The interaction between the electron-positron field and the radiation field is determined by the interaction Hamiltonian density

$$H_I(x) = j^\mu(x)A_\mu(x) - \delta M(x) \tag{8}$$

where

$$j^\mu(x) = -\frac{e}{2}[\bar{\psi}(x)\gamma^\mu,\psi(x)] \tag{9}$$

is the electromagnetic current, $A_\mu(x)$ is the vector potential operator for the radiation field, and

$$\delta M(x) = \frac{\delta m}{2}[\bar{\psi}(x),\psi(x)] \tag{10}$$

is the mass renormalization counter term.

The prescription of Gell-Mann and Low is employed here to obtain an expression for the bound-state level shifts in perturbation theory.[7] Sucher has written an equivalent expression in terms of the S matrix to which the standard renormalization prescription may be applied:[4]

$$\Delta E = \lim_{\epsilon\to 0}\lim_{\lambda\to 1}\frac{i\epsilon}{2}\frac{\frac{\partial}{\partial\lambda}<S_{\epsilon,\lambda}>_c}{<S_{\epsilon,\lambda}>_c} \tag{11}$$

The subscript c denotes the fact that only connected Feynman graphs are included in the matrix element. In (11), $S_{\epsilon,\lambda}$ is the adiabatic S matrix defined by

$$S_{\epsilon,\lambda} = 1 + \sum_{j=1}^{\infty} S_{\epsilon,\lambda}^{(j)} \tag{12}$$

where

$$S_{\epsilon,\lambda}^{(j)} = \frac{(-i\lambda)^j}{j!}\int d^4x_j\cdots\int d^4x_1 e^{-\epsilon|t_j|}\cdots e^{-\epsilon|t_1|}\,T[H_I(x_j)\cdots H_I(x_1)] \tag{13}$$

These equations provide the basic formulation of bound-state quantum electrodynamics.

LEADING CORRECTIONS

To second order in H_I, Eq. (11) yields

$$\Delta E = \lim_{\epsilon \to 0} \frac{i\epsilon}{2}[<S_\epsilon^{(1)}>_c + 2<S_\epsilon^{(2)}>_c - <S_\epsilon^{(1)}>_c^2 + \cdots] \tag{14}$$

where

$$<S_\epsilon^{(j)}>_c = <S_{\epsilon,\lambda}^{(j)}>_c\Big|_{\lambda=1} \tag{15}$$

To order α, only the first two terms in (14) are relevant. The first term gives the mass renormalization term, and the second term gives the main radiative term

$$<S_\epsilon^{(2)}>_c = -\frac{1}{2}\int d^4x_2 \int d^4x_1 e^{-\epsilon|t_2|}\, e^{-\epsilon|t_1|} \tag{16}$$
$$\times <T[j^{\mu_2}(x_2)A_{\mu_2}(x_2)j^{\mu_1}(x_1)A_{\mu_1}(x_1)]>_c + O(\alpha^2)$$

The matrix element in (16) can be evaluated by Wick's theorem, where the electron-positron contractions are given by the propagation function

$$S_F(x_2,x_1) = <0|T[\psi(x_2)\overline{\psi}(x_1)]|0> \tag{17}$$
$$= \frac{1}{2\pi i}\int_{-\infty}^{\infty} dz \sum_n \frac{\phi_n(\bar{x}_2)\overline{\phi}_n(\bar{x}_1)}{E_n - z(1+i\delta)} e^{-iz(t_2-t_1)}$$

The bound-state case differs from the free-particle case in that the current in (9) is not equal to the normal ordered current, and in particular

$$j^\mu(x) = :j^\mu(x): + eTr[\gamma^\mu S_F(x,x)] \tag{18}$$

The second term of the right-hand side corresponds to a vacuum polarization loop with one vertex which vanishes for free particles, but not for bound particles. The contraction of the two vector potentials in Eq. (16) gives the standard photon propagator $D_F(x_2-x_1)$. In Eq. (16), the leading terms in powers of ϵ are identified after carrying out the integrations over time. This is conveniently done by noting that the time dependence of both the electron propagator and the photon propagator is of the form

$$e^{-iy(t_2-t_1)} \tag{19}$$

A typical integration is

$$\int_{-\infty}^{\infty} dt_2 \int_{-\infty}^{\infty} dt_1 e^{-\epsilon|t_2|}\, e^{-\epsilon|t_1|}\, e^{-iy(t_2-t_1)}\, e^{iE_n t_2}\, e^{-iE_m t_1} \tag{20}$$
$$= \frac{2\epsilon}{\epsilon^2+(y-E_n)^2}\frac{2\epsilon}{\epsilon^2+(y-E_m)^2}$$
$$= \delta(E_n,E_m)\frac{2\pi}{\epsilon}\delta(y-E_n) + O(1)$$
$$= \delta(E_n,E_m)\frac{1}{\epsilon}\int_{-\infty}^{\infty} d(t_2-t_1)\, e^{-iy(t_2-t_1)}\, e^{iE_n t_2}\, e^{-iE_m t_1}$$
$$+ O(1)$$

where

$$\delta(a,b) = \begin{cases} 1 & if\, a = b \\ 0 & otherwise \end{cases} \tag{21}$$

With the aid of this and similar expansions for small ϵ, the limit indicated in Eq. (14) is readily carried out, and one obtains the following expression for the level shift of order α

$$\begin{aligned} E^{(2)} = &-4\pi i\alpha \int d(t_2-t_1) \int d\bar{x}_2 \int d\bar{x}_1 D_F(x_2-x_1) \\ &\times \Big[\frac{1}{2} \sum_{nm} \bar{\phi}_n(x_2)\gamma_\mu \phi_m(x_2) \sum_{kl} \bar{\phi}_k(x_1)\gamma^\mu \phi_l(x_1) \\ &\qquad \times \delta(E_n+E_k, E_l+E_m) < a_n{}^* a_k{}^* a_l a_m > \\ &+ \sum_{nm} \bar{\phi}_n(x_2)\gamma_\mu S_F(x_2,x_1)\gamma^\mu \phi_m(x_1) \delta(E_n,E_m) < a_n{}^* a_m > \\ &- Tr[\gamma_\mu S_F(x_2,x_2)] \sum_{nm} \bar{\phi}_n(x_1)\gamma^\mu \phi_m(x_1) \delta(E_n,E_m) < a_n{}^* a_m > \Big] \\ &- \delta m \sum_{nm} \int d\bar{x}\, \bar{\phi}_n(x) \phi_m(x) \delta(E_n,E_m) < a_n{}^* a_m > \end{aligned} \tag{22}$$

This expression for the level shift applies to any zero-order basis state consisting of a well-defined number of electrons. The first three terms in (22) correspond to the one exchanged photon, the self energy, and the vacuum polarization corrections, respectively, represented in Fig. 1.

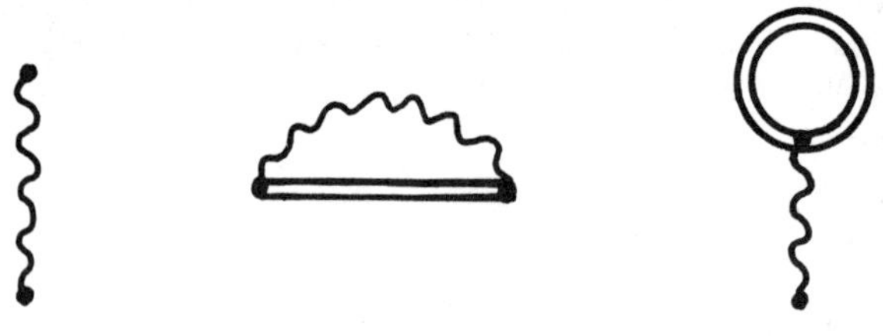

Figure 1: Feynman diagrams for the operators in Eq. (22)

The fourth term in (22) is the mass renormalization term associated with the self-energy correction.

HIGHER-ORDER CORRECTIONS

The extension of the formalism to higher orders is straightforward in principle, but is complicated in practice by the proliferation of diagrams and the appearance of higher inverse powers of ϵ in each order. These terms must, of course, add up to zero in order for the theory to be valid, and it is not obvious in the general case how the cancellations occur. In this section, the cancellation is illustrated in fourth order (in e) for the particular case of the second-order contribution of vacuum polarization.

The terms of interest in the fourth-order level shift are given by

$$E^{(4)}_{\mathrm{VV}} = \lim_{\epsilon \to 0} \frac{i\epsilon}{2} [4 < S^{(4)}_{\epsilon} >_{\mathrm{cVV}} - 2 < S^{(2)}_{\epsilon} >^{2}_{\mathrm{cV}} + \cdots] \tag{23}$$

where the relevant terms of the fourth-order S matrix element are

$$\begin{aligned} < S^{(4)}_{\epsilon} >_{\mathrm{cVV}} = & \int d^4x_4 \int d^4x_3 \int d^4x_2 \int d^4x_1 e^{-\epsilon|t_4|} e^{-\epsilon|t_3|} \\ & \times e^{-\epsilon|t_2|} e^{-\epsilon|t_1|} e^4 Tr[\gamma_\mu S_{\mathrm{F}}(x_4,x_4)] Tr[\gamma_\nu S_{\mathrm{F}}(x_3,x_3)] \\ & \times \Big[\frac{1}{2} \sum_{\mathrm{nmkl}} \overline{\phi}_{\mathrm{n}}(x_2) \gamma^\mu \phi_{\mathrm{m}}(x_2) \overline{\phi}_{\mathrm{k}}(x_1) \gamma^\nu \phi_{\mathrm{l}}(x_1) < a_{\mathrm{n}}{}^* a_{\mathrm{k}}{}^* a_{\mathrm{l}} a_{\mathrm{m}} > \\ & + \sum_{\mathrm{nm}} \overline{\phi}_{\mathrm{n}}(x_2) \gamma^\mu S_{\mathrm{F}}(x_2,x_1) \gamma^\nu \phi_{\mathrm{m}}(x_1) < a_{\mathrm{n}}{}^* a_{\mathrm{m}} > \Big] \\ & \times D_{\mathrm{F}}(x_4 - x_2) D_{\mathrm{F}}(x_3 - x_1) \end{aligned} \tag{24}$$

The two terms in (24) correspond to the two- and one-particle operators represented by the Feynman diagrams in Fig. 2.

Figure 2: Feynman diagrams for the two-particle and one-particle second-order vacuum polarization.

Carrying out the integrations over time in (24) in the limit of small ϵ we obtain

$$4\langle S_\epsilon^{(4)}\rangle_{cVV} - 2\langle S_\epsilon^{(2)}\rangle_{cV}^2 \tag{25}$$

$$= -\frac{2}{\epsilon^2}\sum_{nmkl} U_{nm}U_{kl}\delta(E_n,E_m)\delta(E_k,E_l)\langle a_n{}^*a_k{}^*a_l a_m\rangle$$

$$-\frac{2}{\epsilon^2}\sum_{nm,E_j=E_m} U_{nj}U_{jm}\delta(E_n,E_m)\langle a_n{}^*a_m\rangle$$

$$+\frac{2}{i\epsilon}\sum_{nm,E_j\neq E_m} U_{nj}\frac{1}{E_m-E_j}U_{jm}\delta(E_n,E_m)\langle a_n{}^*a_m\rangle$$

$$+\frac{2}{\epsilon^2}\sum_{nmkl} U_{nm}U_{kl}\delta(E_n,E_m)\delta(E_k,E_l)\langle a_n{}^*a_m\rangle\langle a_k{}^*a_l\rangle$$

$$+O(1)$$

where U is the vacuum polarization operator

$$U_{nm} = 4\pi i\alpha\int d(t_2-t_1)\int d\bar{x}_2\int d\bar{x}_1 D_F(x_2-x_1) \tag{26}$$

$$\times\, Tr[\gamma_\mu S_F(x_2,x_2)]\overline{\phi}_n(\bar{x}_1)\gamma^\mu\phi_m(\bar{x}_1)$$

The first term in (25) arises from the two-particle operator in Eq. (24), and the second and third terms are from the one-particle operator. The second term corresponds to intermediate states in the electron propagator in (24) that are degenerate with the initial state, and the third term corresponds to the rest of the intermediate states. The fourth term in (25) is the square of the vacuum polarization term of order e^2. Substitution of (25) into (23) yields the following expression for the energy shift

$$E_{VV}^{(4)} = \sum_{E_j\neq E_m} U_{nj}\frac{1}{E_m-E_j}U_{jm}\delta(E_n,E_m)\langle a_n{}^*a_m\rangle \tag{27}$$

provided the terms of order ϵ^{-2} in (25) add up to zero. This cancellation is explicitly demonstrated as follows. First, we note the identity

$$\langle a_n{}^*a_k{}^*a_l a_m\rangle = -\langle a_n{}^*a_k{}^*a_m a_l\rangle \tag{28}$$

$$= -\delta_{km}\langle a_n{}^*a_l\rangle + \langle a_n{}^*a_m a_k{}^*a_l\rangle$$

When the right hand side of this identity is substituted into the first term of (25), the first term of (28) cancels the second term of (25), and the remaining term of order ϵ^{-2} is proportional to

$$\sum_{nmkl} U_{nm}U_{kl}\delta(E_n,E_m)\delta(E_k,E_l) \tag{29}$$

$$\times\,[\langle a_n{}^*a_m a_k{}^*a_l\rangle - \langle a_n{}^*a_m\rangle\langle a_k{}^*a_l\rangle]$$

$$= \langle Q^2\rangle - \langle Q\rangle^2$$

where

$$Q = \sum_{nm} U_{nm}\delta(E_n,E_m)a_n{}^*a_m \tag{30}$$

The expression in (29) clearly vanishes for eigenstates of Q, which is the only case that we consider here. In particular, we consider only a spherically symmetric potential for the zero-order one-electron solutions in Eq. (3), and label the solutions by the standard principal, parity, angular momentum, and magnetic quantum numbers. In this case, the vacuum polarization operator in Eq. (26) is spherically symmetric with even parity, and thus diagonal in the parity, angular momentum, and magnetic quantum numbers. Further, the delta function of energy assures that $U_{nm}\delta(E_n,E_m)$ is diagonal in the principal quantum number as well, so that Q can be written as

$$Q = \sum_n U_{nn} a_n{}^* a_n \tag{31}$$

Product states of the form

$$a_i{}^* a_j{}^* \cdots |0\rangle \tag{32}$$

will be eigenfunctions of Q. We are interested in linear combinations of products that form states of well defined total angular momentum. Since such states are formed by summing over magnetic quantum numbers, and the spherical symmetry of the vacuum polarization operator insures that all magnetic substates will have the same value of U_{nn}, it is clear that the states of interest are eigenstates of Q, and hence the term of order ϵ^{-2} vanishes.

ONE-ELECTRON ATOMS

For the case of one-electron atoms, the state vector is given by

$$|i\rangle = a_i{}^*|0\rangle \tag{33}$$

so that

$$\begin{aligned} &\langle a_n{}^* a_k{}^* a_l a_m \rangle = 0 \\ &\langle a_n{}^* a_m \rangle = \delta_{ni}\delta_{mi} \end{aligned} \tag{34}$$

The second-order corrections are given in this case by the well known one-electron expressions

$$E^{(2)} = \Sigma_{ii}(E_i) + U_{ii} \tag{35}$$

where

$$\begin{aligned} \Sigma_{nm}(\omega) = &-4\pi i\alpha \int d(t_2-t_1) \int d\bar{x}_2 \int d\bar{x}_1 D_F(x_2-x_1) \\ &\times \overline{\phi}_n(\bar{x}_2)\gamma_\mu S_F(x_2,x_1)\gamma^\mu \phi_m(\bar{x}_1) e^{i\omega(t_2-t_1)} \\ &- \delta m \int d\bar{x}\overline{\phi}_n(\bar{x})\phi_m(\bar{x}) \end{aligned} \tag{36}$$

is the self energy, and the vacuum polarization U_{nm} is defined in Eq. (26). For one electron, the second-order (fourth-order in e) vacuum polarization correction is

$$E^{(4)}_{VV} = \sum_{E_j \neq E_i} U_{ij} \frac{1}{E_i - E_j} U_{ji} \tag{37}$$

which is the same as the result of ordinary second-order perturbation theory for the perturbation U. There are many other fourth-order corrections that are not discussed here.

TWO-ELECTRON ATOMS

Two-electron states are written as

$$\sum_{ij} C_{ij} a_i{}^* a_j{}^* |0\rangle \tag{38}$$

and for such states

$$<a_n{}^*a_k{}^*a_l a_m> = (C_{kn}{}^* - C_{nk}{}^*)(C_{lm} - C_{ml}) \tag{39}$$
$$<a_n{}^*a_m> = \sum_k (C_{kn}{}^* - C_{nk}{}^*)(C_{km} - C_{mk})$$

The one exchanged photon contribution to the second-order level shift is thus given by

$$E^{(2)}_{pe} = -4\pi i\alpha \int d(t_2-t_1) \int d\bar{x}_2 \int d\bar{x}_1 D_F(x_2-x_1) \tag{40}$$
$$\times \sum_{kn} C_{kn}{}^* \frac{1}{\sqrt{2}}[\bar{\phi}_k(x_2)\bar{\phi}_n(x_1) - \bar{\phi}_n(x_2)\bar{\phi}_k(x_1)]\gamma^{(2)}_\mu \gamma^{\mu(1)}$$
$$\times \sum_{lm} C_{lm} \frac{1}{\sqrt{2}}[\phi_l(x_2)\phi_m(x_1) - \phi_m(x_2)\phi_l(x_1)]\delta(E_n+E_k, E_l+E_m)$$

Eq. (40) is made more explicit by noting that

$$\int d(t_2-t_1) D_F(x_2-x_1) e^{i\eta(t_2-t_1)} = \frac{i}{4\pi} \frac{e^{i|\eta|x_{21}}}{x_{21}} \tag{41}$$

The level shift, understood to mean the real part, is conveniently expressed as

$$E^{(2)}_{pe} = \alpha(Z\alpha)P(Z\alpha)\, mc^2 \tag{42}$$

where P can be expanded as

$$P(Z\alpha) = p_1 + p_3(Z\alpha)^2 + p_5(Z\alpha)^4 + p_7(Z\alpha)^6 + \cdots \tag{43}$$

In (43) the coefficients through p_5 are known exactly for some states, and the next coefficient p_7 is known approximately[8]

$$1^1S_0:\ P(Z\alpha) = \frac{5}{8} + 0.480140(Z\alpha)^2 \tag{44}$$
$$+ 0.219653(Z\alpha)^4 + 0.1507(Z\alpha)^6 + \cdots$$

$$2^3S_1:\ P(Z\alpha) = \frac{137}{729} + 0.076935(Z\alpha)^2$$
$$+ 0.043223(Z\alpha)^4 + 0.0281(Z\alpha)^6 + \cdots$$

$$2^3P_0:\ P(Z\alpha) = \frac{1481}{6561} + 0.219768(Z\alpha)^2$$
$$+ 0.142891(Z\alpha)^4 + 0.1070(Z\alpha)^6 + \cdots$$

$$2^3P_2:\ P(Z\alpha) = \frac{1481}{6561} + 0.040639(Z\alpha)^2$$
$$- 0.001882(Z\alpha)^4 + 0.0037(Z\alpha)^6 + \cdots$$

These expressions provide accurate values for P for Z not too large. For larger Z, the function is readily evaluated by numerical integration.[8]

For a two-electron state composed of the standard one-electron basis states labelled by the principal, parity, angular momentum, and magnetic quantum numbers, summed over magnetic quantum numbers, the lowest-order self-energy correction is

$$E^{(2)}_{SE} = \sum_{nm} \Sigma_{nm}(E_n)\delta(E_n,E_m)<a_n{}^* a_m> = \sum_{orbitals} \Sigma_{nn}(E_n) \tag{45}$$

Similarly, the vacuum polarization correction is

$$E^{(2)}_{VP} = \sum_{orbitals} U_{nn} \tag{46}$$

The total lowest-order two-electron correction is the sum

$$E^{(2)} = E^{(2)}_{pe} + E^{(2)}_{SE} + E^{(2)}_{VP} \tag{47}$$

MANY-ELECTRON ATOMS

The perturbation theory approach described here has a natural extension to the case of many-electron atoms that is a generalization of nonrelativistic many-body perturbation theory. A correction to the Coulomb potential δV is added to the Coulomb potential V to provide a zero-order potential that accounts for the average effect of the electron distribution. Basis states for the expansion of the electron-positron field in (5) are taken as the eigenfunctions of the Dirac equation for this effective potential

$$[-i\vec{\alpha}\cdot\vec{\nabla} + V(x) + \delta V(x) + \beta - E_n]\phi_n(\vec{x}) = 0 \tag{48}$$

The extra term in the potential is compensated by a subtraction in the interaction term

$$H_I(x) = j^\mu(x)A_\mu(x) - j^0(x)A_0{}'(x) - \delta M(x) \tag{49}$$

where

$$\delta V(x) = -\,eA_0{}'(x) \tag{50}$$

This formulation provides a rigorous basis for taking radiative corrections into account in the theory of many-electron atoms. Its practical value will depend on the feasibility of carrying out calculations and the rate of convergence of the perturbation expansion.

REFERENCES

1. D. Layzer and J. Bahcall, Ann. Phys. (NY) 17, 177 (1962).
2. A. Dalgarno and A. L. Stewart, Proc. Phys. Soc. 75, 441 (1960).
3. H. A. Bethe and E. E. Salpeter, *Quantum Mechanics of One- and Two-Electron Atoms*, Springer, Berlin, 1957.
4. J. Sucher, Phys. Rev. 107, 1448 (1957).
5. L. N. Ivanov, E. P. Ivanova, and U. I. Safronova, J. Quant. Spectrosc. Radiat. Trans. 15, 553 (1975).
6. W. H. Furry, Phys. Rev. 81, 115 (1951).
7. M. Gell-Mann and F. Low, Phys. Rev. 84, 350 (1951).
8. P. J. Mohr, Phys. Rev. A , to be published (1985).

SUMMARY OF DISCUSSIONS CONCERNING QED THEORY

G.W.F. Drake
Department of Physics, University of Windsor
Windsor, Ontario, Canada N9B 3P4

ABSTRACT

The discussions following the papers by Sapirstein and Mohr are summarized. The main sections are: (1) a discussion of methods for performing energy level calculations in many-electron atoms, (2) a finite basis set method for calculating hydrogenic Bethe logarithms by John D. Morgan III and (3) calculations of relativistic energy coefficients arising from single photon exchange.

ENERGY LEVEL CALCULATIONS

Following Sapirstein's paper on QED theory, much of the discussion centered around the question posed by him concerning how far many body perturbation theory (MBPT) can be pushed in obtaining an accurate description of atomic properties. Although MBPT has been employed with great success by Kelly[1] and many other workers to a wide range of atomic problems, Stephan Younger pointed out that it may be necessary to take the perturbation expansion to much higher order if a significant improvement in accuracy is to be achieved. Several people expressed the view that the alternative method of configuration interaction (CI) may hold greater promise. The CI method has the advantage that systematic patterns of convergence are obtained when the configurations are expressed in terms of natural orbitals.[2,3] A classic example is provided by the calculations of Carroll et al.[4] for the $1s^2$ state of helium. With a wave function containing 118 NRO's, their final non-relativistic energy of -2.90370 a.u. differs from the exact value by only 0.00002 a.u. A large number of accurate CI calculations for other few-electron systems has been done by Carlos Bunge and co-workers.[5] Other interesting variations on the CI theme include a numerical multiconfiguration Hartree-Fock technique developed by Fischer and Saxena,[6] and combined CI-Hylleraas type expansions studied by Sims and Hagstrom.[7]

For relativistic calculations, multiconfiguration Dirac-Fock techniques and their variants are extensively discussed by Sucher, Dietz and Desclaux in these conference proceedings. However, in addition to this work, Sapirstein pointed out that Caswell and Lepage[8] have made important progress in reformulating the Bethe-Salpeter equation in such a way as to make it more amenable for calculations. Their relativistic two-body formalism reduces to a nonrelativistic Schrödinger theory for a single effective particle with a systematic perturbation series for higher order corrections.

FINITE BASIS SET METHODS FOR BETHE LOGARITHMS

Following the above discussion, John D. Morgan III gave a brief presentation on his finite basis set method of calculting hydrogenic Bethe logarithms. Starting from the definition

$$\ln k_0 = \langle\psi|\vec{p}\cdot(H-E)\ \ln\ (H-E)\ \vec{p}|\psi\rangle$$

$$\equiv \langle\vec{p}\psi|\ (H-E)\ \ln\ (H-E)\ |\vec{p}\psi\rangle \qquad (1)$$

where $(H-E)\psi = 0$, the basic idea is to let H_N be the NxN matrix truncation of H and to work in a basis of its eigenstates, where $\vec{p}\psi$ in included in the original over complete non-orthogonal basis. The only error arises from the fact that $((H-E)\ln(H-E))_N \neq (H_N-E)\ \ln\ (H_N-E)$. In particular, for the 1s state, the basis set $\{\psi_n\}$, n=1,... N is

$$\psi_1 \sim \vec{p}\ \psi \sim \hat{r}\ e^{-r} \qquad (2)$$

$$\psi_n \sim \lambda\vec{r}\ L_{n-2}^{(4)}\ (\lambda r)e^{-\lambda r/2} \quad \text{for } 2 \leq n \leq N \qquad (3)$$

where the $L_{n-2}^{(2\ell+2)}$ are Laguerre polynomials and λ is an adjustable scale factor set equal to unity. One then constructs

$$G_{ij} = (\psi_i,\ \psi_j) \qquad (4)$$

$$H_{ij} = (\psi_i,\ H\psi_j) \qquad (5)$$

and solves the generalized eigenvalue problem for the eigenvalues $E_K^{(N)}$ and orthonormal eigenfunctions $\phi_K^{(N)}$. The exact ψ_1 can be expressed in the form

$$\psi_1 = \sum_{K=1}^{N} d_K^{(N)}\ \phi_K^{(N)} \qquad (6)$$

to obtain the approximation

$$\ln k_0 \simeq \sum_{K=1}^{N}\ (d_K^{(N)})^2\ (\varepsilon_K^{(N)} - E)\ \ln\ (\varepsilon_K^{(N)} - E). \qquad (7)$$

The problem is that the results converge only as ~10/N with increasing N. Much more accurate results can be obtained by use of Neville-Richardson extrapolation. Suppose that X_N has the form

$$X_N = c_0 + \frac{c_1}{N} + \frac{c_2}{N^2} + \frac{c_3}{N^3} + \ldots. \qquad (8)$$

The c_1 contribution can be eliminated by constructing

$$X_N^{(1)} = (N+1)\ X_{N+1} - NX_N = c_0 - \frac{c_2}{N(N+1)} + O(1/N^3) \qquad (9)$$

This can be iterated to eliminate the contribution from all terms up to and including c_M by constructing

$$X_N^{(M)} = \frac{1}{M}\left[(N+M)X_{N+1}^{(M-1)} - NX_N^{(M-1)}\right] \qquad (10)$$

Using N = 120 and M = 6, Morgan obtains a final extrapolated value of

2.9841285557655 (1) Ryd for 1s

and

2.8117698931205 (2) Ryd for 2s.

These results contain three more significant figures than the calculations of Klarsfeld and Maquet.[9] It will be interesting to see if the same ideas can be extended to the two-electron case in a calculation analogous to that of Goldmand and Drake.[10]

RELATIVISTIC ENERGY COEFFICIENTS

The subsequent paper by Peter Mohr on the quantum electrodynamics of high-Z few electron atoms presented results for the coefficients in the energy expansion

$$\begin{aligned} E = {} & E_0^0 Z^2 + E_0^2 \alpha^2 Z^4 + E_0^4 \alpha^4 Z^6 + \ldots\ldots \\ & + E_1^0 Z + E_1^2 \alpha^2 Z^3 + E_1^4 \alpha^4 Z^5 + \ldots\ldots \\ & + \ldots\ldots \end{aligned} \qquad (11)$$

in atomic units. The sum of terms in the first line gives the Dirac one-electron energy, while the second line gives the electron-electron interaction correction due to one exchanged photon. Mohr's paper particularly discusses the terms in the second line, including new exact analytic results for the term E_1^4 (p_5 in his notation) for the two-electron states $1s^2\ {}^1S_0$, $1s2s\ {}^3S_1$, $1s2p\ {}^3P_0$ and $1s2p\ {}^3P_2$. Table I of the present work[11] gives a more extended tabulation of E_0^4 and less accurate numerical estimates of E_1^4 for all states up to and including the n = 3 level in LS-coupling. Also included are the off-diagonal matrix elements responsible for singlet-triplet mixing. The results were obtained by calculating the matrix elements[12]

$$\beta_{n\ell,n'\ell'} = \iint d\vec{r}_1 d\vec{r}_2 \psi_n^*(\vec{r}_1)\psi_\ell^*(\vec{r}_2)\,(V_{12} + B)\psi_{n'}^*(\vec{r}_1)\psi_{\ell'}^*(\vec{r}_2) \qquad (12)$$

over one electron Dirac spinors, where, in the Coulomb gauge,

$$V_{12} + B = (e^2/r_{12})\,[F(r_{12}) - \vec{\alpha}_1 \cdot \vec{\alpha}_2\, G(r_{12})]$$

with

$$F(r_{12}) = 1 + (E_{\ell,\ell'}/2E_{n,n'})(1 - \cos\Omega_{n,n'})$$

$$+ (E_{n,n'}/2E_{\ell,\ell'})(1 - \cos\Omega_{\ell,\ell'})$$

$$G(r_{12}) = \frac{1}{2}(\cos\Omega_{n,n'} + \cos\Omega_{\ell,\ell'})$$

$$E_{n,n'} = E_n - E_{n'}$$

and

$$\Omega_{n,n'} = E_{n,n'}r_{12}/hc.$$

Since Mohr's calculations were done in the Lorentz gauge, it is satisfying to see that the results agree to the figures quoted for the four states he studied. A similar tabulation for the coefficients $E_0^{\,2}$ and $E_1^{\,2}$ for all states up to and including the n = 3 level is given by Drake.[13]

REFERENCES

1. H.P. Kelly, in Atomic Inner Shell Processes, edited by B. Craseman (Academic Press, New York, 1975).
2. P.-O. Löwdin, Phys. Rev. 97, 1474 (1955); P.-O. Löwdin and H. Shull, Phys. Rev. 101, 1730 (1956).
3. For a useful review, see E.R. Davidson, Rev. Mod. Phys. 44, 451 (1972).
4. D.P. Carroll, H.G. Silverstone and R.M. Metzger, J. Chem. Phys. 71, 4142 (1979).
5. See for example C.F. Bunge, Phys. Rev. A 14, 1965 (1976); A.V. Bunge and C.F. Bunge, Phys. Rev. A 30, 2179 (1984).
6. C. Froese Fischer and K.M.S. Saxena, Phys. Rev. A 9, 1498 (1974).
7. J.S. Sims and S. Hagstrom, Phys. Reva. A 4, 908 (1971).
8. W.E. Caswell and G.P. Lepage, Phys. Rev. A 18, 810 (1978).
9. S. Klarsfeld and A. Maquet, Phys. Lett. B 43, 201 (1973).
10. S.P. Goldman and G.W.F. Drake, J. Phys. B 17, L197 (1984); S.P. Goldman, Phys. Rev. A 30, 1219 (1984).
11. G.W.F. Drake, Nucl. Instr. and Method. in press (1985).
12. G.W.F. Drake, Phys. Rev. A 19, 1387 (1979).
13. G.W.F. Drake, Nucl. Instr. and Meth. 202, 273 (1982).

Table I Values of the Coefficients $E_0^{\ 4}$ and $E_1^{\ 4}$ in the Relativistic Energy Expansion (11).

State	$E_0^{\ 4}$	$E_1^{\ 4}$
$1s^2\ ^1S_0$	$-1/8$	0.2197
$1s2s\ ^1S_0$	$-85/1024$	0.1052
$1s2s\ ^3S_1$	$-85/1024$	0.04323
$1s2p\ ^3P_0$	$-85/1024$	0.1430
$1s2p\ ^3P_1$	$-235/3072$	0.05693
$1s2p\ ^1P_1$	$-215/3072$	0.03874
$2^3P_1 - 2^1P_1$	$-5\sqrt{2}/768$	0.03695
$1s2p\ ^3P_2$	$-65/1024$	-0.001877
$1s3s\ ^1S_0$	$-3224/6^6$	0.0324
$1s3s\ ^3S_1$	$-3224/6^6$	0.0172
$1s3p\ ^3P_0$	$-3224/6^6$	0.0408
$1s3p\ ^3P_1$	$-3131/6^6$	0.0190
$1s3p\ ^1P_1$	$-3038/6^6$	0.0141
$3^3P_1 - 3^1P_1$	$-93\sqrt{2}/6^6$	0.0106
$1s3p\ ^3P_2$	$-2945/6^6$	0.00189
$1s3d\ ^3D_1$	$-2945/6^6$	0.00431
$1s3d\ ^3D_2$	$-2935/6^6$	0.00227
$1s3d\ ^1D_2$	$-2930/6^6$	0.00122
$3^3D_2 - 3^1D_2$	$-5\sqrt{6}/6^6$	0.00181
$1s3d\ ^3D_3$	$-2920/6^6$	0.00118
$3^3S_1 - 3^3D_1$	0	6.4×10^{-4}

LAMB SHIFT IN TWO-ELECTRON ATOMS: I. THE LOW-LYING S STATES

A.M.Ermolaev

Department of Physics, University of Durham,
Science Laboratories, Durham DH1 3LE, England,U.K.

ABSTRACT

Revised values of the Lamb shift in $n^{1,3}S$ states , $n \leqslant 5$, of two-electron atoms with $Z \leqslant 10$ are obtained, which are based on a simple and effective approximation for the screening correction to the Bethe logarithm for a two-electron atom. Recent experimental values of the shifts due to Martin are in excellent agreement with the calculations reported here.

1. INTRODUCTION

Recent high precision measurements of some spectral lines in helium and other members of the isoelectronic sequence (1,2,3) have lead to a renewed interest in the improved theoretical values of relativistic and radiative corrections to energy levels in two-electron atoms. Following earlier calculations, particularly by Dalgarno (4) and this author (5), new theoretical work has recently been done, notably by Drake and co-workers and by Hata and Grant. Nevertheless, even a cursory glance at the newly available data for the helium atom reveals that the theory for He is lagging behind the experiment.

For low-Z members of the series, the atomic energy level E of a |nLSJ> state of the atom can be written as an expansion in terms of the parameters αZ, α, and m/M, with the individual terms being expanded in terms of 1/Z to account for correlation between the two atomic electrons. Thus

$$E = E_{nr} - \varepsilon_M + E_r + E_{L,2} + E_{hr} + E_{hQED} + \dots, \tag{1}$$

where E_{nr} is the nonrelativistic (electrostatic Coulomb) energy level of order Z^2 ry in the two-electron atom with an infinitely heavy nucleus of charge Z, and the second term $-\varepsilon_M$ is the non-trivial part of the lowest order correction due to the finite mass of the nucleus (of order Z^2 m/M ry), known sometimes as "mass polarisation" correction. Very accurate numerical values of these two quantities E_{nr} and $-\varepsilon_M$ as well as that of the Breit-Pauli relativistic correction E_r which is of order $(\alpha Z)^2 Z^2$ ry, were obtained by Pekeris and co-workers (6). In (1), $E_{L,2}$ is the main radiative correction (of orders $\alpha(\alpha Z)^2 Z^2$, $\alpha(\alpha Z)^2 Z \ell n \alpha^{1,2}$ and $\alpha(\alpha Z)^2 Z$ ry). Higher order corrections E_{hr} and E_{hQED} contain among them cross terms depending on the parameter m/M.

As well known, the problem of an accurate estimate of the total shift $\Delta E_L = E_{L,2} + E_{hr} + E_{hQED}$ has two different aspects; firstly, it requires an adequate calculation of the main term $E_{L,2}$ which is a task for the nonrelativistic atomic physics. Secondly, a reasonably good estimate of higher-order terms in ΔE_L is also required in view of the high precision of the current experiments. A convergence analysis is also required for the nonrelativistic quantities E_{nr} and $-\varepsilon_M$ as well as for the relativistic correction E_r since these quantities have been obtained with an approximate solution of the two-electron Schrodinger equation, using a finite variational basis. Following Pekeris [6], these variational values have to be replaced by values extrapolated to an infinitely long variational basis set.

In the present paper, I shall concentrate on the $E_{L,2}$ radiative correction. Preliminary results of an analysis which I am carring out at the moment, will be reported here only for the S-states of the two-electron atoms. In this case, the radiative correction $E_{L,2}$ to the atomic level can be written in the following way [7,8,9] :

$$E_{L,2} = E'_{L,2} + E''_{L,2} , \tag{2}$$

where

$$E'_{L,2} = (8/3) Z\, \alpha^3 \langle \delta(\vec{r}_1) + \delta(\vec{r}_2) \rangle \left\{ 2\ln(1/\alpha) + (19/30) - \ln\frac{k}{ry} \right\} \tag{3}$$

and

$$E''_{L,2} = \alpha^3 \langle \delta(\vec{r}_{12}) \rangle \left\{ (28/3)\ln\alpha + (178/15) - (40/3)\, \vec{s}_1\vec{s}_2 \right\} + \alpha^3 Q . \tag{4}$$

In the formulae above, <A> means the expectation value of A in the $|nOSO\rangle$ state of the atom, $\vec{s}_1\vec{s}_2 = -3/4$ for S=0, and 1/4 for S=1, and the quantity Q gives the principal part of the logarithmically diverging $\langle 1/r_{12}^3 \rangle$. In (3), k denotes the average excitation energy of the $|nOSO\rangle$ state of the atom and the corresponding logarithmic quantity (Bethe logarithm) will be a special subject of the subsequent discussion.

It is convenient to introduce, alongside with $E_{L,2}$, the radiative correction for the ionisation energy of the atom. This is achieved by multiplying all one-body terms depending on the elctron density at the nucleus, by the correcting factor $\Delta(Z) = 1 - 2\pi\langle\delta(\vec{r}_1)\rangle/Z^3$, differencing the logarithmic terms, and changing the sign of all two-body terms in the formulae above. In this way, we shall have

$$\Delta_L = E'_{L,1} - E'_{L,2} - E''_{L,2} , \tag{5}$$

where $E'_{L,1}$ is the radiative correction for the hydrogenic ion with nuclear charge Z, of the corresponding order $(\alpha Z)^2 \alpha Z^2$ ry. In these notations, the shift Δ_L will be in correspondence with the experimentally determined "Lamb shift" as it appears in Martin's paper [3]. The theoretical value of the shift will be improved, particularly for excited states, by a treatment of the Bethe logarithm described below and by adding to it higher-order corrections for the hydrogenic ion multiplied by the correcting function $\Delta(Z)$.

2. THE BETHE LOGARITHM ℓnk.

The Bethe logarithm ℓnk in (3) is defined thus:

$$\ell nk = \frac{\sum_{q'} f_{qq'} (E_{q'} - E_q)^2 \, \ell n(E_{q'} - E_q)/ry}{\sum_{q'} f_{qq'} (E_{q'} - E_q)^2} , \qquad (6)$$

where $f_{qq'}$ are oscillator strengths for electric dipole transitions from the initial state $|q>$ ($|nOSO>$, in the present case) to all other states $|q'>$ of the atom including those of the one- and two-electron continuum. Alternatively, ℓnk can be written as

$$\ell nk = \frac{d\, \ell nS(k)}{dk} , \text{ at } k = 2, \qquad (7)$$

where S(k) is the oscillator sum, i.e.

$$S(k) = \sum_{q'} f_{qq'} (E_{q'} - E_q)^k . \qquad (8)$$

I shall now briefly review three different approaches which are often used to calculate ℓnk. They are the pertubation theory in terms of 1/Z applied to (6), direct evaluation of the sums in (6), and semi-empirical methods related to the formula (7).

The perturbational method is the most recent one and it has been suggested by Ermolaev and Swainson [10] and independently by Goldman and Drake [11]. According to this method, the Bethe logarithm is given, to the first order in 1/Z, thus:

$$\ell nk = (\ell nk)_o + \Delta_1 \ell nk , \qquad (9)$$

where the second term $\Delta_1 \ell nk$ is of order O(1/Z). The exact expression for the zeroth-order term $(\ell nk)_o$ is given in [10] where formulae for the screening correction $\Delta_1 \ell nk^o$ can also be obtained from.

$$(\ell nk)_o = \frac{\Delta(n_1,0)\,\ell nK(1) + \Delta(n_2,0)\,\ell nK(2)}{\Delta(n_1,\ell_1) + \Delta(n_2,\ell_2)} , \qquad (10)$$

where $\ell nK(i)$ is the Bethe logarithm for the hydrogenic (one-electron) state $|i> = |n_i \ell_i m_i>$, and $\Delta(n_i,\ell_i) = (\pi n_i^3)^{-1} \delta_{n_i o}$, i=1,2. Therefore the zeroth-order approximation to the two-electron Bethe logarithm is given by a weighted sum of the one-electron Bethe logarithms for the two atomic electrons. This formula is called by some authors "modified Bethe logarithm" or even "*ad hoc*" expression which is hardly justified in view of a clear relation to the perturbation theory.

Recently Goldman and Drake (12) developed a version of the L^2 method to deal with infinite sums in $\Delta_1 \ell nk$ and reported 50 -basis function calculations of the screening corrections in the 1^1S, $2^{1,3}S$, and $2^{1,3}P$ helium-like sequences. In particular, their result for the ground state Bethe logarithm is

$$\ell n\{k(1^1S)/ry\} = \ell n\{19.7693(Z-0.00615)^2\}. \qquad (11)$$

This result confirms earlier findings (7,14) that the Bethe logarithm for the ground state of helium is very close to that of the hydrogenic ion with the same nuclear charge Z = 2. A usual conclusion drawn from (11) is that in nearly all cases (with an exception of the n = 2 states of He where very accurate experimental data is now available), the "hydrogenic" approximation is sufficient to be used in the Lamb shift (3). It is interesting to notice that in the current literature "hydrogenic approximation" to (6) usually means the use of the zeroth-order formula (10) and I shall return to this point later on in this paper.

Kabir and Salpeter (13) were first to apply the method of direct summation to compute (6) for He. They and later Dalgarno (14) established the asymptotic form for the oscillator strenghts for transitions to high energy continuum states. For large $\varepsilon > 0$,

$$\frac{df}{d\varepsilon} \sim C\,\varepsilon^{-7/2} . \qquad (12)$$

For the ground state of helium, singly ionised transitions

$$1s^2 \rightarrow (1s,\varepsilon p) \qquad (13)$$

contribute some 87 per cent to the total sum in (6), with additional 9 per cent coming from simulteneously excited and ionised transitions such as

$$1s^2 \rightarrow (ns,\varepsilon p) \qquad (14)$$

and

$$1s^2 \rightarrow (np,\varepsilon s) . \qquad (15)$$

The coefficient C in the asymptotic form (12) is different for these three series. For (13),

$$C = \frac{512}{3} \left| \int \psi_o(\vec{r}_1,0)\, u_{1s}(\vec{r}_1)\, d\vec{r}_1 \right|^2 , \qquad (16)$$

for (14) it is

$$C = \frac{512}{3} \int |\psi_o(\vec{r}_1,0)|^2 d\vec{r}_1 , \qquad (17)$$

provided that (14) accounts for <u>all</u> final ns states including those of the continuum, and, finally, for the series (15), C = 0. According to (12), transitions (14) cause the sum S(k) to diverge at k = 5/2, whereas the similar singularity due to (15) occurs only at k = 7/2. Behaviour of S(k) at k = 2 is strongly influenced by the singular point at k = 5/2 and therefore the asymptotic form (12) has to be taken into account in direct computations of the sums in the Bethe logarithm (6). Earlier calculations by Dalgarno [14] gave for $\ell nk(1^1S)$ of He the numerical value of 4.3696 with an estimated error lying between ± 0.0300. This result is remarkably good because it differs only by +0.0043 from the perturbational value of Goldman and Drake [12] and, in fact, predicts a small negative screening defect $\Delta_1 \ell nk$. These calculations were extended some time ago to higer members of the ground state sequence up to Z = 10 in a paper by Aashamar and Austvik [15]. It has already been pointed out [10] that their computed values gradually diverge from the hydrogenic values as Z increases from 2 to 10 and, therefore, unreliable.

A similar method applied by Suh and Zaidi [16] to the $2^{1,3}S$ states of helium, gives less accurate values of the Bethe logarithm as it can be seen from Table 1 below, than those for the ground state.

TABLE 1. Calculations of the Bethe logarithm for the $2^{1,3}S$ states of helium. Comparison of the direct summation method with the perturbation theory in terms of 1/Z.

State	Direct summation (Ref.16)	Perturbation theory (Ref.12) $(\ell nk)_o$	$+\Delta_1 \ell n$
2^1S	4.345 ± 0.020	4.35127	4.37157
2^3S	4.380 ± 0.020	4.35127	4.36510

A simple formula

$$\delta E = b_o(1 + 1/n^3)\ \delta(\ell nk)\quad cm^{-1} \tag{18}$$

connects the change $\delta(\ell nk)$ in the numerical value of the Bethe logarithm with the corresponding change δE in the Lamb shift (3). For He, $b_o = 0.41$ showing that an uncertainty of ±0.01 in the logarithm produces an uncretainty of ±0.0088, ±0.0051, and ±0.0043 cm^{-1} in the position of the n=1,2,and 3 energy levels,respectively. These bounds are generally wider than the experimental errors for the respective levels in He quoted by Martin thus setting very high requirements on the accuracy of the logarithmic term in the Lamb shift.

Finally, I shall briefly discuss the group of semi-empirical methods based on the formula (7). Pekeris [17] constructed an interpolating polynomial P(k) = 1/S(k) using accurate values of the oscillator sums at points k=-1,0,1,2, and setting P(2.5)=0 according to the asymptotic expression (12). Then the Bethe logarithm is expressed in terms of the logarithmic derivative of P(k) at k=2. This method gives ℓnk = 4.50 for the 1^1S state, and ℓnk = 2.65 for the 2^3S state of the helium atom showing that the method is poor even for the ground state. The main reason is the fast changing S(k) at k=2 due to the nearly situated singular point k=2.5. This singularity influences S'(k=2) much stronger than the behaviour of S(k) at distant points k=0 and k= -1 also used in constructing the interpolating polynomial P(k). The dominancy of k=2.5 increases with Z and the method gives even poorer results for higher members of the sequence. Hata and Grant suggested an improved version [18] of the Pekeris method, by considering a procedure for the screening correction to the Bethe logarithm rather than to the logarithm itself. This is easily achived by noticing that

$$\Delta_1 \ell nk = \frac{d\ell nT(k)}{dk} \quad \text{at} \quad k = 2 , \tag{19}$$

where

$$T(k) = S(k)/S_H(k) , \tag{20}$$

$S_H(k)$ being the oscillator sum for the hydrogenic ion and the quantity $(\ell nk)_H$ is to be given by (10). In order to obtain an approximation to T(k) near k = 2, they used interpolating functions, for both S(k) and $S_H(k)$, of the form

$$\tilde{S}(k) = a + bk + ck^2 + \frac{C}{5/2 - k} E_c^{k-5/2} , \tag{21}$$

and similarly for $\tilde{S}_H(k)$, fitting them to the oscillator sums at k = 0,1, and 2. Then

$$\tilde{T}(k) = \tilde{S}(k)/\tilde{S}_H(k). \tag{22}$$

This method has produced very good results, particularly, for the ground state of He where the screening correction of Hata and Grant is -0.0073 to be compared with the perturbational value of -0.00615 obtained in [12]. For the 2^3S and 2^3P states, their results for He are 0.0082 and 0.0061 to be compared with the perturbational values of 0.0134 and 0.0048, respectively. However, as Goldman and Drake have pointed out [12], the sreeening correction of Hata and Grant is too small for higher members of the sequence since it decreases as $1/Z^3$ instead of the expected 1/Z dependence.

Though the use of (21) is a great improvement comparing with the original method of Pekeris, there is an uncertainty with regards to the numerical values of E_c and C to be used in such a procedure. E_c is the energy of the ionised electron at which the asymptotic form (7) becomes valid, and Hata and Grant used C as defined by (17). According to calculations of Dalgarno and Stewart [14], the main contribution of nearly 90 per cent comes, to the Bethe logarithm, from single ionisation, a half of this contrubution being from the energy range $E_c > 400$ ry where the asymptotic form is assumed to be applicable. Theferore, the effective value of C may be slightly different from that given by (17) and closer to the value (16).

3. A SEMI-EMPIRICAL 1/Z-EXPANSION METHOD

In view of a considerable difficulty to extend the rigorous pertubational calculations, particularly, to higher excited states of the sequence, I have decided to look into a possibility of formulating an effective semi-empirical method for the Bethe logarithm, free of some problems discussed above. Let us consider, alongside with S(k), the 1/Z expansion of this quantity, i.e.

$$S(k,Z) = \{S_o(k) + S_1(k)/Z + \dots\}Z^{m(k)} . \qquad (23)$$

In the formula above, the zeroth-order term $S_o(k)$ is generally different from a sum of the two hydrogenic terms though in the case of S-states this property holds. We shall now define the quantity T(k,Z) as follows

$$T(k,Z) = S(k,Z)/S_o(k)\, Z^m . \qquad (24)$$

Then

$$T(k,Z) = 1 + T_1(k)/Z + .. \qquad (25)$$

and the screening correction to the Bethe logarithm is given by

$$\Delta_1 \ell nk = \frac{dT_1(k)}{Zdk} \qquad (26)$$

at k=2, where $T_1 = S_1/S_o$. We can also introduce an effective screening correction $\Delta\ell nk$ to the logarithm according to

$$\Delta\ell nk = (dT/dk)/T, \qquad (27)$$

for any given Z. In order to simplify the notations, I shall use $\Delta_1\ell nk$ in all cases below though there is, of course, a difference between (26) and (27) of order $O(1/Z^2)$. A typical behaviour of T as a function of k and Z is displayed in Figs. 1 and 2.

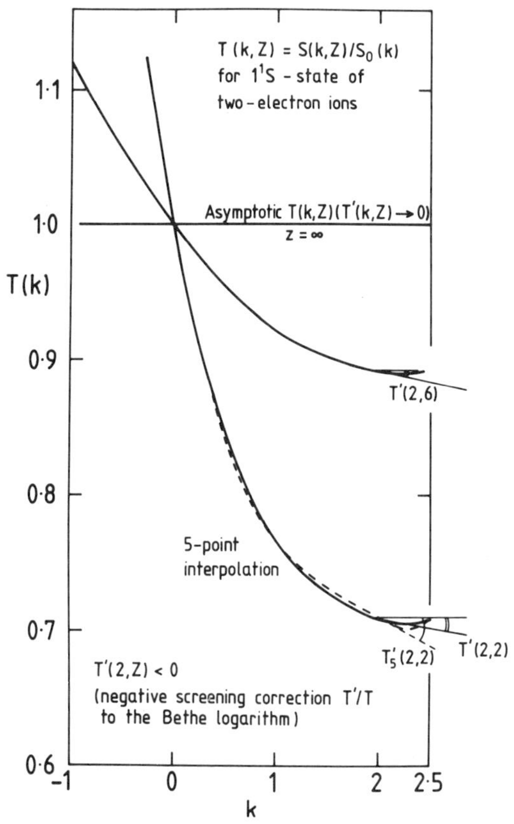

Fig.1. The screening correction $\Delta\ \ell nk = T'(2,Z)/T(2,Z)$ in the ground state of the two-electron atoms.

Two important features of this theory are to be noted: firstly, the property

$$T(2.5,Z) = T(2,Z) \qquad (28)$$

is the exact one and the parameters C and E_c of the asymptotic form (7) do not enter in any explicit way. Secondly, the strong dependence on Z has been completely eliminated, T being a smooth, slow-depending function on both k and Z. Therefore interpolation procedures are expected to work for T better than for the original quantities S(k,Z) and $S_H(k,Z)$.

A simple classification of the screening corrections to the Bethe logarithm can be introduced, based on a 3-point (parabolic) approximation to T as a function of k, in the interval

$$1.0 < k < 2.5. \qquad (29)$$

The two possible cases for the derivative T'(2) are

$$T'(2) > 0 \quad \text{if } T(1) < T(2), \qquad (30)$$

and

$$T'(2) < 0 \quad \text{if } T(1) > T(2). \qquad (31)$$

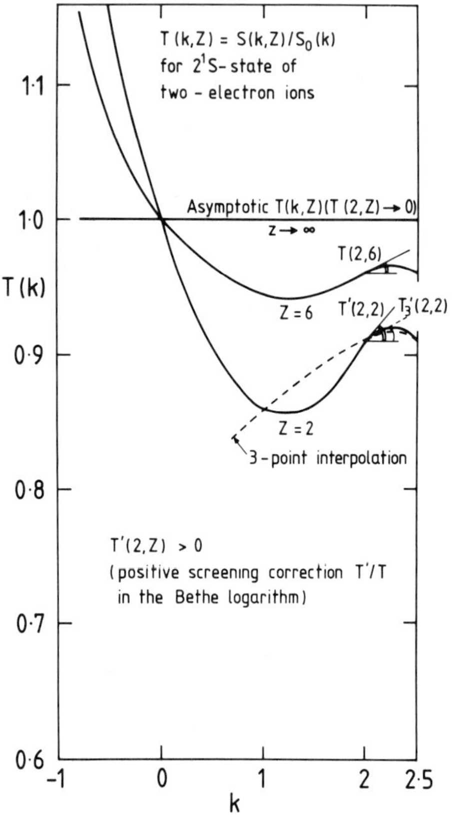

Fig.2. The screening correction $\Delta \ln k = T'(2,Z)/T(2,Z)$ in the excited 2^1S state of the two-electron atoms. Note the reversed sign of the derivative at k = 2, comparing to the case displayed in Fig.1.

As Figs. 1 and 2 show, both possibilities are realised in the two-electron atoms: the present theory explains the small negative correction in the ground state as well as positive corrections in excited S states. I would like to emphasise that, with the points k = 2.5, 2, 1, and two additional points k = 0, and -1, being fixed (by the availability of the accurate numerical data), we have, in fact, very little control over the quality of the prediction by the theory. From this point of view it would have been more advantageous to have the k values spaced more tightly around the point k = 2. An application of this method to S-states is considered below, where numerical values [6] are used for the oscillator sums S(k).

For the 1^1S state, T(1) > T(2) and the screening correction is negative. For He, the method gives the following results:

3-point interpolation (k=2.5,2.0, and 1.0)	-0.020
4-point interpolation (k=2.5,2.0,1.0, and 0.0)	-0.010
5-point interpolation (k=2.5,2.0,1.0,0.0, and -1.0)	-0.008

to be compared with -0.006 from the perturbation theory. The 5-point method applied to ions with Z up to 10, gives corrections which have the correct 1/Z behaviour, with the asymptotic coefficient larger than that of Goldman and Drake by a factor of 3. These results as well as the results of previous calculations by Aashamar and Austvik and by Hata and Grant are presented in Table 2.

For the excited S-states, T(1) < T(2) and the screening correction is positive. The n=2 states have been computed before by Hata and Grant and by Goldman and Drake . The present calculations for 2^1S and 2^3S states are given in Tables 3 and 4. Agreement between the present results and those of Goldman and Drake are very good both in the numerical value of the correction for He and in the value of the asymptotic coefficient. It is interesting to notice that in the case of T(1) < T(2), the 3-point method gives the best agreement with the perturbational results.

In view of the very encouraging results for the $2^{1,3}S$ states, I have extended calculations to obtain the values of the Bethe logarithm in $3^{1,3}S$, $4^{1,3}S$, and $5^{1,3}S$ states of the two-electron atoms with Z < 10. Because the numerical values of the one-electron Bethe logarithm are known to a great precision [19], the uncertainty in ℓnk given in Tables 5-7, comes from $\Delta_1 \ell nk$ and is probably no more than some 20 per cent of the correction. According to formula (18), the corresponding uncertainty of 0.0001 cm^{-1} in the Lamb shift is far to small comparing with higher order relativistic and QED corrections neglected in (3) and (4).

It follows from (10) that the zeroth-order approximation to the two-electron Bethe logarithm, $(\ell nk)_0$ coincides numerically with the hydrogenic logarithm $\ell nK(1s)$ if n=1 or n=∞. The largest deviation of $(\ell nk)_0$ from $\ell nK(1s)$ occurs at n = 2. Tables 2-7 show that the screening correction $\Delta_1 \ell nk$ for n > 2 always shifts the corrected value towards $\ell nK(1s)$, with the resulting value ℓnk being generally closer to the hydrogenic $\ell nK(1s)$ rather than to the zeroth-order logarithm $(\ell nk)_0$. Therefore the recommended choice of the numerical value of ℓnk in the Lamb shift (3) is either $\ell nK(1s)$ or $(\ell nk)_0 + \Delta_1 \ell nk$, depending on the accuracy required, but not $(\ell nk)_0$. A simple formula

$$\Delta_1 \ell nk \sim 0.09/(nZ) \qquad (32)$$

allows to obtain a good estimate of the screening correction for ions with Z > 4 in the n = 3,4, and 5 states. For instance, for Z=5 and n=3, (32) gives 0.0060 to be compared with 0.0064 (3^1S) and 0.0061 (3^3S) from Table 5.

4. THE SHIFT Δ_L OF IONISATION ENERGY.

For an S-state of a two-electron atom with nuclear charge Z, the shift (5) of ionisation energy will be written as follows:

$$\Delta_L = I^A_{L,2} + I^B_{L,2} + I'_{L,2} + I^{HO}_{L,2} + O . \qquad (33)$$

The first two terms are of order $\alpha^3 Z^4$ ry and they give the main contribution to the total shift.

$$I^A_{L,2} = (8\alpha^3 Z^4/3\pi)\ \{\ell n(1/\alpha)+19/30\}\ \Delta(Z)\ \text{ry} , \qquad (34)$$

$$I^B_{L,2} = -(8\alpha^3 Z^4/3\pi)\ \{\ell n K(1s)/\text{ry} - (1-\Delta(Z))\ell n k/\text{ry}\}\ \text{ry} . \qquad (35)$$

The leading two-body QED terms in (33) are given by

$$I'_{L,2} = I^{A'}_{L,2} + I^{B'}_{L,2} , \qquad (36)$$

where

$$I^{A'}_{L,2} = -(28/3)\alpha^3 \ell n\alpha \ \langle\delta(\underline{r}_{12})\rangle\ \text{ry} , \qquad (37)$$

and

$$I^{B'}_{L,2} = -(328/15)\alpha^3\ \langle\delta(\underline{r}_{12})\rangle + (28/3)\alpha^3 Q\ \text{ry} , \qquad (38)$$

with the Q term being defined thus:

$$Q = -(1/4\pi)\lim_{a\to o} \langle r^{-3}_{12}(a) + 4\pi(\gamma + \ell n a)\delta(\underline{r}_{12})\rangle , \qquad (39)$$

and

$$r^{-3}_{12}(a) = \begin{cases} r^{-3}_{12}, & |\underline{r}_1 - \underline{r}_2| > a \\ o, & |\underline{r}_1 - \underline{r}_2| < a \end{cases} \qquad (40)$$

For low Z, these terms give the principal contribution to the total Δ_L. The next-order terms $I^{HO}_{L,2}$ in (33) are usually smaller than the experimental uncertainy, with an important exception of the neutral helium atom being discussed below.

$$I^{HO}_{L,2} = (8\alpha^3 Z^4/3\pi)\{3\pi\alpha Z(427/384-\ell n2/2) - 3(\alpha Z)^2\ \ell n(\alpha Z) + 0.4042(\alpha/\pi)\}\ \Delta(Z)\ \text{ry} . \qquad (41)$$

The last term O in (33) contains all terms which have not been included into the preceding (34) - (41). Among those omitted from the considereation at the present time, are higher-order relativistic corrections of orders $(\alpha^4 Z^6)$ and $(\alpha^4 Z^5)$, higher-order QED corrections, corrections due to finite nuclear size and mass (relativistic recoil). For the helium atom, these corrections are expected to be smaller than the experimental uncertainty [3]. A more detailed analysis will be given later.

The terms in Δ_L as given above, contain three atomic parameters depending on the nonrelativistic wavefunction of the two-electron atom: they are $\Delta(Z)$, ℓnk, and Q.

The function $\Delta(Z)$ introduced in [20],

$$\Delta(Z) \;=\; 1 \;-\; (2\pi<\delta(\underline{r}_1)>/Z^3) \tag{42}$$

gives the departure of the electron density at the nucleus of a two-electron atom, from that in the ground state of the hydrogenic ion with the same nuclear charge Z. Among the one-body corrections in (33), many are proportional to the electron density at the nucleus. The shift of ionisation energy due to these terms is readily found by multiplying the corresponding terms, in the hydrogenic ion, by the factor of $\Delta(Z)$. The expected error in $\Delta(Z)$ is less than 10^{-5} if Pekeris' data for the average values of the delta-functions are used.

The Bethe logarithm ℓnk has been discussed above and its values given in Tables 2-7 are expected to be correct, in terms of the shift (35), within some 0.0010 cm^{-1} or, perhaps, even 0.0001 cm^{-1}, for He.

For the ground state, the quantity Q first introduced by Sucher [9], have been computed by the author in [5] using 20-term variational functions of Hart and Herzberg [21]. In Table 8 the calculations are compared with recent calculations of Q by Drake and Makowski [22] who used 50-term wavefunctions, and with calculations of Hata [23] who used wavefunctions with up to 120 terms. All three calculations show very small magnitude of Q in low-Z atoms as well as a changing sign in Q between Z=3 and 4 [24]. Comparison of the Q values for n=1 with those for n=2 computed in [22] shows that a reasonably accurate estimate for n=2 can be obtained from the n=1 data by a simple $1/n^3$ scaling. This has been used to obtain an estimate of the shift due to the Q-term, in the S-states with n =3, and 4 (for n=5, it has been assumed in the present work that Q=0). An error in the total shift emanating from inaccuracies in Q is expected to be negligible small.

The total shifts Δ_L which include all corrections listed in (34)-(41), are given in Tables 9 and 10. The main uncertainty in the value of the total shift is due to omitting higher-order corrections from (41). Usually, this uncertainty is taken to be of the same magnitude as that of the corrections of the last order included in the estimate of the total shift. For He, this would give $\pm$ 0.023 cm^{-1} in the 1^1S state, and $\pm$ 0.0020 cm^{-1} in the $2^{1,3}S$ states.

Now we are in position to discuss the recent helium data of Martin [3] which is shown in the first column of Table 11. Comparison with the theoretical shifts Δ_L requires first to subtract the nonrelativistic value of the ionisation potential as well as the mass-polarisation correction and relativistic correction (Breit-Pauli) from the experimental value of the potential. The Pekeris' estimate of the probable error in these values is 0.001 cm^{-1} [25]. Table 11 shows that for the $2^{1,3}S$ states of He there exists an agreement between the experimental and theoretical values of Δ_L (columns 2 and 3) within a rather large uncertainty of some 15 per cent of the theoretical Lambshifts computed by Suh and Zaidi [16] and by this author in [20]. The probable theoretical error of ± 0.0160 cm^{-1} of these calculations exceeds by a factor of 3-4 the experimental uncertainty of the new measurements [2,3]. For n >2, no estimates of the Lamb shift were available.

The last columns of Table 11 show that the revised theoretical values of the Lamb shift in the 2^1S and 2^3S states of He are in excellent agreement with the data of Martin. A small difference between the theoretical shifts computed using the values of the Bethe logarithm of the present work and those due to Goldman and Drake, is about 0.0010 - 0.0020 cm^{-1} and smaller than an estimated experimental uncertainty of 0.0040 cm^{-1}. At the same time, the precision of the experimental data is sufficient to conclude that the "modified" Bethe logarithm i.e. the term $(\ell nk)_o$, formula (10), is not good enough and has to be corrected by adding $\Delta_1 \ell nk$.

For the 3^3S and 4^3S terms where experimental data of high precision is also available, good agreement between the experiment and the present theory is also recorded.

The theoretical $1s^2\ ^1S_o$ shift $\Delta_L = -1.381 \pm 0.023$ cm^{-1} in helium is in agreement with the experimental value -1.26 ± 0.15 cm^{-1} cited by Martin [3]. A rather large uncertainty of this 'experimental' value noted there is due to the uncertainty of the position of the ground level relative to the excited levels in the atom. An earlier theoretical value -1.335 ± 0.008 cm^{-1} [15] is likely to underestimate the higher order corrections in the ground state of He and it disagrees with the present result as well as with the value -1.377 cm^{-1} of Hata [23].

5. ACKNOWLEDGMENTS

This work was supported by a research grant from SERC of the United Kingdom. The author wishes to express thanks to N.B.S. for financial support to attend the Workshop on the Atomic Theory at the National Bureau of Standards, U.S.

6. REFERENCES

1. P.Juncar, H.G.Berry, R.Damaschini, and H.T.Duong, J.Phys.B 16, 381 (1983); R.A.Holt, S.D.Rosner,T.D.Gaily, A.G.Adam, Phys.Rev. A 22, 1563 (1980); H.A.Klein, S.Bashkin, B.P.Duval, F.Moskatelli, J.D.Silver, H.F.Beyer, and F.Folkmann, J.Phys.B 15, 4507 (1982); R.D.Deslattes, H.F.Beyer and F.Folkmann, J.Phys. B 17, L689 (1984).
2. C.J.Sansonetti and W.C.Martin, Phys.Rev. A 29, 159 (1984).
3. W.C.Martin, Phys.Rev. A 29, 1833 (1984); ibid. A 30, 651 (1984).
4. A.Dalgarno and A.L.Stewart, Proc.Phys.Soc.London, 75, 441 (1960); also A.Dalgarno and A.L.Stewart, Proc.Phys.Soc.London, 76, 49 (1960).
5. A.M.Ermolaev, Phys.Rev. A8, 1651 (1973); A.M.Ermolaev and M.Jones, J.Phys. B 7, 199 (1974).

6. C.L.Pekeris, Phys.Rev. 112, 1649 (1958); ibid. 115, 1216 (1959); ibid. 126, 143, 1470 (1962); ibid. 127, 509 (1962); see also Y.Accad and C.L.Pekeris, Phys.Rev., A 4, 516 (1971); ibid. A 11, 1479 (1975).

7. P.K.Kabir and E.E-Salpeter, Phys.Rev. 108, 1256 (1957).

8. H.Araki, Progr.Theor.Phys. (Japan), 17, 619 (1957).

9. J.Sucher, Phys.Rev., 109, 1010 (1958).

10. A.M.Ermolaev and R.A.Swainson, J.Phys. B 16, L35 (1983).

11. S.P.Goldman and G.W.F.Drake, J.Phys. B 16, L183 (1983).

12. S.P.Goldman and G.W.F.Drake, J.Phys. B 17, L197 (1984).

13. See Ref.7 where the recommended value of C in formula (12) is 286.

14. A.Dalgarno and A.L.Stewart, Proc.Phys.Soc.London, 76, 49 (1960).

15. K.Aashamar and A.Austvik, Phys.Norv., 8, 229 (1977).

16. K.S.Suh and M.H.Zaidi, Proc.Roy.Soc.London, A 291, 94 (1966).

17. C.L.Pekeris, Phys.Rev., 115, 1216 (1959).

18. J.Hata and I.P.Grant, J.Phys. B 17, 931 (1984).

19. S.Klarsfeld and A.Maquet, Phys.Lett., 43B, 201 (1973).

20. A.M.Ermolaev, Phys.Rev.Lett., 34, 380 (1975).

21. J.F.Hart and G.Herzberg, Phys.Rev., 106, 79 (1957).

22. G.W.F.Drake and A.J.Makowski, J.Phys. B 18, L103 (1985); data for 2^3S states are to be interchanged with those for 2^3P states, for all ions (privite communication).

23. J.Hata, J.Phys. B, 17, L625 (1984).

24. Table 8 shows that data of Ref.5 related to the Kabir-Salpeter terms do agree with those of Ref.22 if a different decomposition of the total correction into two terms, used in these two works, is taken into account.

25. In the nonrelativistic theory where normalisable bound-state solutions of the Schroedinger equation for a two-electron atom exist, the problem of electron-electron correlation is relatively well understood. A failure of CI expansions to compete with the Hylleraas expansions is clearly linked to this problem of correlation since the rate of convergence of CI methods is controlled by the singular point $r_{12}=0$. The method of solution used by Pekeris, provides very accurate and steady-convergent values of the nonrelativistic energy as well as those of other operators including the Breit-Pauli operator. As this author has pointed out in Ref.10, this method is more reliable for computing the Breit-Pauli terms E_r particularly for low Z, rather than using the Dirac-Fock method, an approach suggested by Hata and Grant (J.Phys.B 15, L549, 1982). On the other hand, as Z increases, the relative role of the Breit-Pauli terms recedes and the use of the Dirac-Fock equations becomes more justified, at least from the practical point of view.

TABLE 2. Screening correction $\Delta_1 \ell nk$ to the Bethe logarithm $(\ell nk)_o$ in the ground state of the helium isoelectronic sequence. Comparison of various methods.

Z	$(\ln k)_o$	$\Delta_1^{AA}\ell nk$	$\Delta_1^{HG}\ell nk$	GD		Present work	
				$\Delta_1 \ell nk$	ℓnk	$\Delta_1 \ell nk$	ℓnk
2	4.37042	-0.0005 (100)	-0.00725	-0.00616	4.3643	-0.00822	4.3622
3	5.18135	0.029 (10)	-0.00725	-0.00410	5.1772	-0.00780	5.1735
4	5.75672	0.020 (3)	-0.00125	-0.00308	5.7536	-0.00686	5.7499
5	6.20300	0.011 (3)	-0.00066	-0.00246	6.2005	-0.00590	6.1971
6	6.56765	-0.003 (2)	-0.00039	-0.00205	6.5656	-0.00542	6.5622
7	6.87595	-0.012 (2)	-0.00025	-0.00176	6.8742	-0.00448	6.8715
8	7.14301	-0.028 (2)	-0.00017	-0.00154	7.1415	-0.00397	7.1390
9	7.37858	-0.045 (2)	-0.00012	-0.00137	7.3772	-0.00356	7.3750
10	7.58930	-0.064 (2)	-0.00009	-0.00123	7.5881	-0.00323	7.5861

AA - Ref.15
HG - Ref.18
GD - Ref.12

TABLE 3. Screening correction $\Delta_1 \ell nk$ to the Bethe logarithm $(\ell nk)_o$ in the 2^1S state[1] of the helium isoelectronic sequence. Comparison of different methods.

Z	$(\ell nk)_o$	GD		Present work	
		$\Delta_1 \ell nk$	ℓnk	$\Delta_1 \ell nk$	ℓnk
2	4.35127	0.02030	4.3715	0.01895	4.3702
3	5.16220	0.01355	5.1758	0.01238	5.1746
4	5.73757	0.01017	5.7477	0.00911	5.7467
5	6.18385	0.00814	6.1920	0.00720	6.1911
6	6.54850	0.00679	6.5553	0.00594	6.5544
7	6.85680	0.00582	6.8626	0.00506	6.8619
8	7.12386	0.00509	7.1290	0.00440	7.1283
9	7.35943	0.00453	7.3640	0.00390	7.3633
10	7.57015	0.00408	7.5742	0.00349	7.5736

GD - Ref.12

TABLE 4. Screening correction $\Delta_1 \ell nk$ to the Bethe logarithm $(\ell nk)_o$ in the 2^3S state of the helium isoelectronic sequence. Comparison of various methods.

Z	$(\ell nk)_o$	$\Delta_1^{HG} \ell nk$	GD		Present work	
			$\Delta_1 \ell nk$	ℓnk	$\Delta_1 \ell nk$	ℓnk
2	4.35127	0.00822	0.01383	4.3651	0.01762	4.3689
3	5.16220	0.00224	0.00923	5.1714	0.01142	5.1736
4	5.73757	0.00091	0.00693	5.7445	0.00842	5.7460
5	6.18385	0.00045	0.00554	6.1894	0.00665	6.1905
6	6.54850	0.00026	0.00462	6.5531	0.00551	6.5540
7	6.85680	0.00016	0.00396	6.8608	0.00470	6.8615
8	7.12386	0.00011	0.00347	7.1273	0.00409	7.1280
9	7.35943	0.00008	0.00308	7.3625	0.00325	7.3631
10	7.57015	0.00005	0.00278	7.5729	0.00325	7.5734

HG - Ref.18
GD - Ref.12

TABLE 5. Screening correction $\Delta_1 \ell nk$ to the Bethe logarithm $(\ell nk)_o$ in the 3^1S and 3^3S states of the helium isoelectronic sequence.

Z	$(\ell nk)_o$	3^1S		3^3S	
		$\Delta_1 \ell nk$	ℓnk	$\Delta_1 \ell nk$	ℓnk
2	4.36269	0.01475	4.3774	0.01420	4.3769
3	5.17362	0.01034	5.1840	0.00986	5.1835
4	5.74899	0.00791	5.7569	0.00752	5.7565
5	6.19527	0.00639	6.2017	0.00607	6.2018
6	6.55992	0.00536	6.5653	0.00508	6.5650
7	6.86822	0.00461	6.8728	0.00437	6.8726
8	7.13528	0.00405	7.1393	0.00384	7.1391
9	7.37085	0.00360	7.3745	0.00342	7.3743
10	7.58157	0.00325	7.5848	0.00308	7.5847

TABLE 6. Screening correction $\Delta_1 \ell nk$ to the Bethe logarithm $(\ell nk)_o$ in the 4^1S and 4^3S states of the helium isoelectronic sequence.

Z	$(\ell nk)_o$	4^1S		4^3S	
		$\Delta_1 \ell nk$	ℓnk	$\Delta_1 \ell nk$	ℓnk
2	4.36682	0.01019	4.3770	0.00994	4.3768
3	5.17775	0.00729	5.1850	0.00705	5.1848
4	5.75311	0.00565	5.7588	0.00543	5.7585
5	6.19940	0.00459	6.2040	0.00442	6.2038
6	6.56404	0.00386	6.5679	0.00372	6.5678
7	6.87234	0.00334	6.8756	0.00321	6.8755
8	7.13941	0.00293	7.1423	0.00282	7.1422
9	7.37497	0.00262	7.3776	0.00252	7.3775
10	7.58569	0.00236	7.5881	0.00227	7.5880

TABLE 7. Screening correction $\Delta_1 \ell nk$ to the Bethe logarithm $(\ell nk)_o$ in the 5^1S and 5^3S states of the helium isoelectronic sequence.

Z	$(\ell nk)_o$	5^1S		5^3S	
		$\Delta_1 \ell nk$	ℓnk	$\Delta_1 \ell nk$	ℓnk
2	4.36849	0.00723	4.37572	0.00712	4.37561
3	5.17942	0.00523	5.18466	0.00511	5.18453
4	5.75479	0.00407	5.75886	0.00396	5.75874
5	6.20107	0.00332	6.20439	0.00322	6.20430
6	6.56572	0.00281	6.56852	0.00272	6.56844
7	6.87402	0.00243	6.87644	0.00235	6.87637
8	7.14108	0.00213	7.14321	0.00207	7.14315
9	7.37665	0.00190	7.37855	0.00185	7.37850
10	7.58737	0.00171	7.58908	0.00167	7.58904

TABLE 8. Computed values of the two-electron QED correction $I^{B'}_{L,2}(Z)$ and the Q-correction (see definitions in the text) to the ionisation energy of the 1^1S state of the helium isoelectronic sequence (all shifts are given in cm^{-1}).

Z	$I^{B'}_{L,2}$			$(28/3)\alpha^3Q$		
	a	b	c	a	b	c
2	0.0715	0.0683	0.0681	-0.0276	-0.0308	-0.03106
3	0.493	0.494	0.4930	-0.0046	-0.0036	-0.00462
4	1.52	1.665	1.663	+0.100	+0.245	+0.2427
5	3.64	4.020	4.010	0.552	0.932	0.9218
6	7.23	8.010	7.998	1.50	2.284	2.273
7	12.8	14.13	14.113	3.25	4.586	4.559
8	20.6	22.88	22.858	5.80	8.080	8.061
9	31.4	34.77	34.750	9.70	13.10	13.075
10	45.5	50.35	50.322	15.1	19.94	19.909

a - Ref.5.
b - Ref.23.
c - Ref.22.

TABLE 9. Lambshifts Δ_L of the ionization energies for the singlet S-states (n = 1,2,3,4, and 5) of the helium isoelectronic sequence. (All shifts are given in cm^{-1}).

Z	1^1S	2^1S	3^1S	4^1S	5^1S
2	-1.381	-0.0912	-0.0205	-0.0061	-0.0014
3	-8.947	-0.778	-0.206	-0.0773	-0.0337
4	-30.01	-2.931	-0.811	-0.318	-0.152
5	-73.64	-7.678	-2.170	-0.874	-0.416
6	-150.3	-16.36	-4.678	-1.878	-0.906
7	-271.2	-30.45	-8.777	-3.582	-1.710
8	-449.0	-51.60	-14.96	-6.137	-2.951
9	-696.6	-81.53	-23.74	-9.782	-4.697
10	-1028.0	-122.1	-35.70	-14.74	-7.063

TABLE 10. Lambshifts Δ_L of the ionization energies for the triplet S-states (n = 2,3,4, and 5) of the helium isoelectronic sequence. (All shifts are given in cm^{-1}).

Z	2^3S	3^3S	4^3S	5^3S
2	-0.1322	-0.0291	-0.0091	-0.0033
3	-1.002	-0.257	-0.0956	-0.0440
4	-3.537	-0.950	-0.370	-0.177
5	-8.906	-2.453	-0.975	-0.475
6	-18.46	-5.164	-2.076	-1.024
7	-33.71	-9.531	-3.862	-1.916
8	-56.28	-16.04	-6.536	-3.258
9	-87.94	-25.21	-10.32	-5.160
10	-130.5	-37.60	-15.44	-7.741

TABLE 11. Lambshifts for some S terms of the neutral helium atom (in cm^{-1}).

Term	T_{expt} [a]	$T_{expt} - T_r$ [a]	Δ_L [a]	Present work using ℓnk	Present work using $(\ell nk)_o$
2^1S	32033.2325 (50)	-0.0887 (54)	-0.103 [b] -0.106(16) [c]	-0.0912 [d] -0.0904 [e]	-0.1025
2^3S	38454.6969(40)	-0.1328 (40)	-0.135 [b] -0.129(11) [c]	-0.1322 [d] -0.1345 [e]	-0.1428
3^1S				-0.0205 [d]	-0.0298
3^3S	15073.8839 (40)	-0.0257 (45)		-0.0290 [d]	-0.0374
4^1S				-0.0061 [d]	-0.0121
4^3S	8012.5580 (40)	-0.0107 (45)		-0.0091 [d]	-0.0149
5^1S				-0.0014 [d]	-0.0056
5^3S				-0.0033 [d]	-0.0076

a - from Ref.3.
b - from Ref.20.
c - from Ref.16.
d - present work.
e - from Ref.12.

Calculation of P-Violating and CP-Violating Matrix Elements for Heavy Atoms

W.R. Johnson, Department of Physics,
University of Notre Dame,
Notre Dame IN 46556

ABSTRACT

Calculations of P-violating and CP-violating dipole matrix elements for the heavy atoms Rb,Cs,Au,and Tl are described. These calculations, which are made starting from a variety of model potentials including the relativistic Hartree-Fock potential in zeroth order, are extended to account for first-order correlation corrections. Excitation energies, hyperfine constants, and allowed transition matrix elements are determined and compared with measurement as a control on the calculations. The spread in values of the P-violating and CP-violating matrix elements calculated starting from different potentials, which is a measure of the accuracy of the results, is found to be about 20% in the present work. We expect to be able to reduce this spread by including higher-order correlation correction.

INTRODUCTION

In this work we are concerned with two effects, each of which has to do with weak interaction corrections to electric dipole matrix elements. The first effect is the induction of a nonvanishing off-diagonal matrix element between two states of the same nominal parity caused by the exchange of Z_0 bosons between atomic electrons and the nucleus. The second is the induction of an electric dipole moment in the atomic ground state by an (assumed) intrinsic electron electric dipole moment.

The exchange of Z_0 bosons, which leads to P violation is described by Weinberg-Salam theory.[1] It leads to phenomena such as circular dichroism which have been used to detect P violation in Cs and Tl. The P-violating interaction is characterized by two parameters: G, the weak interaction coupling constant, and Θ_W, the Weinberg angle. This interaction can be expressed in the form of an effective one-particle Hamiltonian[2]:

$$h_w = -\frac{G}{\sqrt{8}} Q_w \gamma_5 \rho_{nuc}(r) \tag{1}$$

where $\rho_{nuc}(r)$ is the nuclear charge density, and where $Q_W = Z(1 - 4 \sin^2\Theta_W) - N$; Z and N being the proton and neutron numbers of the nucleus. Precise atomic measurements of P violation make it possible to determine precise values of the fundamental parameters G and Θ_W, provided only that the associated

0094-243X/85/1360150-12$3.00

atomic calculations can be carried out reliably. During the past few years atomic measurements of P violation at the 10-20% level have been made.[3-5] The corresponding calculations[6-9] have an uncertainty of about 20%, judging by the spread between the more recently published values. Our goal, which has not yet been realized, is to improve the atomic calculations to the point where they could be used to determine values of basic weak interaction parameters at the level of several percent.

The CP-violating effects considered here arise from the coupling of an intrinsic electron electric dipole moment, d_e, with the electromagnetic field. Again the interaction can be reduced to the form of a one-electron Hamiltonian[10]

$$h_w = 2\,i\,\frac{d_e}{e}\,\gamma_4\gamma_5\,\vec{p}^{\,2} \tag{2}$$

where $\vec{p}$ is the electron momentum. The present experimental upper limit on d_e, determined by atomic measurements of of the linear Stark effect in heavy elements is $d_e < 2\times10^{-24}$ e cm.[11,12] Given the potential for improving the measurements by several orders of magnitude, it should be possible to improve the present limit significantly, and ideally to detect and measure d_e. This would have important consequences for super symmetric gauge theories, where predictions[13] presently range from $d_e = 10^{-25}$ to 10^{-30} e cm. The dipole moment D_e induced in the atom by the electron moment is proportional to d_e. Our purpose is to determine accurate theoretical values of the enhancement factor D_e/d_e to help interpret future measurements.

The main difficulty encountered in atomic calculations of P or CP violation is not with the weak interactions, which can be treated in lowest-order perturbation theory, but with the electron-electron Coulomb interaction. The later interaction requires the use of higher-order perturbation theory. Further, there is the problem of judging the reliability of calculations carried out in perturbation theory, since there is no small expansion parameter associated with the residual electron-electron interaction. In the present calculations we start with several different central potentials V(r),and use the spread in values determined in a given order of perturbation theory to measure the accuracy of the calculated atomic parameters. Quantities calculated starting from one potential model may be more sensitive to higher-order effects than another, and thus may be misleading. However, if all evaluations of a quantity starting from different potentials give answers within say 10% of one another then we could assign that 10% to be a theoretical error. The present calculations are carried out to first order in the difference between the two electron Coulomb potential and the model potential. The spread in the calculated weak interaction parameters is found to be closer to 20%. We hope that we will be able to reduce this spread to 1 or 2% by extending the calculations to sufficiently high order.

ZEROTH-ORDER CALCULATIONS

We employ four different potential models in the present studies:

a) Tietz potential[14]

$$V(r) = -\frac{\alpha}{r}\left(1 + \frac{(Z-1)}{(1+tr)^2}\, e^{-\gamma r}\right)$$

b) Green potential[15]

$$V(r) = -\frac{\alpha}{r}\left(1 + \frac{Z-1}{H(e^{r/d}-1)+1}\right)$$

c) Norcross potential[16]

$$V(r) = V_{TF}(\lambda r) - \frac{\alpha}{2r^4}\,\alpha_d\left(1 - e^{-(r/r_c)^6}\right) - \frac{\alpha}{2r^6}\,(\alpha q - 3\beta a_0)\left(1 - e^{-(r/r_c)^{10}}\right)$$

d) Hartree-Fock (HF) potential

$$V_{hf}u(r) = \alpha \sum_a \int \frac{d^3r\,d^3r'}{|\vec{r} - \vec{r}'|}\left[(u_a^\dagger u_a)'u - (u_a^\dagger u)'u_a\right]$$

where u(r) is an arbitrary one electron orbital and where $u_a(r)$ is a core orbital, satisfying the relativistic V(N-1) frozen-core HF equations

$$(h_0 + V_{hf})u_a(r) = \varepsilon_a u_a(r)$$

Table I. Comparison of the theoretical excitation spectrum for Cs determined from the model potentials employed in this work with measured values. Energies in a.u.

State	Tietz	Green	Norcross	HF	Measurement[a]
6s	0.14343	0.14309	0.14301	0.12737	0.14310
$6p_{1/2}$	0.09247	0.09223	0.09250	0.08562	0.09217
$6p_{3/2}$	0.08892	0.08915	0.08928	0.08379	0.08964
7s	0.05827	0.05901	0.05883	0.05519	0.05865
$7p_{1/2}$	0.04379	0.04424	0.04410	0.04202	0.04393
$7p_{3/2}$	0.04270	0.04323	0.04307	0.04137	0.04310
8s	0.03213	0.03251	0.03240	0.03095	0.03230

a) C.E.Moore,Atomic Energy Levels, Natl. Bur. Stand. Ref. Data Ser., Natl. Bur. Stand. (U.S.) Circ. No. 35 (U.S. GPO, Washington, D.C., 1971), Vol.III.

All four of these potentials have been employed by others in calculations of P violation. For the first three potentials the parameters are chosen to fit the low-lying spectra of the atoms being considered. The resulting fit for the case of Cs is shown in Table I. Similar fits were obtained for the other atoms studied.

In calculating the values listed in the table we solve the Dirac equation in each of the potentials considered; nuclear finite size effects, which are important for the weak interaction matrix element calculations are included by appropriately modifying the nuclear Coulomb potential.

One test of the quality of the lowest-order wave functions calculated using these potentials is provided by comparing the values of the calculated ground-state hyperfine intervals with measurements. For the atoms considered, which all have j = 1/2 ground states, the hyperfine interval is given in lowest order by

$$\Delta\nu_{hfs} = \pm 13074.7 \; g_I (I + \tfrac{1}{2}) \cdot \frac{4}{3} \int \frac{2fg}{r^2} \, dr \quad \text{MHz} \qquad (3)$$

where g and f are the large and small component radial wave functions for the valence electron. The values of the integral in Eqn.(3) are compared with experiment in Table II.

Table II. Values of the hyperfine integral in Eqn.(3) from lowest-order calculations compared with experiment.

State	Tietz	Green	Norcross	HF	Measurement
Rb 5s	0.1162	0.1131	0.1114	0.0683	0.1076
Cs 6s	0.1870	0.1872	0.1824	0.1115	0.1799
Au 6s	1.4813	1.6016	1.6254	1.2644	1.8287
Tl $6p_{1/2}$	-0.3772	-0.3749	-0.3653	-0.3092	-0.3730

To include the effects of the weak interactions in our calculations we add the weak Hamiltonian, which for both of the cases considered is described by a one-electron operators h_w to the unperturbed single particle Hamiltonian h_0, which includes the electron's kinetic energy and it's interaction with the nucleus:

$$h_0 \rightarrow h_0 + h_w$$

In the presence of h_w, a small admixture of states with the same angular momentum but opposite parity, is induced in the valence orbital $u_v(r)$:

$$u_v(r) \rightarrow u_v(r) + w_v(r)$$

where the perturbed orbital $w_v(r)$ satisfies

$$(h_0 + V)\, w_v(r) = - h_w u_v(r) \tag{4}$$

The electric dipole matrix element between two valence states of the same nominal parity 1 and 2 becomes :

$$E_0 = \langle w_2 | ez | u_1 \rangle + \langle u_2 | ez | w_1 \rangle \tag{5}$$

which can be easily evaluated once Eqn.(4) is solved for the perturbed orbitals w_1 and w_2. For the CP-violating interaction we must evaluate the diagonal matrix element in Eqn.(5). In the HF case care must be taken in the lowest-order calculations, since the weak interaction modifies the core orbitals leading to a weak correction to the HF potential:

$$V_{hf} \rightarrow V_{hf} + V^{(1)}{}_{hf}$$

The weakly perturbed HF potential, $V^{(1)}{}_{hf}$, gives significant corrections to the dipole matrix element as illustrated in Table III for the case of Cs. In the first row of the table values of the matrix element calculated neglecting $V^{(1)}{}_{hf}$ are listed, while in the second row values calculated using a first-order approximation to $V^{(1)}_{hf}$ are given. In the third row the remaining higher order corrections, determined by using the fully self-consistent value of $V^{(1)}_{hf}$ are given. Comparisons are also shown in this table with other previous calculations. The disagreement seen between the various calculations at the 2% level is due to different choices of nuclear charge densities and illustrates the sensitivity of the P-violating matrix elements to the nuclear size and shape parameters. Results of our lowest-order calculations of P-violating matrix elements for the cases of Cs and Tl are compared with other theoretical and experimental values in Table IV. The spread between the lowest-order theoretical values for these elements is seen to be about 15%, which we assign as a theoretical error to our lowest-order calculations.

Table III. Comparison of lowest order Hartree-Fock values for the parity violating 7s --> 6s matrix element in Cs from the present calculation with other Hartree-Fock calculations. Units : ie a_0 $(Q_w/N)10^{-11}$.

Term	Present a	Present b	Martensson -Pendrill c	Dzuba, et al. d
$V^{(1)}_{hf} = 0$	0.7279	0.7395	0.7361	0.740
first	0.1373	0.1399	0.1393	0.132
higher	0.0467	0.0476	0.0474	0.048
total	0.9119	0.9270	0.9228	0.920

a) Uniform nuclear charge distribution : Rnuc = 6.206 fm.
b) Fermi charge distribution : c = 5.674 fm, t = 2.3 fm.
c) A-M. Martensson-Pendrill (unpublished) : c = 6.125 fm, t=2.4 fm.
d) V. A. Dzuba et al.table8. : c = 5.61 fm, t = 2.5 fm.

The atomic dipole moment induced by the CP-violating interaction given in Eqn.(2) is proportional to the intrinsic moment d_e, the proportionality constant being given by R. As shown by Sandars[18], R is much larger than 1 for heavy, polarizable atoms such as those considered here. The values of the enhancement factor R determined in lowest-order are given in Table V, where they are compared with the previous calculations by Sandars[18] and by Sandars and Sternheimer[12]. The very large value of R from the HF calculations for Tl is due principally to the $6s_{1/2}$ -->$6p_{1/2}$ excitation induced by $V^{(1)}_{hf}$, and indicates that the HF potential may be an unsuitable starting point for many-body perturbation calculations of CP violating effects in this atom.

Table IV. P-violating dipole matrix elements for Cs and Tl in units $i\, e\, a_0\, (Q_W/N)\, 10^{-11}$ calculated in various potential models.

State	Tietz	Green	Norcross	HF	Exp.	Others
Cs (6s->7s)	1.04	1.04	1.02	0.912	0.93(16)[a] 1.00(8)[b]	0.880[c] 0.972[d] 0.996[e] 0.886[f]
Tl ($7p_{1/2}$->$6p_{1/2}$)	- 9.54	-8.49	-9.20	-9.75	7.2(1.4)[g]	-8.73[e] -6.83[h]

a) M.A.Bouchiat et al. Ref.3 . Here Q_W = -68.6.
b) C.E.Wieman et al.(unpublished)
c) V.A.Dzuba et al. Ref.8 .
d) C.Bouchiat,C.A.Piketty,and D.Pignon,Ref.6 .
e) D.V.Neuffer and E.D.Commins,Ref.7 .
f) A-M Martensson-Pendrill (unpublished).
g) P.Drell and E.D.Commins, Ref.5 . Here Q_W = -112.7.
h) B.P.Das et al., Ref.17.

Table V. Ground-stste electric dipole moment enhancement factor, $R = D_{atom}/d_e$, calculated using various model potentials.

Atom	Tietz	Green	Norcross	HF	Other
Rb	34.0	29.7	30.0	26.7	24[a]
Cs	158	139	144	127	119[a]
Au	393	397	370	340	
Tl	-681	-675	-685	-1910	-700[b]

a) P.G.H.Sandars, Ref.18 (with shielding).
b) P.G.H.Sandars and R.M,Sternheimer, Ref. 12.

FIRST-ORDER CALCULATIONS

Starting with any of the potentials described in Sec.II one can easily determine the corrections to the atomic wave function due to the difference between the electron-electron Coulomb potential and the model potential in first-order perturbation theory. Two types of correction to the dipole matrix element in Eqn.(5) occur in first order: corrections to the valence orbitals caused by the change in potential seen by the valence electron, and corrections associated with polarization of the atomic core. In the HF case the first type of correction vanishes.

For the parametric model potentials, zeroth-order dipole matrix elements are known to be independent of the gauge of the electro magnetic field; consequently, matrix elements of ez are identical to matrix elements of e v/ω_0 ,where v is the z component of the velocity, and ω_0 is the zeroth approximation to the transition frequency:

$$E_0 = \langle 2 \mid ez \mid 1 \rangle_0 = \langle 2 \mid \frac{eV}{\omega_0} \mid 1 \rangle_0$$

In first order one can show

$$E_1 = \langle 2 \mid ez \mid 1 \rangle_1 = \langle 2 \mid \frac{eV}{\omega_0} \mid 1 \rangle_1 - \frac{\omega_1}{\omega_0} \langle 2 \mid \frac{eV}{\omega_0} \mid 1 \rangle_0$$

where ω_1 is the first order correction to ω. The sum of zeroth and first order terms is therefore seen to be gauge independent to first order since the second term on the right hand side of Eqn.(9) is just what is required to account for the first-order change in the transition frequency. In addition to being an important physical constraint on the matrix elements, the property of gauge independence provides a useful check on our numerical codes. In Table VI we give a detailed example of the contributions to ez and ev/ω matrix elements for an allowed transition matrix element in Cs illustrating the accuracy of our numerical calculations. It can be seen that although the individual contributions to the first-order matrix elements in length and velocity forms are quite different, the sums of these contributions are precisely equal.

Table VI. Comparison of length and velocity forms of the reduced matrix element for the 6s --> $6p_{1/2}$ transition in Cs. Tietz potential: ω_0 = 0.054507 ω_1 = -0.003292.

Term	$\langle 6p_{1/2} \| z \| 6s_{1/2} \rangle$	$\langle 6p_{1/2} \| v/\omega \| 6s_{1/2} \rangle$
E_0	-6.7705191	-6.7705191
E_1(val)	-0.1844816	0.9331966
E_1(core)	1.1880296	0.4792892
$-\omega 1/\omega 0\ E_0$		-0.4089378
E_1(tot)	1.0035480	1.0035480
$E_0 + E_1$	-5.7669711	-5.7669711

Since the zeroth-order potential in the HF case is nonlocal the corresponding zeroth-order and first-order matrix elements are not gauge independent. This difficulty is remedied in the present calculations by replacing the first-order core-polarization corrections by corrections calculated in the random-phase approximation (RPA). A discussion of this procedure from the point of view of field theory has been given by Feldman and Fulton.[19] From the point of view of many-body perturbation theory, the RPA adds those terms in second and higher order that result from iteration of the first-order core polarization correlation corrections, and restore guage independence.

In Table VII we compare values of the allowed transition matrix elements for the case of Cs, calculated to first order starting from the four potentials considered, with precise experimental values.

Table VII. Reduced matrix elements for the $6s_{1/2}$ --> $6p_{1/2}$,$6p_{3/2}$ transition in Cs calculated in first order using various starting potentials.

Transition	Tietz	Green	Norcross	HF	Experiment [a]
6s --> $6p_{1/2}$					
E_0	-6.711	-6.644	-6.722	-7.462	
E_1	1.004	-0.296	0.227	0.412	
E_0+E_1	-5.767	-6.940	-6.496	-7.014	-6.361(18)
6s --> $6p_{3/2}$					
E_0	4.850	4.743	4.805	5.278	
E_1	-0.815	0.161	-0.223	-0.303	
E_0+E_1	4.035	4.904	4.581	4.975	4.414(20)

a) L.N.Shabanova,et al. Opt.Spectrosk. 47,3 (1979).

The magnitude and sign of the first order corrections are seen to vary from one potential to another in the table, resulting in an even larger spread in values in first order results than in zeroth order. This large fluxuation in the calculations is partially due to the fact that the model potentials were constructed by fitting only the valence electron spectrum, with no regard to properties of the core. Since first-order perturbation theory couples the resulting poorly described core orbitals with the valence orbitals, the corresponding first-order corrections fluxuate from case to case. We hope, and indeed expect, that these fluxations will be reduced in second order, since the first-order corrections already account partially for the poor initial description of the core.

Weak interactions are included in our first-order calculations just as they were in the lowest order; viz., each orbital occuring in the first-order calculation is replaced by a parity mixed orbital. The counting problems associated with this replacement become complex; e.g., the 10 individual terms (Bruckner-Goldstone diagrams) occuring in first-order increase to 36 when the weak interaction is included. We reduce these terms to radial differential equations which are solved numerically. Results from our first-order calculations of P-violation for the $6s_{1/2}$ --> $7p_{1/2}$ transition in Cs are shown in Table VIII. The row labeled E_0 gives the zeroth-order values calculated using a Fermi distribution to describe the nucleus and are thus different from those given in Table IV. The rows labled E_v and $E_{v'}$ give results for two distinct types of valence orbital correlation corrections, the term E_v gives all matrix elements of one weakly-perturbed valence orbital with the correlation correction to the other, while $E_{v'}$ gives matrix elements of one unperturbed orbital with the weakly perturbed valence correlation corrections to the other. The rows E_a and $E_{a'}$ give the corresponding core polarization corrections.

Table VIII. First-order P-violating dipole matrix element for the $75_{1/2}$ -->$6s_{1/2}$ transition in Cs. Units: ie a_0 (Q_w/N) 10^{-11}. Fermi charge distribution parameters : c = 5.674 fm, t = 2.3 fm.

Term	Tietz	Green	Norcross	HF	Exp.
E_0	1.079	1.073	1.062	0.927	
E_v	0.009	1.244	0.822		
$E_{v'}$	-0.174	-1.339	-0.907		
E_a	-0.344	-0.465	-0.369		
$E_{a'}$	0.247	0.363	0.247		
E_1	-0.325	-0.197	-0.207	-0.037[a]	
E_0+E_1	0.754	0.867	0.855	0.890	0.93(16)[b] 1.00(8)[c]

a) A-M Martensson-Pendrill (unpublished).
b) M.A.Bouchiat et al. Ref.3 . Here Qw = -68.6.
c) C.E.Wieman et al.(unpublished).

As in the case of allowed transitions first-order results fluctuate from one potential to another for reasons given above. Our experience after applying first-order perturbation corrections to each of the atoms considered is that these corrections can be as large as 50% and tend to worsen the agreement between values calculated from different initial potentials.

Despite this uncomfortable result, we plan to continue the perturbation theory approach at least through second order with the expectation that when the important pair-correlation effects, occuring for the first time in second order, are accounted for the resulting matrix elements will be under better control. To extend our calculations through second-order we must carry out sums involving pairs of excited orbitals. Such sums can be reduced to partial differential equations in two radial variables. While such "geminal" equations have been studied nonrelativistically, the corresponding relativistic calculations have not as yet been treated and pose a challenge to be met before second-order values of P-violating matrix elements can be given.

ACKNOLEDGEMENTS

This work was part of a collaboration on P and CP violation between the author, J. Sapirstein, D.S. Guo, M. Idrees, Z.W. Liu and J. Liu. Support was provided by NSF grant #PHY83-08136.

REFERENCES

1) S. Weinberg, Phys. Rev. Lett. 19,1264 (1967); A. Salam, in Elementary Particle Theory, edited by N. Svartholm (Almquist and Forlag, Stockholm, 1968).
2) M. Bouchiat and C. Bouchiat, J. Phys. (Paris) 35, 111 (1974); 36, 493 (1975).
3) M. A. Bouchiat, J. Guena, L. Pottier, and L. Hunter, Phys. Lett. 134B, 463 (1984).
4) P. H. Bucksbaum, E. D. Commins, and L. R. Hunter, Phys. Rev. D24, 1134 (1981).
5) P. Drell and E. D. Commins, Phys. Rev. Lett. 53, 968, (1984).
6) C. Bouchiat, C. A. Piketty, and D. Pignon, Nucl. Phys. B221, 68 (1983).
7) D.Neuffer and E. D. Commins, Phys. Rev. A16,1760 (1977); 16, 844 (1977).
8) V. A. Dzuba, V. V. Flambaum, P. G. Silverstrov and O. P. Sushkov,J. Phys. B18, 597 (1985).
9) A.-M. Martensson-Pendrill (unpublised).
10) This one-body form for CP-violation can be inferred from P. G. H Sanders, J. Phys. B1, 511 (1968).
11) H. Gould, Phys. Rev. Lett. 24, 1091 (1970).
12) P. G. H. Sandars and R. M. Sternheimer, Phys. Rev. A11, 473 (1975).
13) M. B. Gavela, in Theoretical Symposium on Intense Medium Energy Sources of Strangeness, University of California at Santa Cruz, 1983, edited by T. Goldman, H. E. Haber, and H. F. W. Sadrozinski: (American Institute of Physics, New York, 1983), Vol. 102.
14) T. Tietz, J. Chem. Phys. 22, 2094 (1954).
15) A.E.S.Green, D. L.Sellin and A. S. Zachor, Phys. Rev. 184, 1 (1969).
16) D. W. Norcross, Phys. Rev. A7, 606 (1973).

17) B. P. Das, J.Andriessen, M. Vajed-Samii, S.N.Ray, and T. P. Das, Phys. Rev. Lett. 49, 32 (1982).
18) P. G. H. Sandars, Phys. Lett. 22,290 (1966).
19) G. Feldman and T. Fulton, Annals of Phys. 152, 376 (1984).

RELATIVISTIC CALCULATIONS FOR MANY ELECTRON ATOMS

J.P. Desclaux
Centre d'Etudes Nucléaires de Grenoble
DRF/Service de Physique
85X, 38041 Grenoble Cédex, France

ABSTRACT

Many improvements have now been introduced in ab-initio methods for relativistic atomic structure calculations. After a short description of the different methods, we review the various contributions to energy levels and compare the most recent theoretical and experimental results for few electron heavy ions.

INTRODUCTION

Beginning of atomic physics started with one or few electrons systems, then and until few years ago, the research in this field focussed mainly on neutral atoms or slightly ionized ions with few vacanies either in the valence or inner shells. Now the area of few electron systems has regained an important vitality due to the development of ion sources and the use of high energy accelerators and plasma sources which make possible to study very heavy and highly ionized atoms. The experimental results emerging from these new facilities have challenged the theorists since for these systems no simple extrapolation from the low Z region is adequate and that because the velocity of the electrons is high enough to make relativistic effects a major contribution to the energy levels or other observables.

The relativistic formulation of atomic theory is provided by the Bethe and Salpeter equation which included both the relativistic effects and the electron-electron correlation, but no exact solution is known for this equation. Thus all the practical calculations start from a zero order approximation generally taken as a sum of Dirac one-electron operators associated to the Breit interaction. Then quantum electrodynamics (QED) effects are added to correct this no-pair approximation. For more than one-electron systems these QED corrections are only estimated and comparison with experiment is used to assess the validity of the approximations.

In calculating energy levels many various terms are involved: electron-electron interaction, magnetic interaction, retardation in the interaction and QED corrections to name only the most important ones. As these terms are not of the same sign, agreement with experiment is not by itself a definite proof of the validity of the assumptions. The aim of this paper is thus to review in detail each of these contributions and consequently to try to reach some conclusion concerning the accuracy of nowdays ab initio calculations for few electron systems. Section I summarizes some of the methods use to perform relativistic calculations, section II concentrates on the non QED many-body problem in the relativistic case, section III deals with the retardation in the electron-electron interaction, section IV with

0094-243X/85/1360162-14$3.00

QED corrections and the last section is devoted to comparison with experimental results.

I. RELATIVISTIC METHODS FOR MANY ELECTRON SYSTEMS

The choice of a relativistic many-body Hamiltonian is discussed in an other contribution to these proceedings[1] and we shall not consider it here. Let us assume that a correct description is given by:

$$H = \sum_{i} h_D(i) + \sum_{i}\sum_{j} g(i,j) \qquad (1)$$

where h_D is the Dirac one-electron operator:

$$h_D = c\vec{\alpha}\cdot\vec{p} + \beta c^2 + V_N(r) \qquad (2)$$

with c the velocity of light, p the electron momentum, V_N the nuclear potential and the fourth order Dirac matrices given by:

$$\vec{\alpha} = \begin{bmatrix} 0 & \vec{\sigma} \\ \vec{\sigma} & 0 \end{bmatrix} \qquad \beta = \alpha_4 - 1 \qquad \alpha_4 = \begin{bmatrix} I & 0 \\ 0 & -I \end{bmatrix} \qquad (3)$$

In equation (1) g(i,j) is the electron-electron interaction to be discussed later. This Hamiltonian does not allow for pair production and occurence of negative energy states is avoided in an ad hoc way (for example in numerical calculations by imposing strictly electron and not positron boundary conditions).

The next step after the choice of the Hamiltonian is that of the many-electron wavefunction. At this stage many different methods have been used and we shall, for simplicity, divide them in two categories. In the first one, the non relativistic correlation energy is determined with high accuracy and then relativistic corrections are added either in the Breit Pauli approximation or fully relativistically. The second category is characterized by an attempt to include many-body and relativistic effects on an equal footing. In both cases QED corrections are added afterwards as a perturbation on a more or less sophisticated level to be discussed in section IV. Most representative examples of the first category of methods are:

the Z-expansion technique introduced by Layzer and Bahcall[2] and Doyle[3]and extensively used by Safronova and coworkers[4]. In this method the total energy is assumed to be expanded in the double serie:

$$E = Z^2 \sum_{n}\sum_{m} A_{nm} Z^{-n} (\alpha^2 Z^2)^m \qquad (4)$$

where the A_{nm} coefficients are independant of Z and are associated with the non-relativistic energy ($n \geq 0$, $m=0$), the relativistic Dirac corrections ($m \geq 0$, $n=0$) and the other terms correspond to the Breit interaction and various other corrections because of the non additivity of correlation and relativistic effects. The non-relativistic part of the energy is calculated using the S matrix perturbation theory with hydrogenic basis functions. Relativistic corrections are then taken into account by including expansion of the Sommerfeld formula in the Pauli approximation, Breit interaction and intermediate coupling mixing.

The unified relativistic theory of G.W.F. Drake[5] in which the total Hamiltonian is writen as the sum of three terms:

$$H = (H^{o}+B_{p})^{LS} + R(H^{1})R^{-1} - \Delta \qquad (5)$$

In the first term H^{o} is the non-relativistic Hamiltonian and B_p a sum of operators representing the spin-orbit, spin-spin, orbit-orbit, mass-velocity and Darwin corrections. The second term is equivalent to the Hamiltonian given in equation (1) but includes the recoupling transformation R from jj to LS couplings. The third term corrects for double counted contributions in the first two. The first term is calculated, as indicated by the subscript LS, with best available non-relativistic wave functions. For the second one hydrogenic products of Dirac spinors are used.

The MCDF-EAL method has been used by I.P. Grant and coworkers[6] to obtain, through a limited configuration interaction calculation, the relativistic corrections to the energy levels while the non-relativistic part is again obtained from highly correlated non-relativistic wave functions.

The point in common with all the methods of the second category is that they all use configuration state functions (CSF) of the type:

$$| \nu JM\pi\rangle = A\ \phi_1(1)\phi_2(2)\ldots\ldots\ldots\phi_n(n) \qquad (6)$$

that are eigenstates of the total angular momemtum J, its projection M and the parity π. And ν stands for all other quantum numbers needed to fully specify the CSF. This CSF is an antisymmetric product of Dirac spinors:

$$\phi_{n,\kappa,m} = \frac{1}{r}\begin{bmatrix} P_{n,\kappa}(r)\chi_{\kappa,m}(\theta,\psi) \\ iQ_{n,\kappa}(r)\chi_{-\kappa,m}(\theta,\psi)\end{bmatrix} \qquad (7)$$

and the total wavefunction is generally written as a linear combination of these CSF's, i.e.:

$$\Psi(JM\pi) = \sum_n C_n\ | \nu_n JM\pi\rangle \qquad (8)$$

The Multiconfiguration Dirac Fock Method (MCDF)[7] uses the variationnal principle applied to the Hamiltonian of equation (1) to optimize simultaneously the radial functions P and Q of the Dirac spinors (7) and the weights C_n of the various CSF's. The most crucial problem with this method is certainly its rather slow convergence with respect to the number of CSF included.

The Relativistic Random Phase Approximation (RRPA) developpe by W.R. Johnson and coworkers[8] is discussed somewhere else in these proceedings. It has the advantage of summing to all order the so-called bubble diagrams of the many-body perturbation theory and the drawback of neglecting other types of diagrams. To overcome this problem it has been extended to handle multiconfiguration reference states.[9]

The model potential methods[10–11] differ from the MCDF method in that they use a central potential with adjustable parameters (fitted to reproduce experimental results or to satisfy some criteria) to generate a basis of Dirac spinors for the building of the CSF's that are then used to perform a configuration interaction calculation. **The g-Hartree method**[12] recently introduced takes advantage of the fact that it is possible to construct an independant single particle basis by solving Hartree-Fock type equations depending upon a parameter g and that this parameter can be determined such that the correlation energy with respect to this basis vanishes to a given order of the perturbation theory. This method has up to now only been used up to second order and its convergence has to be verified.

This outlook of the various methods available to handle many-body relativistic calculations show that none is really completly satisfactory since some are not really self-consistent in mixing non-relativistic and relativistic basis sets while for the other convergence is very difficult to assess and that without saying anything concerning higher order relativistic corrections (beyond the Dirac-Breit Hamiltonian) nor QED. This workshop may be a good opportunity to clarify all these problems.

II. NON-QED MANY-BODY PROBLEM.

In the pure non-relativistic limit, the two particle interaction reduces to the instantaneous Coulomb repulsion between the electrons while, in the relativistic case, this interaction involves other contributions, the so-called magnetic and retardation terms. Let us assume here that the g(1,2) term of equation (1) is given by the zero frequency limit of the interaction reduced to the magnetic term. This term is not gauge dependant and so we can postpone the discussion of the choice of the gauge with that of the retardation term to be considered in the next section. In that case we have to deal with the operator:

$$\frac{1}{R} - \frac{\vec{\alpha}_1 . \vec{\alpha}_2}{R} \qquad (9)$$

in which R is the interelectronic distance and the α's are the Dirac matrices already defined. The first term is thus the Coulomb repulsion and the second the magnetic interaction.

Table I: Convergence of the Coulomb correlation energy (in eV) for the $1s^2$ state as a function of the basis size.

Configurations	n=2	n=3	n=4
Z=18	-0.999	-1.163	-1.210
Z=36	-1.007	-1.165	-1.208

To obtain the relative contribution of these two terms to the correlation energy in two-electron systems, we have performed a set of MCDF calculations using all configurations, with the correct symmetry and parity, that can be constructed from 1s up to 4f one-electron orbitals. First we include only the Coulomb interaction in the self-consistent process (MCDF-OL one) and then add the magnetic interaction as a first order perturbation correction. Convergence in the correlation energy of the $1s^2\ {}^1S_0$ ground state is illustrated in Table I for argon and krypton ions. The basis we are using appears to be good enough to reproduce at least 95% of the correlation energy obtained by Safronova[4] which gives $-1.26+0.27/Z$ eV as the non-relativistic correlation energy of this $1s^2$ configuration. This rather good convergence gives us confidence to compare Coulomb and magnetic contributions that are displayed in Table II for the $1s^2$ ground state and for the $1s2p\ {}^3P_2$ excited level of He-like ions.

Table II: Coulomb and magnetic correlation energies (in eV) for the $1s^2\ {}^1S_0$ and $1s2p\ {}^3P_2$ levels of He-like ions (all values are negative).

	Z=18		Z=26		Z=36		Z=54	
	1s2	1s2p	1s2	1s2p	1s2	1s2p	1s2	1s2p
Coul.	1.210	0.103	1.210	0.105	1.208	0.105	1.212	0.104
Magn.	0.153	0.002	0.330	0.005	0.649	0.009	1.412	0.021
Total	1.363	0.105	1.540	0.110	1.857	0.114	2.624	0.125

It is immediatly observed that while the Coulomb contribution remains almost constant over the entire range of Z values considered in the Table (small variations may be due to the different nuclear radius used for the various ions), the magnetic contribution is a rapidly increasing function of the nuclear charge. To understand these results we make again use of the double series expansion of the total energy as given by equation (4). The dominant terms are easily seen to correspond to:

A_{00}: hydrogenic non-relativistic energy
A_{10}: intra-shell Coulomb interaction
A_{n0}: with n>1, inter-shell Coulomb interaction
A_{0m}: expansion of the Dirac single particle energy for the two electrons
A_{11}: intra-shell Breit interaction
A_{21}: inter-shell Breit interaction.

In doing MCDF calculations for the various states of the two-electron systems considered here we are calculating exactly both the single particle and intra-shell interaction, consequently correlation energy is expected to behaves as: $a+b/Z$ for the Coulomb contribution and cZ^2+dZ for the magnetic one. Thus it is not unexpected that the results given in Table II show indeed such a behaviour. What cannot be predicted without doing the calculations is the magnitude of the various numerical coefficients and the important conclusion for very heavy stripped ions is that the magnetic correlation energy dominates the Coulomb one in the ground state (the crossing between the two terms arises at Z=48).

Owing to the importance of the magnetic contribution to the total correlation energy, the neglect of this interaction in the self-consistent process may be questionned. We have thus started preliminary calculations that include both the Coulomb and the magnetic interactions in the self-consistent field. Because these calculations are computer time consumming we have up to now restricted them to the most important configurations (i.e. up to n=2 for the ground state and up to n=3 for the excited level $^3P_2$). The results are reported in Table III.

Table III: Correlation energies (in eV) for two electron systems obtained with the magnetic interaction as a perturbation (a) or included in the self-consistent field (b).

	Z=36 (1s2)	Z=54 (1s2)	Z=36 (1s2p,$^3P_2$)
(a)	−1.447	−2.052	−0.113
(b)	−1.413	−1.983	−0.110
(a) − (b)	−0.034	−0.169	−0.003

The variations are certainly of second order with respect to the total correlation energy but become nevertheless sizeable for high Z ions.

III. RETARDATION IN THE ELECTRON-ELECTRON INTERACTION.

Up to now we have only discussed the Dirac-Breit theory without taking into account the finite velocity of light. The correct expression for the electron-electron interaction must obviously be obtained from QED and then, for practical purpose be reduced to an effective interaction between one-electron states, i.e. put in the form:

$$\iint \phi_A^+(1)\phi_B^+(2)g(1,2)\phi_C(1)\phi_D(2)\; d^3x_1 d^3x_2 \qquad (10)$$

As usual a gauge has to be choosen to obtain an explicit form of the operator g(1,2). It turns out that the result in the Coulomb gauge is given by:

$$g^C = \frac{1}{R} - \frac{\vec{\alpha}_1 . \vec{\alpha}_2}{R} \exp(i\omega R) + \frac{(\vec{\alpha}_1 . \vec{\nabla})(\vec{\alpha}_2 . \vec{\nabla})}{\omega^2 R} [\exp(i\omega R) - 1] \qquad (11)$$

while in the Lorentz gauge one obtains:

$$g^L = \frac{(1 - \vec{\alpha}_1.\vec{\alpha}_2)}{R} \exp(i\omega R) \quad (12)$$

in the above equations ω is the frequency of the virtual exchange photon that, because of energy conservation, is given in terms of the one-particle ϵ by:

$$\omega_{AC} = | \epsilon_A - \epsilon_C | /c = \omega_{BD} = | \epsilon_B - \epsilon_D | /c \quad (13)$$

As discussed recently by Hata and Grant[13] these two forms are equivalent when a local potential is used to define the single particle states, this can be shown with the help of the double commutator:

$$\left[h_1, \left[h_2, \frac{e^{i\omega R} - 1}{\omega^2 R}\right]\right] \quad (14)$$

When the integral defined by equation (10) is neither a direct nor an exchange one, the energy conservation is not fullfilled. To overcome this problem Mittleman[14] has extended the definition of the operator g and suggested to use half the sum of two identical operators, the first associated with $\omega=\omega_{AC}$ and the second with $\omega=\omega_{BD}$.

Table IV: Retardation contribution to the fine structure splitting along the Boron isoelectronic sequence (in cm^{-1}).

Z	Lorentz gauge	Coulomb* gauge	% of the splitting
10	-0.3	0.	0.02
18	-9.	-9.	0.04
26	-66.	-66.	0.06
34	-274.	-265.	0.07
42	-833.	-806.	0.09
54	-3 109.		0.11
92	-52 203.		0.16

* Reference 13.

Hata and Grant[13] have studied the influence of the non zero frequency part of the retardation correction on the fine structure splitting between the $^2P_{3/2}$ and $^2P_{1/2}$ levels in the ground state of boron-like ions. They performed a MCDF-EAL calculation taking into account the three $1s^22s^22p$, $1s^22s2p^2$ and $1s^22p^3$ configurations and used the Coulomb gauge form of the operator. To study the possible influence of the choice of the gauge (this influence is obviously completely unphysical but may arise only because of the use of approximate wave functions) we have done a similar calculation but using the Lorentz form of the operator. The results are given in Table IV that also shows the contribution of this term (in percent) to the fine structure splitting. If the agreement between the two gauges is excellent for low Z ions, it seems that a small discrepancy starts to show up between the two sets of results in the case of the highest Z value consider by Hata and Grant. This point deserves further study to make sure that no numerical problem occurs at this level of accuracy between the two codes used to perform the calculations.

To understand why retardation may be important for the fine structure, let us consider the $\omega \to 0$ limit of equation (12), i.e.:

$$g \to \frac{(1 - \vec{\alpha}_1 . \vec{\alpha}_2)}{R} + \frac{1}{2} (\vec{\alpha}_1 . \vec{\alpha}_2) \omega^2 R \qquad (15)$$

The most important contribution comes from integrals involving electrons with the same principal quantum number (maximum overlap between the wave functions). Then the Z dependance originates mainly from the ω^2 energy term. For the $^2P_{1/2}$ level, the energies of the 2s and $2p_{1/2}$ electrons are degenerate in the Dirac hydrogenic limit and consequently the difference between them is proportional to Z only because of intra-shell Coulomb interaction. On the other hand the spin-orbit interaction in the p level will result in a Z^4 dependance of the energy separation between the 2s and $2p_{3/2}$ electrons of the $^2P_{3/2}$ level. Thus we expect that retardation will be more important for this later level and consequently contributes to the fine structure splitting. This simple minded explanation is obviously too crude to give the correct Z dependance since interaction with 1s electrons cannot be completely neglected.

The contribution of the retardation to the electron-electron interaction is illustrated in Table V for the various levels of the 1s2p configuration and appear to be quite substantial as well as having different signs between triplet and singlet levels.

Table V: Retardation contribution to the energy levels of the 1s2p configuration (in eV).

	$^3P_2$	$^3P_1$	$^3P_0$	$^1P_1$
Z=18	-0.078	-0.074	-0.076	+0.071
Z=26	-0.246	-0.203	-0.235	+0.198
Z=36	-0.693	-0.411	-0.634	+0.420
Z=54	-2.641	-0.973	-2.171	+1.134

Beside the choice of the gauge, an other problem in estimating this contribution arises in the definition of the one-particle energies (the ϵ's in the above equations). For single particle models these quantities are quite well defined and in Hartree-Fock type of calculations, Koopmans theorem can be used to approximate the binding energies. But in other kind of calculations, like correlated wave functions or MCDF, no single particle parameter is easily identified as an approximation to the ϵ's. This simple comment questions the use that has been made (including by myself) in some calculations of the Breit operator. Let me be more explicit, taking the $\omega \to 0$ limit of the Coulomb operator of equation (11), we recover the Breit interaction:

$$g^C = \frac{1}{R} - \frac{\vec{\alpha}_1 . \vec{\alpha}_2}{R} + \frac{(\vec{\alpha}_1 . \vec{\nabla})(\vec{\alpha}_2 . \vec{\nabla})}{2} R \qquad (16)$$

that indeed includes contribution of order ω^2. But, as was just pointed out, there is no simple way of estimating the value of the ω's in say a MCDF-OL calculation. Should have we been using the Lorentz gauge we would not have been able to estimate this ω^2 contribution. This not all satisfactory situation certainly deserves more theoretical studies. For the time being my own prescription is not to use the full operator of equation (16) when single particle energies cannot be defined unambiguously.

IV. QED CORRECTIONS.

As this subject will be extensively discussed in other contributions to this workshop we just summarize here how these corrections, the most important ones after the electron-electron interaction, are introduced in accurate calculations of energy levels. Two term contribute to the Lamb shift: i) the self-energy that is the dominant one for electrons and arises from the emission and reabsorption of virtual photons and ii) the vacuum polarization associated with the creation of virtual electron-positron pairs. As usual in a many electron system, we shall distinguish between one-body and many-body contributions. In the case of the self-energy, very accurate results have been obtained for the one-electron terms either in the pure hydrogenic case[15] or for the 1s level of some very heavy neutral atoms[16]. These calculations solve numerically the exact Dirac Green's function either in the Coulomb case or in some screened potential, so doing they avoid the convergence problem of series expansion in powers of αZ. These calculations have recently been extended to take into account the finite size of the nucleus.[17] Vacuum polarization is generally obtained as the expectation value of the Uehling potential[18] that is known to give the correct lowest order contribution. When needed, higher order terms can also be introduced as the expectation values of higher order potentials[19]. If it thus appears that the one-electron terms are well under control, the situation is much less satisfactory for the many-electron ones.

For not too heavy two-electron systems, recent calculations[20-23] have evaluated the self-energy contribution starting from the formulation of Kabir and Salpeter[24] that, to lowest order and in the non-relativistic limit, gives the two-electron self-energy correction as:

$$\Delta E^{KS} = \frac{4}{3} Z\alpha^3 \langle \delta(\vec{r}_1) + \delta(\vec{r}_2) \rangle \left[\frac{19}{30} - 2\ln(\alpha Z) + \ln(Z^2 \mathrm{Ryd}/k_0) \right] \qquad (17)$$

where k_0 is the two-electron Bethe logarithm calculated using various prescriptions certainly to be discussed during the workshop. This lowest order term is further corrected through essentially higher order one-electron hydrogenic terms but taking into account the screened electron density at the nucleus. Under these assumptions, quite good agreement was obtained with experiment up to argon He-like ions. Nevertheless questions remain as to how far this can be extended to much higher Z values for two reasons. The first one is related to the fact that this formulation relies on an expansion in powers in αZ. As these series expansions are known to have convergence problems for one-electron systems and it is more than likely that they will suffer from the same sickness in the many-electron case. Next these corrections involve the electron density at the nucleus that is known to be largely enhanced by relativity for high Z values. How this enhancement can be incorporated coherently is less than obvious since, as already pointed out, the Kabir and Salpeter formula holds at the non-relativistic limit only. A coherent and numerically tractable method to compute these many-body QED corrections for high Z atoms or ions is still a challenge for the theoreticians.

V. COMPARISON WITH EXPERIMENT.

High precision absolute measurements of $1s2p \rightarrow 1s^2$ transition energies[25-28] have been made during the last two years and are certainly the most severe test of relativistic many-body calculations. In Table VI we display, for the argon He-like ion, the various contributions to the transition energies as obtained by the MCDF-OL method. These values are slightly different from the ones of reference 25 since: (a) the full retardation in the Lorentz gauge has been included, (b) configuration interaction was extended to include n=4 orbitals, (c) only the magnetic part of the Breit interaction is used to define the non Coulomb correlation energy in agreement with our discussion of this problem in section III and (d) hydrogenic QED corrections are used. This last point deserves some explanation. In our previous calculation we tried to take into account the screening of the QED corrections by interpolating the hydrogenic results of Mohr[15] for a given effective nuclear charge instead of the bare one. The effective charge was defined so that Dirac-Fock orbitals have the same mean expextation value of r that hydrogenic orbitals for this effective charge. This prescription, we suggested a long time ago, has no fundamental background and relies only on the empirical observation that, in so doing, it was possible, for heavy neutral atoms, to obtain good agreement with screened self energy results.[16] But, according to the latest calculations of two-electron QED,[23] it appears that we were overestimating the screening for low Z ions. The various corrections to the Dirac-Fock values are: the magnetic interaction, the Coulomb and magnetic correlation energies, the retardation in the electron-electron interaction, the mass polarization and the hydrogenic QED values that are given in lines 2 to 6 respectively of Table VI. All these corrections, excepted for the mass polarization term, are much greater than the experimental uncertainty. In this table we also make comparison of various theoretical results and it can be observed that all of them agree within the experimental uncertainty even for our own $^1P_1$ calculation that is certainly the less accurate value because no orthogonality constraint to the $^3P_1$ level was required.

Table VI: Transition energies (in eV) from 1s2p to $1s^2$ in He-like argon.

	$^3P_2 - {}^1S_0$	$^3P_1 - {}^1S_0$	$^1P_1 - {}^1S_0$
Dirac-Fock	3 128.1259	3 125.2078	3 141.2989
Magnetic int.	-1.9004	-1.7232	-2.1067
Coulomb corr.	1.1074	1.1236	1.1045
Magnetic corr.	0.1509	0.1507	0.1493
Retardation	-0.0778	-0.0747	-0.0708
Mass polariz.	-0.0093	-0.0093	0.0075
Hydrog. QED	-1.1601	-1.1682	-1.1635
Total	3 126.2366	3 123.5067	3 139.3608
Hata*	3 126.2796	3 123.5309	3 139.5632
Drake*	3 126.2512	3 123.5112	3 139.5274
Experiment*	3 126.251(36)	3 123.516(36)	3 139.549(36)

* see Reference 28.

Table VII gives analogous results for Z=36 but, for this high Z value, the screening of the hydrogenic QED corrections cannot be neglected. According to the results of Hata[25] the various corrections scaled like $Z^{3.5}$ or Z^4 when going from helium to argon. Thus if we defined the screening for Z=18 as the difference between the hydrogenic value and that of Hata we get 0.077 eV for the $^3P_2$ level and 0.069 eV for the $^3P_1$ one. The extrapolation of these values, according to the above scalling laws, amounts to 0.8 to 1.2 eV for Z=36, a result very close, certainly fortuitously, from our own estimate of 0.8 eV obtained from the effective nuclear charge. Experimental results published up to now have too large uncertainties to draw a definite conclusion. A new experiment has been done at GANIL for this argon He-like ion for which the accuracy is expected to be much higher but the results are not yet available.

Table VII: Transition energies (in eV) from 1s2p to $1s^2$ in He-like krypton.

	$^3P_2 \rightarrow {}^1S_0$	$^3P_1 \rightarrow {}^1S_0$
Dirac-Fock	13 116.879	13 051.877
Magnetic int.	-15.840	-15.630
Coulomb corr.	1.103	1.103
Magnetic corr.	0.640	0.639
Retardation	-0.693	-0.411
Mass polariz.	-0.085	-0.085
Hydrog. QED	-11.482	-11.634
Screening	0.797	0.822
Total	13 091.319	13 026.681
Experiment*	13 091.2(±1.5)	13 023.8(±2.2)

* see Reference 27.

Again for He-like krypton we have just completed a calculation of the 1s2p $^3P_2$ to 1s2s $^3S_1$ transition wavelength and obtained 111.147 angstroms to be compared with the experimental value of 111.16(±0.08) recently obtained at GANIL by Desesquelles et al.[29].

CONCLUSION

The above comparison with experimental results shows that for not too heavy ions the theoretical predictions are quite accurate, although many problems remain in extending some of these calculations to much higher Z values. First, highly accurate non-relativistic correlation energies are not available beyond Z=10 and it is not obvious how far their extrapolation is reliable. Furthermore, as pointed out in section II, correlation and relativistic effects are not additive and at some point it will become necessary to handle the correlation problem in a fully relativistic way. At present the MCDF method is the only fully relativistic available one that has been shown to be competetive at low Z with methods using non-relativistic highly correlated wavefunctions. Nevertheless it suffers from some drawbacks: slow convergence and more seriously the ambiguity to define single particle energies needed in the calculation of the retardation term (in fact this last point is indeed not specific to the MCDF method). The next main problem to be solved is that of the many-body QED effects in non-perturbative αZ expansion way and this one will certainly require a tremendous numerical effort.

ACKNOWLEDGMENTS

Some of the MCDF results reported here have been obtained in collaboration with O. Gorceix and P. Indelicato from the Laboratoire de Physique Atomique et Nucléaire (Institut Curie, Paris). The author would like to thank them and J.P. Briand for useful discussions as well as J. Desesquelles for communicating his experimental results prior to publication.

REFERENCES

1. J. Sucher in these proceedings.
2. D. Layzer and J. Bahcall, Ann. Phys., NY17, 177 (1962)
3. H.T. Doyle, Adv. At. Mol. Phys. 5, 337 (1969)
4. U.I. Safronova, Phys. Scr. 23, 241 (1981) see also E.P. Ivanova and U.I. Safronova, J. Phys. B 8, 1591 (1975)
5. G.W.F. Drake, Phys. Rev. A 19, 1387 (1979)
6. J. Hata and I.P. Grant, J. Phys. B 16, 507 (1983)
7. J.P. Desclaux, Comput. Phys. Comm.,9, 31 (1975)
8. W.R. Johnson and C.D. Lin, Phys. Rev. A 20, 964 (1979)
9. W.R. Johnson and K.N. Huang, Phys. Rev. Lett. 48, 315 (1982)
10. M. Aymar and E. Luc-Koenig, Phys. Rev. A 15, 821 (1977)
11. L.N. Ivanov and L.I. Podobedova, J. Phys. B 10, 1001 (1977)
12. K. Dietz, O. Lechtenfeld and G. Weymans, J. Phys. B 13, 4301 (1982)
13. J. Hata and I.P. Grant, J. Phys. B 17, L107 (1984)
14. M.H. Mittleman, Phys. Rev. A24, 1167 (1981)
15. P.J. Mohr, At. Data Nucl. Data Tables 29, 453 (1983)
16. K.T. Cheng and W.R. Johnson, Phys. Rev. A 14, 1943 (1976)
17. W.R. Johnson and G. Soff, GSI Preprint 84-72 (1984)
18. E.A. Uehling, Phys. rev. 48, 55 (1935)
19. G. Kallen and A. Sabry, K. Dan. Vidensk. Selks. Mat. Fys. Medd. 29, 17 (1955)
20. S.P. Goldman and G.W.F. Drake, J. Phys. B 17, 2187 (1984)
21. G.W.F. Drake and A.J. Makowski, J. Phys. B 18, L103 (1985)
22. J. Hata and I.P. Grant, J. Phys. B 17, 931 (1984)
23. J. Hata, J. Phys. B 17, L625 (1984)
24. P.K. Kabir and E.E. Salpeter, Phys. Rev. 108, 1256 (1957)
25. J.P. Briand, J.P. Mossé, P. Indelicato, P. Chevallier, D. Girard-Vernhet, A. Chetioui, M.T. Ramos and J.P. Desclaux, Phys. Rev. A 28, 1413 (1983)
26. J.P. Briand, M. Tavernier, R. Marrus and J.P. Desclaux,

Phys. Rev. A 29, 3143 (1984)
27. J.P. Briand, P. Indelicato, M. Tavernier, O. Gorceix, D. Liesen, H.F. Beyer, B. Liu, A. Warczak and J.P. Desclaux, Z. Phys. A 318, 1 (1984)
28. R.D. Deslattes, H.F. Beyer and F. Folkmann, J. Phys. B 17, L689 (1984)
29. J. Desesquelles et al. to be published.
Phys. Rev. A 29, 3143 (1984)
27. J.P. Briand, P. Indelicato, M. Tavernier, O. Gorceix, D. Liesen, H.F. Beyer, B. Liu, A. Warczak and J.P. Desclaux, Z. Phys. A 318, 1 (1984)
28. R.D. Deslattes, H.F. Beyer and F. Folkmann, J. Phys. B 17, L689 (1984)
29. J. Desesquelles et al. to be published.

THE ANALYSIS OF THE HIGH-RESOLUTION X-RAY SPECTRA EMITTED FROM A LASER-IRRADIATED GOLD PLASMA

S. Kiyokawa T. Yabe, N. Miyanaga, K. Okada, H. Hasegawa,
T. Mochizuki, T. Yamanaka and C. Yamanaka
Institute of Laser Engineering, Osaka University,
Suita, Osaka 565, Japan

T. Kagawa
Department of Physics, Nara Women's University,
Nara 630, Japan

ABSTRACT

The fifteen lines in the x-ray spectra between 3.1 and 4.5 keV emitted from a laser-irradiated Au plasma were observed and identified with the relativistic Hartree-Fock-Slater(RHFS) calculation. It is shown that these lines are emitted from Au ions with the charge state of 26+∿31+ containing multiple inner-shell vacancies: one in the $3p_{1/2}$ or $3p_{3/2}$ shell and one or two more in the M or N shell. It is seen that the energy separation of the corresponding lines between the two groups in the spectra observed, which is about 400eV, comes from the difference of the binding energy between the $3p_{1/2}$ and $3p_{3/2}$ shell in Au ions due to the spin-orbit interaction. A possible mechanism of creating multiple vacancies in inner-shells of the ions in the plasma is also discussed.

1. INTRODUCTION

The X-ray line spectra emitted from ions immersed in plasmas have so far been used as a diagnostic tool to determine the plasma parameters such as the temperature, density and charge state of ions[1-5]. For example, the existence of very highly charged ions such as Gd^{36+}, W^{45+}, Mo^{13+} and Xe^{25+} in laser-produced and tokamak plasmas has been confirmed from the detailed analysis of the x-ray spectra emitted from Cu- or Zn-like ions[6-9], where the relativistic Hartree-Fock calculations have been carried out to identify the transition energies observed. These ascertained ions have relatively simple electronic structures which consist of one or two valence electrons outside the closed 3d shell. The method of identification of such charge states was developed by Reader and Luther[10]. The x-ray spectrum from plasmas containing heavy ions, however, is very complex because the multiple vacancies in various shells of an ion in the plasma are easily created through many different atomic processes, especially by the ion-ion or electron-ion collisions.

In recent studies on the x-ray spectra in a laser-produced plasmas, there has been much interest in the mechanism how the inner-shell vacancies of an ion in the plasma are created.

Matthews et. al.[12] have reported the M-band x-ray spectra from gold in a laser-irradiated disk. However, the detailed analysis of the spectral lines has not given in their paper.

In this report, we present the measurement of the high-resolution x-ray spectra from a gold plasma produced by a high-power glass laser, and the identification of the spectral lines through the the theoretical ralativistic Hartree-Fock-Slater (RHFS) calculations. A possible mechanism which leads to the x-ray emission observed is also discussed.

2. DETAILS OF EXPERIMENT

In the experiment, twelve laser beams from the high-power glass laser system GEKKO XII were focused onto a glass microballoon (GMB) which is coated with 10μm-thick parylene(CH) and 0.3∼0.5μm thick gold as shown in Fig. 1. The GMBs were 200∼300μm in diameter and 2.1∼3.0μm in wall thickness. The average laser intensity on target was $(1.7\pm0.4)\times10^{15}$ W/cm^2 at 500 psec(FWHM) and 1.053μm. The space- and time-integrated x-ray spectrum was analyzed by a flat KAP diffracting crystal(2d=26.64Å) with a 25μm-thick beryllium entrance window. The x-ray spectra covering the range of 3.1∼4.5keV were recorded on the low-noise, sensitive Fuji X-ray film. The micro densitometer used was a Sakura PDM-5 whose objective was x4 with the numerical aperture of 0.1. The so-called H-D characteristics of this film were independently measured by using the monochromatic x-rays obtained by spectrally resolving the

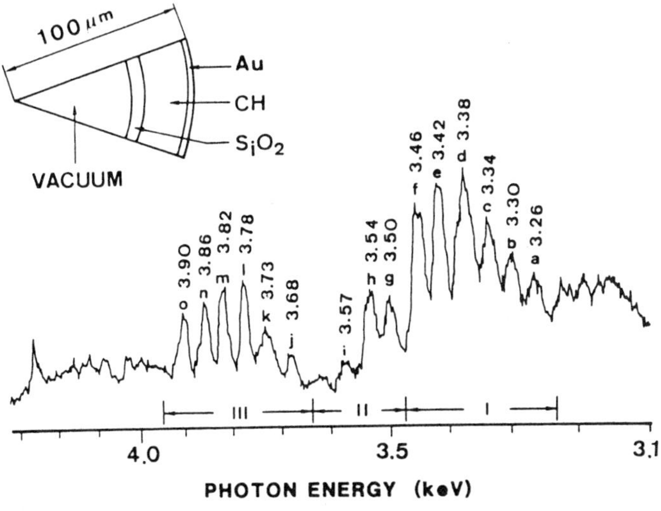

Fig. 1. The target structure (Au 0.3μm, CH 10μm, SiO_2 2.1μm) and the x-ray spectrum observed for the laser-irradiated Au plasma with the photon energy between 3.1 and 4.3 keV.

pulsed emission from a laser-produced plasma. The absolute sensitivity was calibrated by comparing this H-D curve with that of Kodak No-Screen Film[15] which was simultaneously obtained from the same x-ray source. The reliability of this calibration was checked by comparing the other H-D curve with the monochromatic CW x-ray from a standard source.

The photon energy was determined from the geometrical calculation againt the source plasma and crystal spectrometer location. The energy was not calibrated by using known line emission, but the relative spacing should be accurate.

The x-ray spectra observed is shown in Fig. 1. The spectra obtained in all shots are quite reproducible and have a common satellite structure which consists of sequential 6 lines(group I) in 3.2∿3.4keV, 3 lines(group II) in 3.4∿4.5keV and 6 lines(group III) in 3.6∿4.0keV. The intensity of these lines was almost independent of Au thickness, so these lines were emitted from a relatively low density region in the plasma. As apparent from the figure, the groups I and III have a similar structure: The line intensity peaks at the center of the group and the line emission abruptly disappears at the high energy side.

3. COMPARISON OF THE RHFS RESULTS WITH EXPERIMENT

It is expected that in a high-temperature plasma most ions are in an highly-excited state, where there are many vacancies in various shells. Moreover, these excited states consist of a large number of complex multiplet terms. Considering the resolution of the spectrometer used in our experiment, the effect of the multiplet structure on the x-ray spectra for many-electron systems may appear in the broardness of a spectral line rather than its split in most cases. However, the position of the peak of the lines in the spectra can be used to determine a charge state of ions, since each line corresponds to the transition for ions in different charge states. So the RHFS method, which calculates the average transition energy for atomic systems, can effectively be used to identify the lines in the spectra observed here.

A RHFS computer program using the finite difference method was constructed to calculate the transition energy for the system, where the parameter α for the exchange potential is unity and Latter's correction used. This program was checked by comparing the calculated eigen values for various atoms with those of Liberman et. al.[12] and the experimental binding energies for solid gold[19]. The transition energy was obtained by taking the difference of the binding energies between the upper and lower states so called Slater's "transition states". The RHFS energies were always slightly higher than the corresponding experimental values[16]. The identification of the spectra can be performed by taking the calculated values which are slightly higher than the experimental ones.

Before we start calculations, we estimate the states

for the transition from the characteristic peaks in the spectra between the group I and III in Fig 1. As the energy difference of all corresponding peaks between group I and III is about 400eV, it is probable that the appearance of these two groups in the spectra are attributed to the transitions in which a $3p_{1/2}$ or a $3p_{3/2}$ vacancy is filled by an outer-shell electron, since the energy difference of 400eV observed is close to that of orbital energies between the $3p_{1/2}$ and $3p_{3/2}$ electrons for Au due to the relativistic effects such as the spin-orbit interaction[17]. In this case the spectra observed can be assigned to the M x-ray lines emitted from Au ions.

In order to make sure of the estimation, some trial calculations with various configurations for Au ions in a different ionization stage were carried out. The calculations show that the spectra obtained cannot be explained as the transitions between two shells outside the 3p one in the ion. For example, if Au^{51+} or a more highly charged state is considered, the valence shell is 3d and the transition energies above 3keV are obtained as the transitions of $3d^{n+1} \leftarrow 3d^{n}5p$ (n=9,8,7,...). However, the energy spacings for these transitions corresponding to different charge states, are 80-90eV which is too large compared with the experimental ones (30-40eV), namely, the energy difference between neighboring peaks of the group I or III in the spectra.

This discrepancy is removed if a vacancy in the $3p_{1/2}$ or $3p_{3/2}$ shell instead of the 3d shell and 4f electrons as valence ones in the initial configuration for the transitions considered is taken into account. In this case removal of electrons from the valence 4f shell one by one leads to about 40eV as a spacing of the transition energy close to the energy separation between the neibouring lines in the spectra observed. The infered charge state for the ions becomes between 26+ and 31+ according to the change of the number of vacancies in the 4f shell. The charge state for the ions around the 30+ is close to those obtained from the Thomson parabola measurement in a similar experimental condition for the same plasma: The charge states between 30+ and 40+ for the gold ions in the plasma have been observed[18].

So the transition energies from various excited states with a single $3p_{1/2}$ or $3p_{3/2}$ vacancy were extensively calculated by varying the number of electrons in the 3d and 4f shells, and shown in Table I. In the table, the transition energies for the $3p_{3/2} \leftarrow 6d_{3/2}$ transition are slightly larger than those in our experiment, whereas those for the $3p_{1/2} \leftarrow 6d_{3/2}$ transition are too small. On the other hand, in the case of the transition of $3p_{3/2} \leftarrow$ 4d or 4s, we must consider very highly ionized states (60+) inconsistent with the Thomson-parabola data so that this case is discarded.

Coincidence between the theoretical and experimental values can further be improved if double vacancies in the M-shell for the configuration are assumed. In particular, if the vacancies in both $3p_{1/2}$ and $3p_{3/2}$ shells are considered, the calculated transition energies for the $3p \leftarrow 6d$ transitions listed in Table II are in good agreement with the 12 lines both the group I and III in the spectra in Fig. 1. The energy shift for the transition between the states having a vacancy in other M subshells, that is, the 3s or 3d one is about 1∿3eV, which are within the spectral line width. So our calculation indicates that the probability of creating two vacancies in the M shell of the ions is higher than that of a single hole production in the plasma, since 400eV observed as the difference of the binding energy between the $3p_{1/2}$ and $3p_{3/2}$ electrons cannot be obtained with the configuration containing a single vacancy in the 3p shell in the calculation.

Table I. Calculated RHFS transition energies (in keV) for the $3p_{1/2} \leftarrow 6d_{3/2}$ and $3p_{3/2} \leftarrow n\ell j$ (= $6d_{3/2}$, $6s_{1/2}$, $5d_{5/2}$, $5d_{3/2}$ and $5s_{1/2}$) transitions in Au(26+) through Au(42+), where the electronic configuration for the initial state in the ions is $[Kr(3p)^{-1}](4d_{3/2})^{n_1}(4d_{5/2})^{n_2}(4f_{5/2})^{n_3}(4f_{7/2})^{n_4}$ $(n\ell j)$ and $[Kr(3p)^{-1}]$ means the electronic configuration for krypton-like atom containing a vacancy in the $3p_{1/2}$ or $3p_{3/2}$ shell.

Charge state	No. of 4d and 4f elctrons n_1	n_2	n_3	n_4	RHFS transition energy (keV) $6d_{3/2} \downarrow 3p_{1/2}$	$6d_{3/2} \downarrow 3p_{3/2}$	$6s_{1/2} \downarrow 3p_{3/2}$	$5d_{5/2} \downarrow 3p_{3/2}$	$5d_{3/2} \downarrow 3p_{3/2}$	$5s_{1/2} \downarrow 3p_{3/2}$
26	4	6	6	1	3.577	3.226				
27	4	6	6		3.616	3.264	3.182			
28	4	6	5		3.656	3.304	3.219			
29	4	6	4		3.696	3.345	3.258	3.099		
30	4	6	3		3.737	3.386	3.298	3.129		
31	4	6	2		3.780	3.428	3.338	3.161		
32	4	6	1		3.822	3.472	3.380	3.192		
33	4	6			3.867	3.516	3.422	3.225	3.215	
34	4	5			3.908		3.462	3.255	3.245	
35	4	4					3.502	3.285	3.275	
36	4	3					3.543	3.316	3.305	
37	4	2						3.347	3.335	
38	4	1						3.378	3.366	
39	4							3.410	3.398	3.207
40	3							3.442	3.430	3.238
41	2							3.475	3.462	3.269
42	1							3.508	3.495	3.301

Table II. RHFS transition energies (in keV) for the $3p_{3/2} \leftarrow 6d_{3/2}$ transition in Au(26+) through Au(31+) having one more vacancy in the M-shell in addition to a $3p_{3/2}$ hole in the electronic configuration for the initial state. The configuration considered here is written as $[Kr(3p_{3/2})^{-1}, M^{-1}] (4d_{3/2})^4(4d_{5/2})^6(4f_{5/2})^n(4f_{7/2})^{n'} (6d_{3/2})$ for M = $3s_{1/2}$, $3p_{1/2}$, $3p_{3/2}$, $3d_{3/2}$, and $3d_{5/2}$.

Charge state	No. of 4f electrons n	n'	RHFS transition energies (keV) for the $3p_{3/2} \leftarrow 6d_{3/2}$ transition $(3s_{1/2})^{-1}$	$(3p_{1/2})^{-1}$	$(3p_{3/2})^{-1}$	$(3d_{3/2})^{-1}$	$(3d_{5/2})^{-1}$
26	6	2	3.261	3.264	3.260	3.265	3.264
27	6	1	3.300	3.302	3.299	3.304	3.303
28	6		3.339	3.342	3.338	3.343	3.342
29	5		3.380	3.382	3.379	3.384	3.383
30	4		3.421	3.424	3.420	3.426	3.425
31	3		3.464	3.466	3.463	3.468	3.467

The RHFS calculation also indicates a possibility of taking a configuration with triple vacancies in the inner-shells, namely, one in the M shell and two in the N shell. In this case, we considered the two more vacancies in the $4p_{1/2}$ and $4p_{3/2}$ shells in addition to a $3p_{1/2}$ or a $3p_{3/2}$ shell vacancy in the initial configuration for the $3p \leftarrow 6d_{3/2}$ transition. However, it is difficult to uniquely determine the number of vacancies in a particular shell in ions from the RHFS calculation because one can construct some configurations with various combinations for vacancies among inner- and outer-shells to give good transition energies.

Comparison of the calculated results with experiment are given in Table III for the lines of group I and III and in Table IV for those of group II in the specra observed. The calculated results agree well with experiment except the single hole configuration for the $6d_{3/2} \leftarrow 3p_{1/2}$ transition in Table III. Our calculations were carried out so as to obtain not only the energy spacing of corresponding lines between group I and III but also that for neibouring peaks of lines in each groups in the spectra. However, the present calculation for the transition energy does not give an appropriate answer for the question why the sequencial six lines in both group I and III in the spectra were intensively observed in the experiment. It must be another key to diagnose the plasma and still remains as a future problem. This problem will be clarified if the x-ray spectra observed in wider wave-length region for the plasma is analyzed by appropriate relativistic calculations.

Table III. Comparison of the RHFS transition energies with photon energies of the lines for group I and III in the spectra (in keV).

Charge state	Configuration[a)]			Transition energy (keV)			
	No. of holes in M and N shells	No. of 4f electrons		$3p_{3/2} \leftarrow 6d_{3/2}$		$3p_{1/2} \leftarrow 6d_{3/2}$	
		n	n'	RHFS	Exp.	RHFS	Exp.
27	Single	6		3.264		3.616	
26	Double	6	2	3.264	3.26	3.681	3.68
27	Triple	6	2	3.265		3.683	
28	Single	5		3.304		3.656	
27	Double	6	1	3.302	3.30	3.719	3.73
28	Triple	6	1	3.303		3.722	
29	Single	4		3.345		3.696	
28	Double	6		3.342	3.34	3.759	3.78
29	Triple	6		3.343		3.762	
30	Single	3		3.386		3.737	
29	Double	5		3.382	3.38	3.800	3.82
30	Triple	5		3.384		3.803	
31	Single	2		3.428		3.780	
30	Double	4		3.424	3.42	3.842	3.86
31	Triple	4		3.426		3.845	
32	Single	1		3.472		3.822	
31	Double	3		3.464	3.46	3.885	3.90
32	Triple	3		3.468		3.888	

a) The configuration considered is written as [Kr(M, N holes)] $(4d_{3/2})^4(4d_{5/2})^6(4f_{5/2})^n(4f_{7/2})^{n'}(6d_{3/2})$, where [Kr(M, N holes)] means $[Kr(3p_{1/2})^{-1}$ or $(3p_{3/2})^{-1}]$ for single-, $[Kr(3p_{1/2})^{-1}$ and $(3p_{3/2})^{-1}]$ for double- and, $[Kr\ \{(3p_{3/2})^{-1}$ or $(3p_{3/2})^{-1}\}$, $(4p_{1/2})^{-1}$ and $(4p_{3/2})^{-1}]$ for triple-vacancy configurations.

Table IV. Comparison of the RHFS transition energies with the photon energies of the lines for group II in the spectra (in keV).

Charge state	Configuration[a)]		Transition energy (keV)	
	No. of holes in M and N shells	No. of $4f_{5/2}$ electrons	$3p_{1/2} \leftarrow 5d_{3/2}$	
			RHFS	Exp.
29	Double	5	3.509	3.50
30	Double	4	3.539	3.54
31	Double	3	3.570	3.57

a) The configuration considered is $[Kr(3p_{1/2})^{-1}(4s_{1/2})^{-1}]$ $4d_{3/2})^4(4d_{5/2})^6(4f_{5/2})^n(5d_{3/2})$.

4. DISCUSSION

It is possible to construct a process for the creation of the double inner-shell vacancies mentioned above. A single L-shell vacancy is produced by the inelastic collisions of energetic electrons, i.e., hot electrons whose energy should be about 20keV, which is close to the hot electron temperature deduced from the filter-fluorescer x-ray spectrograph. This vacancy decays through a radiationless process such as $L\text{-}M_1M_2$ or $L\text{-}M_2M_3$ or so on. Since the fluorescence yield of a gold L shell is about 0.3[19], a large fraction of the decay can be due to the radiationless process. In some cases, one of the Auger electrons from the M-shell does not jump to a free state, but occupies a P-shell (n=6) orbital, i.e., shake-up occurs. If P-shell electrons in ions could exsit by the condition of thermal equilibrium, more O-shell electrons than the P-shell ones must exist, so that the spectra would be more complex. However, such complex spectra due to the transition in which O-shell electrons jump to innner orbitals are not obtained in our experiment. On the other hand, a P-shell electron can make the $3p \leftarrow 6d$ transition causing line emissions in groups I and III as shown in Table III.

When one of the two M-shell vacancies which may be created by the L-MM Auger process, decays through an Auger process such as the M-NN transition, triple vacancies are created. Since the transition energies for the $3p \leftarrow 6d$ transition with the configuration between the double- and triple-vacancy cases as shown in Table III are almost the same, this type of the triple-vacancy state may contribute to the intensity of the lines in group I and III in the spectra, although the triple-vacancy charge state is higher by one unit than the double-vacancy one.

We can consider another branch of the decay of the double vacancies in the M-shell to explain the 3 lines in the group II observed in the spectra. It is a radiative decay such as the 3p ← 4s transition which makes one vacancy in the M shell and another in the N-shell. For the configuration containing two vacancies in the ions, the 3p ← 5d radiative transition in appropriate ionization stages can give the 3 lines in the group II as shown in Table IV, if it is assumed that a 5d electron is produced as a result of a transition in which a P-shell electron fills a 5d-shell vacancy.

Although some assumptions on atomic processes and conditions for the plasma are needed, it is shown that possible mechanisms to give the x-ray emission by gold ions in the plasma can be selected by use of the RHFS results.

In the calculations reported here, the effect of the shielding of the nuclear charge by free electrons or other ions on the transition energies is not taken into account. Strictly speaking, this effect cannot be neglected in the calculation of energies for ions especially in a high-density plasma because the shielding will reduce the potential due to the nuclear charge especially in inner shells. This means that the magnitude of the energy lowering is not the same for various shells of an ion in a plasma. As has already mensioned in section 2, Au ions emitting the x-ray radiation observed are considered to be in a low-density region of the plasma. So it is said that the energy shift in the ions due to the effect of the shielding of the nuclear charge by surrounding plasma particles is not large and lies within the accuracy of the calculation in the configurations discussed here.

From a diagnostic point of view, the distribution of these lines includes much information about the plasma and will be a useful tool for diagnostics. For an example, as seen from Table III, the peaks of line intensity in the groups I and III imply a peak in the charge state distribution at $Z = 28 \sim 29$. More precisely, the relative abundance of charge states could be deduced by calculating the transition probabilities for each of the lines in the group I or III.

5. CONCLUSION

The line spectra from laser-irradiated gold plasma have been found to originate from the mutiple inner-shell vacancies which were produced through Auger processes: The fifteen lines in the spectra observed have been identified as the satellite lines of the M-shell x-ray emitted from Au ions in a different ionization stages. The relativistic effect, the spin-orbit interaction, has been proved to appear in the M x-ray spectra for Au ions. The high-resolution x-ray spectra for heavy ions can be used as a tool for diagnostics of plasmas: The charge state distribution and the relative abundance of ions in defferent ionization stages could be obtained from the analysis of the line intensity in the spectra if the transition probabilities for the transitions are known.

ACKNOWLEDGEMENT

The authors would like to thank the GEKKO XII laser operation group, the diagnostic group, and the target fabrication group for their contribution to this experiment. We also thank Dr. S. Sakabe for providing experimental data of charge states in typical laser-irradiated gold plasma.

REFERENCES

The material appears in Phys. Rev. Lett. 54, 1999 (1985).
$ Present address is Dept. of Phys., Nara Women's Univ., Nara 630, Japan
1. S. Skupsky, Phys. Rev. A21, 1316 (1980).
2. R. K. Landshoff and J. D. Perez, Phys. Rev.A13, 1619 (1976).
3. W. H. Tucker and R. J. Gould, Astrophys. J. 144, 244 (1966).
4. H. R. Griem, Phys. Rev. A27, 2566 (1983).
5. D. J. Nagel et. al., Phys. Rev. Lett. 33, 743 (1974).
6. M. D. Krause et. al., Phys. Rev. A6, 871 (1972).
7. P. G. Burkhalter et. al., Phys. Rev. A9, 2331 (1974).
8. P. G. Burkhalter et. al., Phys. Rev.A 15, 700 (1977).
9. L. F. Chase et. al., Phys. Rev. A13, 1497 (1976).
10. J. Reader and G. Luther, Phys. Rev. Lett. 25, 609 (1980).
11. V. L. Jacobs et. al., Phys. Rev. A21, 1917 (1980).
12. D. Liberman et. al., Phys. Rev. 137, A27 (1965).
13. D. L. Matthews et. al., J. Appl. Phys. 54(8), 4260 (1983).
14. J. V. Gilfrich, Appl. Spectroscopy 29, 322 (1975).
15. C. M. Dozier et. al., J. Appl. Phys. 47, 3732 (1976).
16. J. A. Bearden, Rev. Mod. Phys. 39, 78 (1967).
17. J. A. Bearden and A. F. Burr, Rev. Mod. Phys. 39, 125 (1967).
18. S.Sakabe, private comunication; a similar measurement with a slightly different laser intensity was given by S.Sakabe and T.Mochizuki, Annual Progress Report on Laser Fusion Program, ILE, Osaka Univ., ILE-APR-79, p56 (1979) (unpublished).
19. E. J. McGuire, Phys. Rev. A3, 587 (1971), W. Bambynek et. al., Rev. Mod. Phys. 44, 716 (1972).

Minimal Quantum Electrodynamics

R. Jáuregui and M. Berrondo
Instituto de Física, UNAM, Cd. Univeristaria
Apdo. Postal 20-364, México, D.F., 01000

Abstract

A simple and coherent formulation of quantum electrodynamics is obtained within the general framework of the LSZ field theory. The commutation relations for the intereacting fields are obtained rather than being postulated a priori and the current densities fulfill the one particle stability conditions. Thus, the inconsistencies which appear in the canonical formalism are avoided. The resulting spectral representations do not have any ambiguities so that we do not have to introduce the "renormalization" concept.

1.INTRODUCTION.

The great sucess of quantum electrodynamics (QED) in providing the necessary means for making quantitative predictions of physical data [1], is partially obscured by the ultimate problems QED faces. The Lagrangian approach and the perturbation method, both in the interaction picture [2] and in the Heisenberg picture [3], have run up against difficulties of principle that have been solved, at most, from a practical point of view. The canonical formalism implicitly assumes that all the results in the theory should follow from the postulated Lagrangian and the field (anti-)commutation relations. However, the product of operators at equal spacetime events is ill-defined [4] so that the equal time commutation relations for the interacting fields cannot be assumed in advance in a consistent fashion, thus questioning not just the existence of the interaction picture but also of a well defined Lagrangian in Heisenberg's picture. As a consequence, QED in the canonical formalism cannot be regarded as a conceptually consistent, physically complete relativistic quantum field theory.

Taking that into account, the axiomatic approaches arose not just in the natural search of generality but,most of all, trying to achieve the neccessary logical coherence in quantum field theories. Thus,very general assumptions which seem to be compatible to one another, are made. These include basic properties of the physical state taken as a Hilbert space and transformations therein, causality, locality, uniqueness of the vacuum state and asymptotic completeness. In the LSZ version [5-7], there is a relationship between the interacting field (or interpolating field) and a corresponding free field (<u>in</u> or <u>out</u>) through the asymptotic boundary conditions. These schemes usually leave open the question of how to build a specific theory, <u>e</u>.<u>g</u>. QED, since their interest lies in defining a general framework.

This work is devoted to the construction of a field

theory of QED with a minimun of assumptions-which can be taken essentially as those of the LSZ formulation- and avoids the in compatibilities which appear in the canonical approach.

In section II, the notation is set and the consequences of the general assumptions that give rise to the inconsistencies of the canonical approach are clearly stated. In order to avoid them, we do not take any a priori form of the equal-time commutation relations for the interacting fields. Neither do we take the usual definition [3] for the current density and the interaction term in the equations of motion. In section III, we do show that they both follow in an unique fashion as results from our general postulates and the first order interaction which defines QED. Finally section IV contains a general discussion and comparasion with other approaches.

II. GENERAL FRAMEWORK.

QED describes the interaction of structureless charged particles(e.g. electrons) via photons. We assume the existence of Heisenberg fields for the photon and the electron $A_\mu(x), \Psi(x)$. These fields tend asymptotically to free incoming fields $a_\mu(x)$, $\phi(x)$ when $t \rightarrow -\infty$ and free outgoing fields $a_\mu^{out}(x), \phi^{out}(x)$ when $t \rightarrow +\infty$. They both fulfill the sourceless equations:

$$K_x a_\mu(x) \equiv \partial_\lambda \partial^\lambda a_\mu(x) = 0 \qquad (2.1)_a$$

$$\mathcal{D}_x \phi(x) \equiv (i\gamma_\lambda \partial^\lambda - m)\phi(x) = 0 \qquad (2.1)_b$$

in the Lorentz gauge, the usual commutation relations for the photon free field:

$$[a_\mu(x), a_\nu(x')] = -i D_{\mu\nu}(x-x'), \qquad (2.2)_a$$

and anticommutation relations for the electron-positron field:

$$\{\bar{\phi}_\alpha(x), \phi_\beta(x')\} = -i S_{\alpha\beta}(x-x'). \qquad (2.2)_b$$

Here D and S are the Jordan-Pauli functions.

The in fields are unitarily related to the out fields by the S-matrix

$$S^{-1} a_\mu(x) S = a_\mu^{out}(x), \qquad (2.3)_a$$

$$S^{-1} \phi(x) S = \phi^{out}(x) \qquad (2.3)_b$$

which, assuming asymptotic completeness in the usual sense[5-9], can be expanded in terms of normal ordered products of the <u>in</u> fields

$$S = 1 + \sum_{m,n=1}^{\infty} \frac{i^n(-)^m}{n!(m!)^2} \int C_{m,n}(x_1,\dots,x_m;y_1,\dots,y_m;z_1,\dots,z_n) : \bar{\phi}(x_1)\cdots\bar{\phi}(x_m)\phi(y_1)\cdots\phi(y_m)a_{\mu_1}(z_1)\cdots a_{\mu_n}(z_n): . \quad (2.4)$$

The expansion coefficients are given by the reduction formulae[5,10,11] in terms of the 2m+n-point propagator

$$C_{m,n}(x;y;z) = \mathcal{D}_1\cdots\mathcal{D}_m K_1\cdots K_n \langle T\bar{\Psi}(x_1)\cdots\Psi(y_1)\cdots A_{\mu_1}(z_1)\cdots\rangle \overleftarrow{\mathcal{D}}_1\cdots\overleftarrow{\mathcal{D}}_m . \quad (2.5)$$

It can be shown[8,9] that defining the current densities $j_\mu(x)$ and $f(x)$ from the S matrix as

$$j^\mu(x) = i\left[\delta S/\delta a_\mu^{out}(x)\right] S^+, \quad (2.6)_a$$

$$f(x) = i\left[\delta S/\delta \phi^{out}(x)\right] S^+ \quad (2.6)_b$$

the Heisenberg fields $A_\mu(x)$, $\Psi(x)$ obey the equations of motion

$$K_x A_\mu(x) = e\, j_\mu(x), \quad (2.7)_a$$

$$\mathcal{D}_x \Psi(x) = e\, f(x) \quad (2.7)_b$$

in the Lorentz gauge. The integral form of these equations in terms of the <u>in</u> fields is[12]

$$A_\mu(x) = a_\mu(x) - e\int D_{\mu\nu}^{ret}(x-x')\, j^\nu(x')\,dx', \quad (2.8)_a$$

$$\Psi(x) = \phi(x) + e\int S^{ret}(x-x')\, f(x')\,dx'. \quad (2.8)_b$$

Now, taking into account the relativistic covariance of the theory as well as the existence of a unique vacuum, it can be shown[4,9] that the current densities must fulfill the stability conditions

$$\langle 0|\, j_\mu(x)\,|1\ \text{ph.}\rangle = 0 \quad (2.9)_a$$

$$\langle 0|\, f(x)\,|1\ \text{el.}\rangle = 0 \quad (2.9)_b$$

Notice that these conditions are not fulfilled by the canonical current densities. We may also observe that a similar relationship, related to the vacuum stability, gives rise to the need of introducing normal ordering in the definition of the free electron current . In fact the free currents

$$e\, j_\mu^{(1)}(x) = e : \bar{\phi}(x)\gamma_\mu \phi(x) :, \qquad (2.10)_a$$

$$e\, f^{(1)}(x) = e\, \not{a}(x)\phi(x) \qquad (2.10)_b$$

satisfy the identities

$$\langle 0 | j_\mu^{(1)}(x) | 0 \rangle = 0 \quad ; \quad \langle 0 | j_\mu^{(1)}(x) | 1ph \rangle = 0 \, , \qquad (2.11)_a$$

$$\langle 0 | f^{(1)}(x) | 0 \rangle = 0 \quad ; \quad \langle 0 | f^{(1)}(x) | 1el. \rangle = 0 \qquad (2.11)_b$$

so that in a coupling constant expansion of the ineracting currents, they can be taken to be the lowest order terms. Equations (2.10) then correspond to chosing the elementary vertex to be such that

$$S^{(1)} = -ie \int : \bar{\phi}(x)\gamma_\mu \phi(x) a^\mu(x) : dx \qquad (2.12)$$

This is, of course, the "minimal coupling" interaction. It is local and yields a gauge inavariant theory. As it is well known, it is also C, P and T invariant.

We have then found that the canonical currents contra dict our general postulates. What about the equal time commutation relations? That they are ill-defined is a well known fact[6,9]. In particular Haag's theorem[9,13] shows that they cannot be assumed to be the canonical ones. These apparently negative result may be seen from a different point of view. We may ask how much the general postulates restrict the commutation rules. Or even better, given the elementary interaction and the general assumptions, can the commutation rules be uniquely determined? As already anticipated in the introduction, the answer is affirmative and part of purpose of the next section is precisely to show it.

III. CONSTRUCTION OF QED.

Given the specific form of the elementary interaction, we can already calculate tree diagrams. It is just necesssary to calculate the appropiate functional derivatives[11,14,15]. The derivative with respect to a_μ, e.g., is obtained by introducing an infinitesimal change $a_\mu \rightarrow a_\mu + \delta a_\mu$, taking δa_μ as an external field.

Consider then the equation of motion $(2.8)_b$ to lowest order we then get[3]

$$\mathcal{D}_x\, \delta S^c(x,y) = e\int \delta \not{a}(\xi)\, S^c(\xi,y)\, d\xi\,, \tag{3.1}$$

hence the first functional derivative yields

$$\frac{\delta S^c(x,y)}{\delta a_\mu(\xi)} = e\, S^c(x,\xi)\,\gamma^\mu S^c(\xi,y). \tag{3.2}$$

This gives an extra elementary vertex with a factor e; functional differentiation increases in one order a perturbative term. Hence, the second derivative will give the lowest order Compton scattering term:

$$\frac{\delta^2 S^c(x,y)}{\delta a_\mu(\xi)\,\delta a_\nu(\eta)} = e^2\left[S^c(x,\xi)\,\gamma^\nu S^c(\xi,\eta)\,\gamma^\mu S^c(\eta,y) + S^c(x,\eta)\,\gamma^\mu S^c(\eta,\xi)\,\gamma^\nu S^c(\xi,y)\right]; \tag{3.3}$$

setting $\delta a_\mu = 0$, and using the reduction formula $(2.5)_b$ together with equation (3. 1) we get

$$S^{(2)}_{\text{Compton}} = -\frac{ie^2}{2}\iint\left[:\bar{\phi}(\xi)\not{a}(\xi)\,S^c(\xi-\eta)\,\not{a}(\eta)\phi(\eta): - :\not{a}(\xi)\phi(\xi)\,S^c(\eta-\xi)\,\bar{\phi}(\eta)\not{a}(\eta):\right] d\xi\, d\eta \tag{3.4}$$

The other tree diagrams can be obtained in a similar way. Additional functional derivatives give rise to higher order terms with additional vertices. Extra electron terms are obtained by taking functional derivatives with respect to ϕ and $\bar{\phi}$, in a well known procedure.

However, since functional differentiation does not create any closed loop, we cannot obtain by this single procedure any radiative corrections. We have thus to study the general behaviour of the two electron propagators, where the loops first appear.

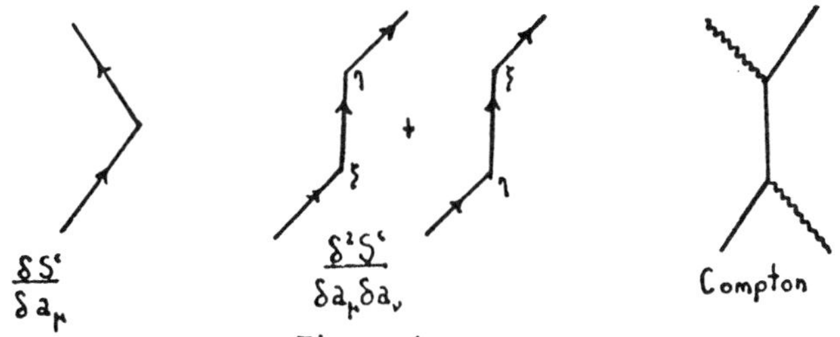

Figure 1

Two point functions.- The spectral representation for the photon and electron propagators can be found[16,17], with the aid of the integral equations (2.8) and the stability condition (2.9). We shall explicitly do it for the photon propagator and, at the end, it will be self evident that an entirely similar result holds for the exact electron propagator.

The exact photon propagator is defined as

$$\mathcal{D}^c_{\mu\nu}(x_1-x_2) = i\,\langle T A_\mu(x_1) A_\nu(x_2)\rangle. \tag{3.5}$$

In order to have a correct relativistic definition of the T product we introduce the scalar function:

$$\mathcal{D}^c(x_1-x_2) = \frac{i}{3}\,\langle T A_\mu(x_1) A^\mu(x_2)\rangle \tag{3.6}$$

and use the transversality property[10,18]:

$$\mathcal{D}^c_{\mu\nu}(k) - D^c_{\mu\nu}(k) = \left(g_{\mu\nu} - \frac{k_\mu k_\nu}{k^2}\right)[\mathcal{D}^c(k) - D^c(k)] \tag{3.7}$$

in momentum space. Substituting now Eq.(2.8)$_a$ into Eq.(3.6)

$$\mathcal{D}^c(x_1-x_2) = D^c(x_1-x_2) + \frac{ie^2}{3}\left\{\theta(\tau)\int D^{ret}(x_1-y_1)D^{ret}(x_2-y_2)\langle j_\mu(y_1) j^\mu(y_2)\rangle + (1\leftrightarrow 2)\right\}; \tag{3.8}$$

$$\tau = t_1 - t_2 .$$

There are several points worth remarking in this expression. The first one is that the cross terms vanish, in view of the one photon stability property (2.9)$_a$. The second one is that the causal character of the propagator follows directly from its definition in terms of the T-product, so it becomes irrelevant whether we use retarded functions in the integrand, or any other photon free propagator. Thirdly, we notice that the argument in the step functions refers to the original variables, and not to the integration variables, appearing as arguments of the current densities. Finally, the current-current correlation function is a well defined function (more precisely a tempered distribution[7,19]) with well known spectral properties[10]. Its Fourier transform has the form:

$$-\frac{1}{3}\int \langle j_\mu(y_1) j^\mu(y_2)\rangle e^{iky}\,dy = J(k^2)\,\theta(k^0) \quad , \quad y = y_1 - y_2 \; , \tag{3.9}$$

where J depends on k^2 only. It can be obviously calculated knowing the current density $j_\mu(y)$.

The last step to obtain the spectral representation of $\mathcal{D}^c(k)$ is to take the Fourier transform of Eq. (3.4), yielding a convolution in ω, the conjugate variable of the time difference τ, using the representation of θ :

$$\theta(\pm\tau) = \frac{i}{2\pi}\int_{-\infty}^{\infty} \frac{e^{-i\omega\tau}}{\pm\omega + i\varepsilon} \; . \tag{3.10}$$

So we finally have:

$$\mathcal{D}^c(k) = D^c(k) - \frac{e^2}{2\pi}\int_{-\infty}^{\infty} d\omega \, \frac{J(q^2)}{q^4}\left[\frac{\theta(\omega)}{k^0-\omega+i\varepsilon} + \frac{\theta(-\omega)}{-k^0+\omega+i\varepsilon}\right] \tag{3.11}$$

where

$$q^\mu = (\omega, \vec{k}) \; . \tag{3.12}$$

As we mentioned above, we see that we do not need to specify the character of the Green function $1/q^2$.

If we now change the integration variable in Eq. (3.11) to the scalar

$$\lambda \equiv q^2 = \vec{k}^2 - k^{0 2} + \omega^2 \tag{3.13}$$

we get

$$\mathcal{D}^c(k) = D^c(k) + \frac{e^2}{2\pi}\int_0^{\infty} \frac{J(\lambda)}{\lambda^2} \, \frac{d\lambda}{\lambda - k^2 - i\varepsilon} \; . \tag{3.14}$$

This spectral representation shows that the correct spectral density for the interacting part of the photon propagator is $J(\lambda)/\lambda^2$.

The photon proper energy function is defined through the relation [18] :

$$\mathcal{D}^c = D^c + D^c \Pi D^c \tag{3.15$_a$}$$

$$\Pi(k^2) = k^2\left[\mathcal{D}^c(k) - D^c(k)\right] k^2 \tag{3.15$_b$}$$

and the transverse projection gives the gauge invariant function:

$$\Pi_{\mu\nu}(k) = \left(g_{\mu\nu} - \frac{k_\mu k_\nu}{k^2}\right)\Pi(k^2) . \tag{3.16}$$

It then follows immediately, from Eq. (3.14) and (3.15), that

$$\Pi(k^2) = \frac{e^2}{2\pi} k^4 \int_0^{\infty} \frac{J(\lambda)}{\lambda^2} \, \frac{1}{\lambda - k^2 - i\varepsilon} \, d\lambda . \tag{3.17}$$

A similar procedure for the electron propagator yields[16]

$$\mathcal{G}^c(p) = S^c(p) - \frac{e^2}{2\pi}\int_0^{\infty} \frac{\Sigma_1(\lambda) + \not{p}\,\Sigma_2(\lambda)}{(\lambda - m^2)^2(\lambda - p^2 - i\varepsilon)} \, d\lambda \; , \tag{3.18}$$

in terms of the functions

$$\Sigma_1(\lambda) = 2\lambda m \, R_1(\lambda) + (\lambda + m^2) R_2(\lambda) \tag{3.19$_a$}$$

$$\Sigma_2(\lambda) = (\lambda + m^2) R_1(\lambda) + 2m \, R_2(\lambda) \tag{3.19$_b$}$$

with
$$\mathcal{F}\langle f(y_1) f(y_2)\rangle = \theta(p^0)\left[\not{p}\, R_1(p^2) + R_2(p^2)\right]. \tag{3.20}$$

The electron proper energy
$$\Sigma(\not{p}) = (\not{p}-m)\left[G^c(p) - S^c(p)\right](\not{p}-m), \tag{3.21}$$
is hence given by :
$$\Sigma(\not{p}) = -\frac{e^2}{2\pi}(\not{p}-m)^2 \int_0^\infty \frac{\Sigma_1(\lambda) + \not{p}\,\Sigma_2(\lambda)}{(\lambda-m^2)^2(\lambda-p^2-i\varepsilon)}\, d\lambda . \tag{3.22}$$

The contribution of the two-point propagators to the S-matrix obtains from the reduction formulae, Eq. (2.5)

$$C_{0,2}(z_1,z_2) = -i\,\Pi(z_1-z_2) \tag{3.23$_a$}$$

and

$$C_{1,0}(x_1;y_1) = i\,\Sigma(x_1-y_1)\,. \tag{3.23$_b$}$$

These expressions automatically fulfill the usually imposed "renormalization conditions"[3,10,18],

$$\Pi(\lambda)\Big|_{\lambda=0} = \Pi'(\lambda)\Big|_{\lambda=0} = 0, \tag{3.24$_a$}$$
$$\Sigma(\not{p})\Big|_{\not{p}=m} = \Sigma'(\not{p})\Big|_{\not{p}=m} = 0. \tag{3.24$_b$}$$

The origin of subtraction terms in Eqs. (3.16) and (3.22) is the appearence of the inverses of K and $\mathcal{D}$ respectively when solving the equations of motion (2.7). This is in agreement with the fact that the interaction implies necessarily substracted dispersion relations[19].

Now, we observe that, due to the e^2 factor in the spectral representations, we can calculate the lowest order radiative corrections for the two point propagators from Eqs. (2.4-5). In the case of vacuum polarization, we need to compute the j-j correlation function. To lowest order the current is
$$j_\mu^{(1)} = :\bar{\phi}\gamma_\mu\phi: \tag{2.10$_a$}$$

so that

$$e^2 \langle j_\mu(x_1)\, j^\mu(x_2)\rangle^{(2)} = e^2 \langle :\bar\psi(x_1)\gamma_\mu \psi(x_1)::\bar\psi(x_2)\gamma^\mu \psi(x_2):\rangle \tag{3.25}$$

and the function J in Eq. (3.9) is , to this order[3] ,

$$e^2 J^{(2)}(\lambda) = -\frac{e^2}{3(2\pi)}\,\theta(\lambda-4m^2)\sqrt{\frac{\lambda-4m^2}{\lambda}}\,(\lambda+2m^2) \tag{3.26}$$

and hence:

$$\Pi^{(2)}(k^2) = -\frac{e^2}{12\pi^2}\,k^4 \int_{4m^2}^{\infty} \frac{d\lambda}{\lambda^2}\sqrt{\frac{\lambda-4m^2}{\lambda}}\,\frac{\lambda+2m^2}{\lambda-k^2-i\varepsilon} \;; \tag{3.27}$$

for small k

$$\Pi^{(2)}(k^2) \xrightarrow[k^2\to 0]{} -\frac{\alpha}{15\pi m^2} \tag{3.28}$$

where $\alpha = e^2/4\pi$ is the fine structure constant.

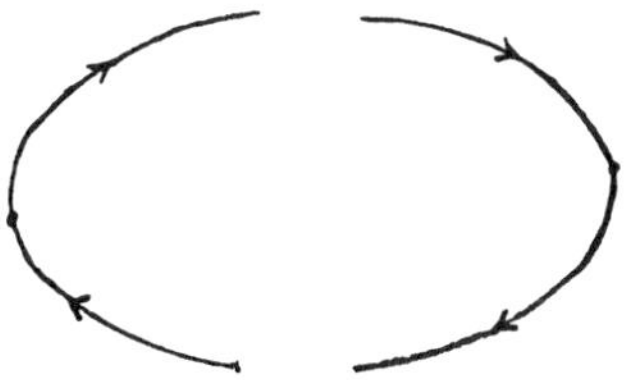

$e^2 \;\langle jj \rangle$

Figure 2

Thus, a loop is formed through the correlation function. However, since there is no multiplication by step functions, The product in Eq. (3.25) is well defined as a product of distributions[7] . The diagram in Fig. 2 is then similar to the one used in S-matrix theory[20]. However, in the present work there are no assumptions about the analytic properties, and the substractions in the dispersion relations , Eq(3.17) are derived, instead of being imposed

A similar calculation gives the radiative correction to the electron propagator[3,21]. Again a loop is formed in the computation of the f-f correlation function. In general, all the new loops result from the contractions appearing in the spectral densities $J(\lambda)$, $\Sigma_1(\lambda)$, $\Sigma_2(\lambda)$. These are not time ordered functions and are well behaved.

$e^2 \langle f\bar{f} \rangle$

Figure 3

Iteration.- We have thus shown that the second order terms can be computed just on the basis of our general assumptions and the elementary vertex. The higher order terms can be computed by iteration: from the two and four-point propagators to second order, we construct the S-matrix, and hence the currents to the same order, using Eqs. (2.4-6). At the same time, we can compute the third order contributions to the vertex by taking the functional derivative $\delta\Sigma/\delta a_\mu$,which yields the S-matrix to third order. Substituing these expressions for the currents in the spectral representatoions for the photon and electron proper energies, Eqs (3.17) and (3.22), gives their correct expression to fourth order, and the process can be iterated once the other n-propagators have been calculated to that order by functional differentiation. The result obtained in this way concide with the usual perturbation expansion, once the latter has been properly renormalized.

Commutation relations.- By the iterative procedure we have just described, QED can be constructed. Thus, the commutation relations can be calculated to the desired order. However, the procedure applied for obtaining the spectral representations of the two-point propagators can be applied to obtain similar relations for the vacuum expectation value of the (anti-)commutation relations.[21] Starting from Eq. (3.14) in configuration space

$$\mathcal{D}^c(x_1-x_2) = D^c(x_1-x_2) - \frac{e^2}{2\pi}\int_0^\infty \frac{J(\lambda)}{\lambda^2} D^c(x_1-x_2;\lambda)\,d\lambda \qquad (3.29)$$

with

$$(\partial^2 + \lambda) D^c(x;\lambda) = -\delta(x). \tag{3.30}$$

Taking the even part of Eq.(3.5), and using Eqs. (3.7) and (3.29), we get

$$\langle [A_\mu(x_1), A_\nu(x_2)] \rangle = -i \left[D_{\mu\nu}(x_1 - x_2) - \frac{e^2}{2\pi} \int_0^\infty \frac{J(\lambda)}{\lambda^2} D_{\mu\nu}(x_1 - x_2;\lambda) d\lambda \right. . \tag{3.31}$$

Taking the time derivative of Eq. (3.31) and evaluating it at equal times $t_1 = t_2$, we get

$$\langle [\partial_t A_\mu(x), A_\nu(x')] \rangle \Big|_{t=t'} = -i\, \delta(\vec{x} - \vec{x}') g_{\mu\nu} \left[1 - \frac{e^2}{2\pi} \int_0^\infty \frac{J(\lambda)}{\lambda^2} d\lambda \right] \tag{3.32}$$

in Loretz gauge.

Similarly, for the electron anticommutation relation, we get

$$\langle \{ \Psi_\alpha(x), \overline{\Psi}_\beta(x') \} \rangle \Big|_{t=t'} = \gamma^0_{\alpha\beta}\, \delta(\vec{x} - \vec{x}') \left[1 + \frac{e^2}{2\pi} \int_0^\infty \frac{\sqrt{\lambda}\, \Sigma_1(\lambda) + \Sigma_2(\lambda)}{(\lambda - m^2)^2} d\lambda \right] \tag{3.33}$$

Just as expected, the commutation relations depend on the interaction. Furthermore, since the delta functions factor out, we are left with a divergent constant multiplying the delta function, in both cases.

IV. CONCLUSION.

A coherent and systematic picture of QED has been obtained by defining current densities which obey the one-particle stability condition, and taking the integral form of the equations of motion as a starting point. The commutation relations for the interacting fields are computed a posteriori, instead of being assumed, as it is done in canonical theory. In this way we work with "renormalized" fields from the very beginning. The resulting expressions for the two-point propagators turn out to be the correct ones. The rest of the n-point functions are computed by taking the appropiate functional derivatives.

In our procedure, there is no need to impose any additional conditions, either on the mass shell [22], or for large momenta [23]. The electron and photon propagators fulfill automatically the "renormalization conditions [2,10], while the other propagators inherit the correct boundary conditions by functional differentiation, and no quasilocal operators remain undetermined

It is also very clear that the starting point for the chain of functional derivatives should not be the vacuum to vacuum amplitude. Instead, we must start from the two-point propagators. The reason is apparently, that in this way we defined the interaction in a unique way[19], and compute the modification suffered by the free propagators due to the interaction.

Bibliography

1.- See, e.g., the interesting contributions of J. Sapirstein and P.J. Mohr to this workshop.

2.- J. Schwinger, "Selected Papers and Quantum Electrodynamics "(Dover Publ., New York, 1979)

3.- G. Källen, "Quantum Electrodynamics" (Springer Verlag, Berlin (1972).

4.- W. Zimmermann, in "Lectures on Elementary Particles and Quantum Field Theory" Bradies University Summer Institute (1980)

5.- M. Lehman, K. Symansik & W. Zimmermann, Nuovo Cimento 1, 1425 (1955); 6, 319 (1964)

6.- R.F. Sthreater & A. S. Wightman, "PCT, Spin and Statistics, and See that "(Benjamin, New York, 1975)

7.- N.N. Bogoliubov, A.A. Logunov & S.T. Todorov, "Axiomatic Field Theory" (Benjamin, New York, 1975)

8.- S. Schrueber, "An Introduction to Relativistic Quantum Field Theory" (Harper & Row, New York, 1961).

9.- P. Roman, "Introduction to Quantum Field Theory" (J. Wiley & Sons,

New York, 1969).

10.- N.N. Bogoliubov & D.U. Shirkov, "Introduction to the Theory of Quantized Fields" (Interscience, New York, 1959)

11.- S. Bialincki - Birula & Z. Bialincka - Birula, "Quantum Electrodynamics" (Pergamon Press, Oxford, 1975)

12.- C.N. Yang I D. Feldman , Phys. Rev. 79, 972 (1950)

13.- R. Haag, Kgl. Dnsk, Vid. Slskb. Mat. Pys. Medd. 29, 12 (1955)

14.- A. Visconti, "Theorie Quantique des Champs" (Gauthiers-Villars, Paris, 1965) Vol. II

15.- H.M. Fried, "Functional Methods and Models in Quantum Field Theory" (MIT Press, Cambridge, Mass., 1972)

16.- M. Berrondo & R. Jáuregui, Lett. Nuovo Cimento 45, 451 (1984)

17.- M. Berrondo & R. Jáuregui, Kinam 5, 41 (1984)

18.- E.M. Lifsithz, L.P. Pitaevskii, "Relativistic Quantum Theory II" (Pergamon Press, Oxford, 1973)

19.- M. Muraskin & K. Nishijima, Phys. Rev. 122, 321 (1961)

20.- T. Chou & M. Dresden, Rev. Mod. Phys. 39, 143 (1968)

21.- M. Berrondo & R. Jáuregui (to be published)

22.- J. Wray, J. Math. Phys. 9, 537 (1968); 9 552, (1968)

23.- R.E. Pugh, Ann. Phys. (NY) 23, 335 (1963)

SESSION ON RELATIVISTIC CALCULATIONS FOR MANY-ELECTRON ATOMS

I.P. Grant
Department of Theoretical Chemistry, 1 South Parks Road,
Oxford, OX1 3TG, England

INTRODUCTION

The session on relativistic calculations for many-electron atoms divided into two distinct parts. The first paper, by Walter Johnson, gave a very careful analysis of the progress of his work with Sapirstein on the calculation of radiative transitions induced by parity-violating electro-weak interactions. This was followed by a very brief discussion centring on the information that atomic physics measurements may provide about electro-weak interactions and its relation to other data coming from high-energy physics and from such experiments as low-energy electron scattering at small momentum transfer. The second paper, by Jean-Paul Desclaux, dealt with the status of relativistic atomic structure calculations for systems with relatively few electrons. This was followed by another brief discussion, and there were a number of short contributions which are summarized below.

DISCUSSION ON WEAK INTERACTIONS IN ATOMS

G.L. Greene (NBS) : Prof. Johnson has correctly noted that experiments on atoms can provide important data concerning T-violation. As he has pointed out, the current best limits on electron dipole moment (e.d.m.) are set by experiments on atoms. However, it is important to note that experiments on atoms (and molecules) are also sensitive to a variety of other T-violating effects. Such experiments provide the most sensitive tests now known for such effects.

The essential experiment is the attempt to detect a permanent electric dipole moment in an atom (or molecule with non-degenerate ground state). Such an atomic e.d.m. may arise from a variety of more fundamental causes. These may be categorized as : i) electron e.d.m.; ii) proton e.d.m.; iii) neutron e.d.m.; iv) induced atomic e.d.m. arising from T-odd interactions between electrons and nucleons; and v) induced nuclear e.d.m. arising from T-odd interactions between nucleons. With the exception of the neutron e.d.m. (iii) whose limit is set by experiments on free neutrons, limits on all of the effects listed are most accurately set by atomic (or molecular) experiments. There are two separate approaches. In the first, one uses a highly polarizable species in which an enhancement of certain T-odd effects may occur; experiments on Cs, $^3P_2$ metastable Xe and TlF. The second approach uses a species which is less polarizable, but on which extremely sensitive measurements can be made. An experiment of this sort has been made on ground state Xe.

The interpretation of results from such experiments is somewhat ambiguous. While any observation of an atomic e.d.m. is certainly *prima facie* evidence for some type of T-violation, the nature of the interaction which gives rise to such T-violation requires a thorough theoretical understanding of the atomic system under consideration. As such, there is a clear need for convincing calculations of the type outlined by Prof. Johnson. I would add that such calculations are useful only when carried out on atomic systems amenable to experimental investigation. Since such experimental systems are rather limited, I would hope that interested theoreticians seek appropriate direction from the experimentalists.

DISCUSSION ON STRUCTURE OF FEW-ELECTRON ATOMS

A major issue raised by Desclaux was the choice of gauge for the covariant Coulomb interaction; it is usually assumed that the conventional choice, the Coulomb gauge, leads to the most satisfactory results but this has never been properly checked. The Breit interaction is normally treated as a first order perturbation correction only, but it would be interesting to make further tests, at higher values of Z, of the effect of including it in the self-consistent part of the calculation in the manner indicated by Desclaux. There is no reason not to do this, despite folklore restricting use of the Breit interaction to first order perturbation theory.

Fulton suggested that it would be easier to study gauge effects in positronium where the answer is known : spurious terms appear in Lorentz gauge which are not present in the Coulomb gauge, but which can be shown to cancel out in higher orders of perturbation theory. Sucher agreed, and pointed out that, by iterating the one-photon interaction, one generates all the ladder diagrams but not those in which photon lines cross. Drake added that, in addition to the one-electron QED terms discussed in the talk, there are specifically two-electron terms proportional to $\langle\delta(\vec{r}_{12})\rangle$ and $\langle 1/r_{12}{}^3\rangle$ which should be included in a comparison with experiment. In particular, the $\langle 1/r_{12}{}^3\rangle$ terms, which have traditionally been neglected for triplet states, are nearly as large for triplet states as for singlet states[1].

The Casimir effect, which is prominent in the g-Hartree method described by Dietz in the first session of the Workshop, was discussed inconclusively. Few of those present were clear what magnitude it would have, nor how it should be interpreted physically; it was agreed to be essential to obtain more information.

Not enough time was available to discuss other issues of importance in many-electron relativistic calculations. The sessions on QED devoted much time to the formulation of relativistic atomic structure theory at the Dirac-Fock or MCDF level, the only exception being Zygelman's brief description of his investigations with Mittleman into three-body interaction terms, which do not appear to be very significant. There was also little discussion of the way in which MCDF codes are applied in practice, but it seems as though most practitioners are coming round to the view that the method can be given "a clean bill of health" for most purposes, although not everyone agrees on how it can be fitted into quantum electrodynamics.

Radiative corrections are at least as important as the Breit interaction in higher-Z atoms, and need further attention. Although Mohr outlined on Thursday a possible approach to self-energy calculations in many-electron systems, his proposal is very far from being a simple prescription which is economical enough to use in a routine fashion. For the foreseeable future, it will be necessary to rely on the prescriptions currently used in computer codes: these are based on interpolation at a screened value of the nuclear charge in Mohr's hydrogenic tables for n = 1,2. Values of the Lamb shift for higher principal quantum numbers must be generated by using $1/n^3$ scaling. Clearly such prescriptions have no theoretical foundation, and should never be expected to give better than an order of magnitude estimate, if that. Experiments to test QED at higher values of Z will soon show how crude these estimates really are.

The range of topics discussed in this session and the vigour of the discussion gave impressive evidence of the strength with which relativistic atomic structure theory is developing. The close interaction between theory and experiment which has brought such rewards in the past will fuel further progress in this field of research in the next few years.

SHORT CONTRIBUTIONS

A. K.T. Lu (NBS) discussed how the nuclear hyperfine interactions can be studied through analysis of high-resolution spectra of Rydberg electrons using a modified "Fermi-Segré-Goudsmit" formula in conjunction with multichannel quantum defect theory. This allows predictions of the splitting of ionization potentials by hyperfine interactions; of level crossing and inversions due to competition between electrostatic, fine structure and hyperfine interactions; non-hydrogenic scaling laws for s, $p_{\frac{1}{2}}$, ... orbitals due to channel interactions; of nuclear shapes; and of volume isotope shifts in the continuum.

B. V.L. Jacobs (NRL) described a unified treatment of angular momentum changing electron collisions and radiative corrections for resolvable satellite spectra in dielectronic recombination done in collaboration with Rogerson (NRL), Chen (Lawrence Livermore National Laboratory) and Cowan (Los Alamos National Laboratory). For a problem with one discrete level and two continuum channels, the branching ratios for autionization and radiative decay rates are modified by replacing the autoionization rate $A_a(a \rightarrow i\varepsilon_i)$ and the radiative decay rate $A_r(a \rightarrow f)$ by quantities $\tilde{A}_a(a \rightarrow i\varepsilon_i)$ and $\tilde{A}_r(a \rightarrow f)$ which depend on two parameters: a continuum-continuum coupling parameter Ψ related to the cross-section for photoionization from the state $|f\rangle$ and a multi-channel Fano line profile parameter Q_f. He discussed applications to satellite spectra of helium-like and lithium-like ions.

C. Rocio Jauregui Renaud (UNAM) discussed the origin of renormalisation in QED.

D. W.E. Baylis (University of Windsor) has collaborated with Migdalek on the treatment of electron correlation in many-electron atoms. Intravalence electron interaction is treated by the multiconfiguration method; only configurations which contain virtual excitations solely from the valence shell are taken into account.

Valence-core and intracore correlations are described approximately by adding a dipole polarization potential to represent the effect of the core on the valence electrons. The polarization interaction also modifies the dipole-moment transition operator. This procedure brings a significant improvement in theoretical estimates of the wavelengths of the $^1S_0$ - $^1P_1$ and $^1S_0$ - $^3P_1$ lines in the Cd and Hg isoelectronic sequences compared with straightforward relativistic Hartree-Fock predictions.

REFERENCES

1. G.W.F. Drake and A.J. Makowski, J. Phys. B. 18, L103 (1985).

IONIZATION AND POSITRON EMISSION IN GIANT QUASIATOMS

G. Soff
Institut für Theoretische Physik, Justus-Liebig-Universität
Heinrich-Buff-Ring 16, D-6300 Gießen, West Germany

U. Müller and P. Schlüter
Gesellschaft für Schwerionenforschung (GSI), Planckstraße 1
Postfach 110 541, D-6100 Darmstadt, West Germany

J. Reinhardt, T. de Reus, K.-H. Wietschorke, A. Schäfer, B. Müller, and W. Greiner
Institut für Theoretische Physik der Johann Wolfgang Goethe-Universität, Robert-Mayer-Straße 8-10, Postfach 111 932
D-6000 Frankfurt am Main, West Germany

ABSTRACT

Electron excitation processes in superheavy quasiatoms are treated within a relativistic framework. Theoretical results on K-hole production rates as well as δ-electron and positron spectra are compared with experimental data. It is demonstrated that the study of heavy ion collisions with nuclear time delay promises a signature for the spontaneous positron formation in overcritical systems. Corresponding experimental results are confronted with our theoretical hypothesis. Recent speculations on the origin of the observed peak structures in positron spectra are critically reviewed. Atomic excitations are also employed to obtain information on the course of a nuclear reaction. Using a semiclassical picture we calculate the emission of δ-electrons and positrons in deep-inelastic nuclear reactions. Furthermore some consequences of conversion processes in giant systems are investigated.

INTRODUCTION

At present atomic physics of superheavy collision systems and quantum electrodynamics of strong fields is subject to widespread theoretical[1-38,92-98] and experimental[39-53] activities. In collisions of very heavy ions with projectile energies close to the nuclear Coulomb barrier superheavy quasiatoms are formed for a short period of time T ~ 10^{-20} s. In these exotic systems binding energies E_b and wavefunctions of electrons are determined by the combined charge of the projectile and target nuclei. Several characteristic features of a supercritical collision system are summarized[45] in fig. 1. The upper part displays a classical Rutherford trajectory, i.e. the internuclear distance R(t) as function of collision time t, for a central $^{238}U + ^{238}U$ collision with a bombarding energy at the nuclear Coulomb barrier (E_{Lab} = 5.9 MeV/u). A corresponding heavy ion collision is schematically indicated. Here the classical impact parameter b is uniquely connected with the ion scattering angle Θ. The lower part gives the energy eigenvalues of the strongest bound 1sσ, $2p_{\frac{1}{2}}\sigma$ and 2sσ states as function of t. The energy eigenvalues in the separated uranium atoms may be deduced from $t = \pm\infty$. For the innermost electrons $E_b(t)$ is comparable to the electron rest mass, the spatial distribution of electrons is localized within the Compton wavelength. Relativistic effects completely dominate the electron excitation processes. The transitions a and b denote electron excitation processes from the negative energy continuum, the Dirac sea, into the positive energy continuum and into vacant bound states, respectively. The created holes are detectable as positrons. The total cross section for this type of electron-positron pair formation exhibits a rather strong dependence[54,55] on the combined nuclear charge $Z = Z_1 + Z_2$, $\sigma_{e^+,e^-} \sim Z^{20}$. The most fascinating feature is that the quasiatomic K-shell enters the negative energy continuum as a resonance state for internuclear separations $R < R_{cr} \simeq 27$ fm. This allows for spontaneous positron production (process c in fig. 1), provided a partial K-shell depletion has occured on the ingoing path of the Rutherford trajectory. The related inner-shell vacancy formation, which is accompanied by δ-electron emission and subsequent molecular orbital radiation (MO x-rays) during the collision or K x-ray emission after the collision is also indicated schematically. However, even for the heaviest colliding system investigated up to now, U + Cm with Z = 188, this overcritical scenario ($E_b > 2\ m_e c^2$) exists only for about $2 \cdot 10^{-21}$ s, whereas the spontaneous decay width corresponds to $\tau \sim 10^{-19}$ s. In consequence this peculiar particle creation process remains undetectable in elastic Rutherford scattering. To overcome this obstacle nuclear reactions were investigated theoretically[1-4,6-9,11,12,26,27,29,56,57], leading to a prolonged overcritical field configuration. As one particular aspect of spectroscopy studies in superheavy atomic systems we summarize in this paper the current information on measured line structures in positron spectra. The relationship of the experimental facts[39,40,45,48,52,53] and our hypothesis of spontaneous positron emission in the strong Coulomb field of giant nuclear molecules is illustrated.

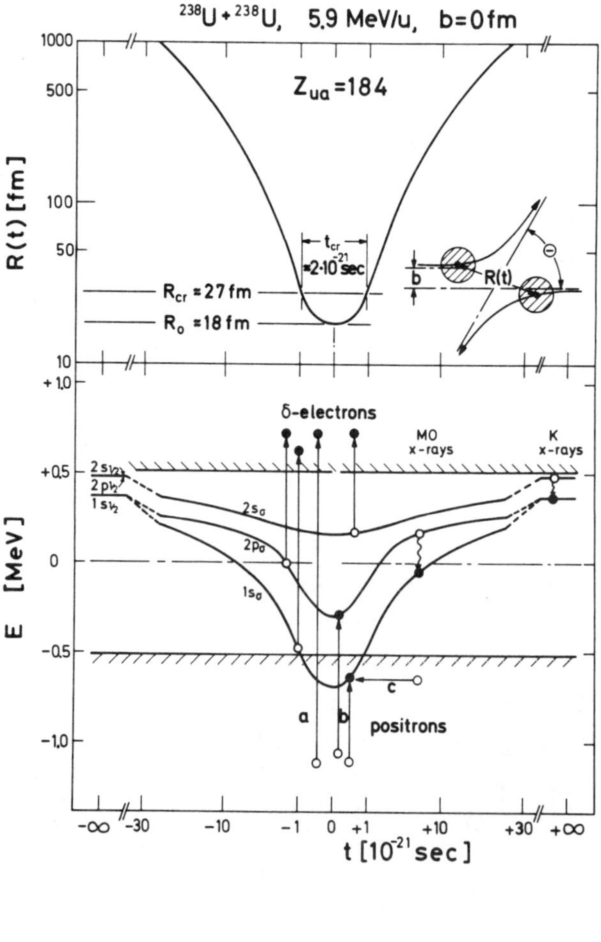

Fig. 1: The dependence of the internuclear separation R(t) and the energy eigenvalues E of strongly bound electrons on the collision time t. Processes a and b denote electron-positron pair production processes induced by the collision dynamics. The spontaneous positron creation is indicated by process c.

A first estimate of the electronic sensitivity on nuclear processes can be obtained by evaluating the electron probability density distribution inside the nucleus. Fig. 2 displays $|\Psi(r=0)|^2$ for the 1s-, 2s- and $2p_{\frac{1}{2}}$-state, respectively versus the nuclear charge number Z in superheavy atoms. A strong increase is visible for total charges below Z = 160, being most pronounced for the $2p_{\frac{1}{2}}$-state. Between Z = 92 (U) and Z = 184 $|\Psi_{2p\frac{1}{2}}(r=0)|^2$ changes by more than four orders of magnitude. For Z > 160 a saturation behaviour appears indicating the sharp localization of the electron wavefunction close to the origin.

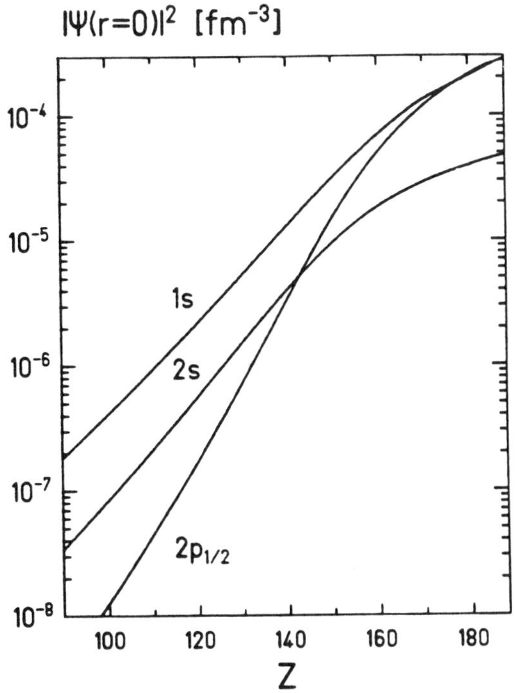

Fig. 2: Radial electron density $|\Psi(r)|^2 = f^2(r) + g^2(r)$ at the nuclear origin (r = 0) for 1s-, 2s- and $2p_{1/2}$-electrons as function of the nuclear charge number Z in superheavy atoms. The relative increase is most pronounced for the $2p_{1/2}$-state.

Fig. 3 shows a corresponding quantity for various states with angular momentum quantum numbers $\kappa = \pm 1$ in the positive and negative energy continuum. The considered total electron energies are indicated. For supercritical systems (Z > 170) the 1s-state leaves the bound state gap and gets imbedded as resonance in the negative energy continuum. This peculiar fact is reflected in the strong enhancement of the probability density at the resonance position. But in summary the Z-dependence of $|\Psi(r=0)|^2$ for continuum states is much weaker compared with that of inner-shell electron states.

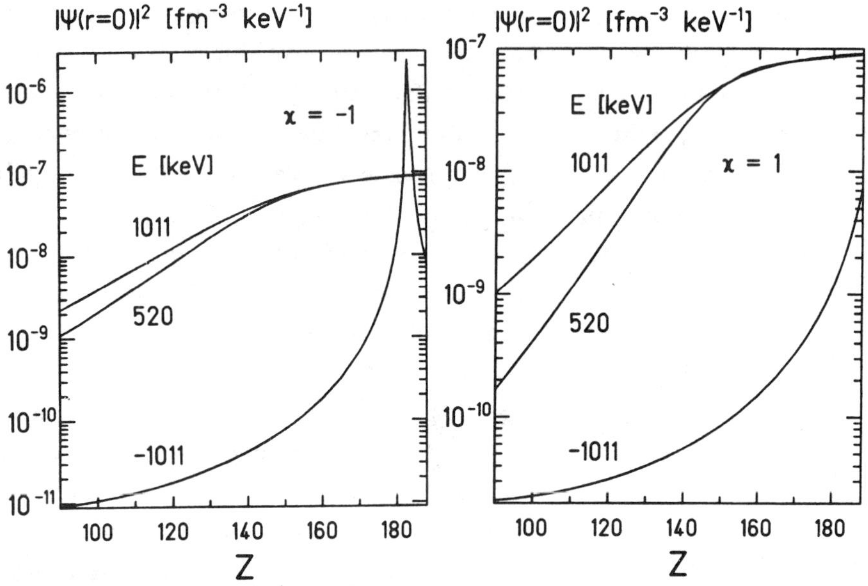

Fig. 3: Radial electron density $|\Psi(r)|^2 = f^2(r) + g^2(r)$ at the nuclear origin ($r = 0$) for continuum state electrons as function of the nuclear charge number Z in superheavy atoms. Electron states with quantum numbers $\kappa = -1$ and $\kappa = 1$ and with total energies E = 520, 1011 and -1011 keV are considered. The cusp for the state with $\kappa = -1$ and E = -1011 keV indicates the K-shell resonance imbedded into the negative energy continuum.

These electron densities are transiently produced in quasimolecular collisions of very heavy ions. For a short period of time $T \sim 10^{-20}$ s a superheavy quasiatomic system is formed, where all electrons experience the combined charge of the projectile and target nuclei. This time interval must be compared with the time scale of $\tau \leq 10^{-14}$ s being typical for radiative nuclear transitions. As demonstrated in fig. 1 innermost electron states exhibit drastic energy variations as function of the internuclear distance R along a classical Rutherford trajectory. This is also examplified in fig. 4 for the quasiatoms Z = 182 (U + Th) and Z = 188 (U + Cm). In the domain $R \leq 100$ fm the binding energies change by several hundred keV. To account for significant screening corrections we calculated the energy eigenvalues within the relativistic Hartree-Fock-Slater formalism[5].

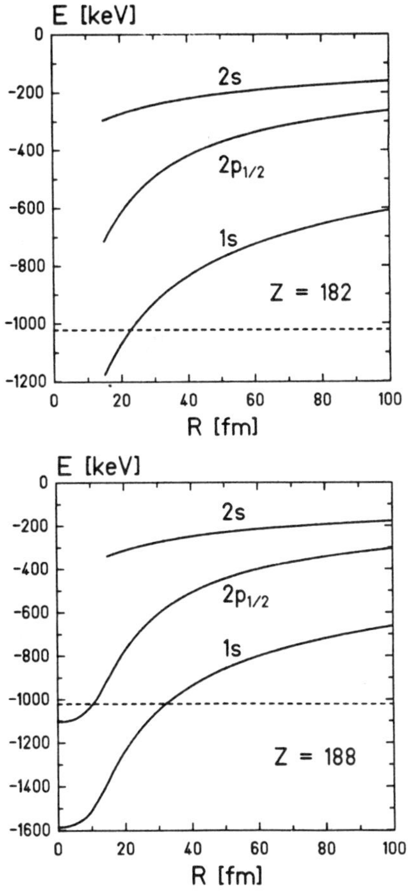

Fig. 4: Binding energies of 1s-, $2p_{1/2}$- and 2s-electrons versus internuclear distance R in the superheavy systems Z = 182 (U + Th) and Z = 188 (U + Cm). Electron screening and nuclear size corrections are taken into account. The border line to the negative energy continuum is indicated by the dashed line.

In consequence sharp electron or positron lines correlated with nuclear conversion processes involving transitions in or to bound states may result only from the separated atoms or from a long-living compound system which is subject of current investigations.

After this introduction we briefly outline the major theoretical ingredients for the calculation of electron excitation processes. Special emphasis is laid on the solution of the two-center Dirac continuum, which was numerically achieved employing finite-element techniques. To verify the correct theoretical treatment of ionization phenomena a few selected examples will be presented comparing theoretical and experimental data for vacancy and δ-electron production rates. After that we turn to the discussion of positron emission in elastic collisions of very heavy ions as well as in deep-inelastic nuclear collisions. It is demonstrated that the energy spectrum of emitted positrons is a sensitive tool to measure rather short nuclear reaction times. In this survey we will focus upon spontaneous positron formation in giant nuclear systems. This transition from a neutral to a charged vacuum ground state[1,2,78,79] represents the center of our theoretical investigations. Theoretical concepts are confronted with recent experimental observations. Various speculations on the origin of the observed peak structures in positron spectra are critically reviewed. Finally we discuss as novel aspects the K-hole coincident positron detection in heavy ion collisions and the positron production in crossed beams of bare uranium nuclei as well as particular studies of conversion processes in superheavy systems.

BASIC THEORETICAL FORMALISM

A semiclassical method is used to describe the electron dynamics during the collision. It is based on the time-dependent two-center Dirac equation ($\hbar = c = 1$):

$$i\,\partial/\partial t\;\Phi_i(\vec{R}(t)) = H_{TCD}(\vec{R}(t))\;\Phi_i(\vec{R}(t)), \tag{1}$$

where

$$H_{TCD} = \vec{\alpha}\cdot\vec{p} + \beta m + V_{TCD}(\vec{r},\vec{R}), \tag{2}$$

$$V_{TCD}(\vec{r},\vec{R}) = -Z_1e^2/|\vec{r}-\vec{R}_1| - Z_2e^2/|\vec{r}-\vec{R}_2|, \qquad \vec{R} = \vec{R}_1 - \vec{R}_2 .$$

The two-center potential may be extended to account for finite nuclear size and electron screening. The electrons are treated relativistically and the time-dependence is parametrized via the internuclear separation R(t). At non-relativistic bombarding energies it is appropriate to expand the wavefunction Φ_i using the molecular eigenstates ϕ_j of H_{TCD}:

$$\Phi_i(R(t)) = \sum_j a_{ij}(t)\;\phi_j(R(t))\;e^{-i\chi_j(t)}, \tag{3}$$

$$\chi_j(t) = \int^t dt' \, E_j \, (t'), \tag{4}$$

summing and integrating over bound states and continuum states of positive and negative energies, respectively. Inserting (3) into (1) and projecting with ϕ_j, we obtain a set of first order coupled differential equations for the amplitudes $a_{ij}(t)$:

$$\dot{a}_{ij}(t) = -\sum_{k \neq j} a_{ik}(t) \langle \phi_j | \partial/\partial t | \phi_k \rangle \, e^{i(\chi_j - \chi_k)}, \tag{5}$$

with the initial condition $a_{ij}(t=-\infty) = \delta_{ij}$. Splitting the time derivative $\partial/\partial t$ into a radial coupling, $\dot{R}\, \partial/\partial R$, and a rotational one, $i\vec{\omega} \cdot \vec{J}$, and neglecting the latter, the coupled equations (5) can be solved numerically. In most calculations the monopole approximation was used, i.e., V_{TCD} was restricted to its spherically symmetric part.

The solutions of the stationary two-center Dirac equation deserves special consideration. The eigenvalue spectrum of the stationary two-center Dirac equation consists of a discrete and a continuous part. The bound states[58] may be found by an application of the methods described in refs. 59-60. In figure 5 the energies of the 1sσ and 2sσ states are shown as a function of the two-center distance R for two lead nuclei. These eigenvalues are compared with results of the monopole approximation (dotted line), where the two-center potential has been replaced by its dominant monopole part.

For continuum states one does not need to solve a boundary value problem. Every solution regular at the origin yields a normalizable wavefunction. Here one has to orthonormalize the solutions, if possible following some guiding principle. The most obvious method seems to look for wavefunctions which are eigensolutions of the angular momentum operator $\hat{K}$ for large distances. However, in general such solutions will not exist for potentials which are not symmetric. In consequence this concept has to be generalized. Utilizing the local representation[59] such a generalization can be found. This procedure has the following properties: (a) the result is independent of the basis functions, which are found by numerical integration and from which the orthonormalized functions are constructed, and (b) for spherically symmetric potentials the orthonormalized functions are eigenfunctions of $\hat{K}$. By these means we constructed for various energies and two-center distances R wavefunctions which for $R \to 0$ transform into $\kappa = -1$ states. In figure 6 the $\partial/\partial R$ matrix elements between these states and the $1s\sigma$ bound state are shown. Again, they are compared with results of the monopole approximation (dotted line). Both methods yield nearly the same value despite the fact that the binding energies display a larger difference.

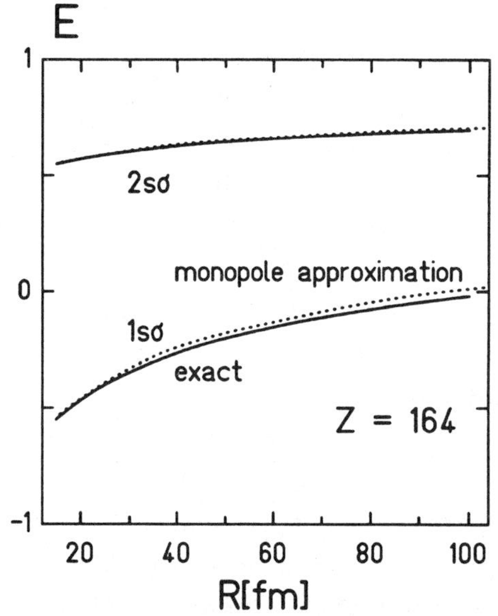

Fig. 5: Energy eigenvalues in natural units ($m_e = 1$) of $1s\sigma$ and $2s\sigma$ electrons versus the internuclear separation R in fermi. The collision system Pb + Pb is considered. Two-center calculations (solid lines) are compared with results of the monopole approximation (dotted lines).

An adequate description of positron production in supercritical collision systems, where Z_T+Z_P exceeds 173, requires a slight modification of the formalism. In a supercritical system the 1s-state is represented as a resonance in the positron s-wave continuum and not by a single eigenstate of the Hamiltonian H_{TCD}. A formalism that avoids those difficulties and moreover has heuristic value for the interpretation of the positron creation process was developed by Reinhardt et al[55]. The method is based on the observation that the continuum wavefunction of the supercritical system at resonance energy $E_p = E_{res}$ is quite similar to the discrete 1s state in the subcritical case except for an oscillating tail, small in amplitude but reaching out to infinity. This structure reflects the occurence of a tunnelling process through the barrier separating the particle and antiparticle solutions of the Dirac equation in a semi-classical picture. Apart from the asymptotic behaviour the 1s wavefunction retains many of its properties, e.g., the strong localization and the radial matrix elements which may be continued smoothly to the supercritical region if the tail of the wavefunction is neglected.

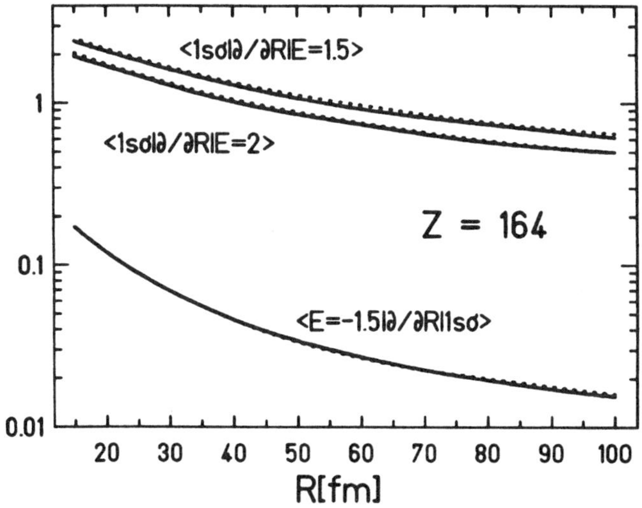

Fig. 6: Radial coupling matrix elements in natural units versus internuclear separation R in fermi in the superheavy system Z = 164. Transitions between the strongest bound 1sσ-state and σ-continuum states with E = 1.5, 2 and -1.5 m_e are considered. The solid lines follow from a full two-center calculation for bound as well as continuum states. The dotted lines give the result of the monopole approximation.

This idea was used within the projection formalism: After having defined a normalizable quasibound wavefunction ϕ_R, a new continuum $\tilde{\phi}_{Ep}$ is constructed which spans a subspace orthogonal to ϕ_R and replaces the old continuum ϕ_{Ep}. The modified continuum states satisfy the original Dirac equation supplemented by an inhomogeneous term that ensures orthogonality with respect to the resonance wavefunction ϕ_R:

$$(H_{TCD} - E_p)\, |\tilde{\phi}_{Ep}\rangle = \langle\phi_R|H_{TCD}|\tilde{\phi}_{Ep}\rangle\, |\phi_R\rangle\, . \qquad (6)$$

If the states ϕ_R and $\tilde{\phi}_{Ep}$ are used as part of the basis in eq. (3) the 1s-state ϕ_R couples to the new positron continuum by two separate coupling operators

$$\dot{R} \langle\tilde{\phi}_{Ep}|\partial/\partial R|\phi_R\rangle + i \langle\tilde{\phi}_{Ep}|H_{TCD}|\phi_R\rangle. \tag{7}$$

The second matrix element arises since ϕ_R and $\tilde{\phi}_{Ep}$ are not exact eigenstates of the two-center Hamiltonian H_{TCD}. It does not depend on the nuclear motion and leads, in the static limit $R(t) = \text{const} < R_{cr}$, to an exponential decay of a hole prepared in ϕ_R with a decay width

$$\Gamma = 2\pi|\langle\tilde{\phi}_{E_{res}}|H_{TCD}|\phi_R\rangle|^2. \tag{8}$$

The formalism thus naturally leads to the emergence of 'induced' and 'spontaneous' positron creation, the latter resulting from the presence of an unstable state ϕ_R in the expansion basis.

To account for the presence of all the other electrons in a realistic heavy ion collision, we evaluated the observable quantities within the framework of the second quantization prescription. Neglecting explicit electron-electron correlations one may express the excitation probabilities in terms of single particle occupation amplitudes. A detailed representation of the employed formalism is presented in refs. 55, 61 and 62.

IONIZATION PHENOMENA

To justify the employed theoretical framework we show in fig. 7 a measurement performed by Liesen et al.[63] of Pb + Cm 1sσ vacancies at 5.9 MeV/u in comparison with computed results. The full curve, being calculated with a relativistic Hartree-Fock-Slater (HFS) potential, is in good agreement with experimental data. We have chosen the initial Fermi surface above the 3sσ state, indicated by the notation F = 3sσ, in short F = 3. The sensitivity of the ionization probability on screening corrections may be deduced from the difference between the full and dashed line. The latter result was obtained employing hydrogen-like wavefunctions. The inset shows a comparison of $P_{1s\sigma}$ for the same system but at 5.4 MeV/u bombarding energy[64]. For this region of impact parameters K-holes produced by atomic excitation could be distinguished from those resulting from internal conversion. Theory[5] und experiment are in fair agreement which also illustrates the importance of using a realistic HFS potential.

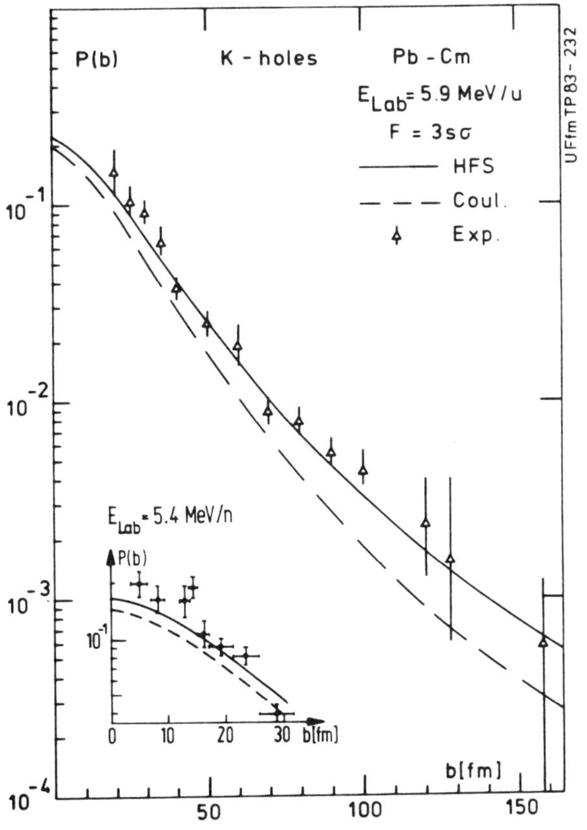

Fig. 7: $P_{1s\sigma}(b)$ is displayed for the system Pb + Cm at E_{Lab} = 5.9 MeV/u. The calculations are compared with experimental data of Liesen and co-workers[63]. In the inset we compare our results with experimental data of Ito et al.[64].

Most striking is the rather high ionization probability of about 10% for central collisions ($b \simeq 0$). This is caused by the sharp localization of the initial wavefunction due to relativistic effects and the associated high momentum components. The ionization rate calculated within the non-relativistic framework would be lower by several orders of magnitude. The fall-off constant in the exponential decline of $P_{1s\sigma}$ as function of the impact parameter is mainly determined by the combined nuclear charge Z in the collision system. It has been demonstrated[65] that a systematic measurement of this constant could lead to a spectroscopy of inner-shell electron states in superheavy quasiatoms.

With decreasing projectile energy the agreement between theoretical and experimental ionzation probabilities becomes less convincing. This is exemplified in fig. 8 where again measurements of Liesen et al[63] for $P_{1s\sigma}$ in the Pb + Cm system are compared with our computed results[5]. For E_{Lab} = 3.6 MeV/u the observed values for $P_{1s\sigma}(b)$ are always above the theoretical expectations. This could be caused by the neglection of purely molecular effects (vacancy sharings, crossings etc.) and by additional inner-shell vacancies already brought into the collision.

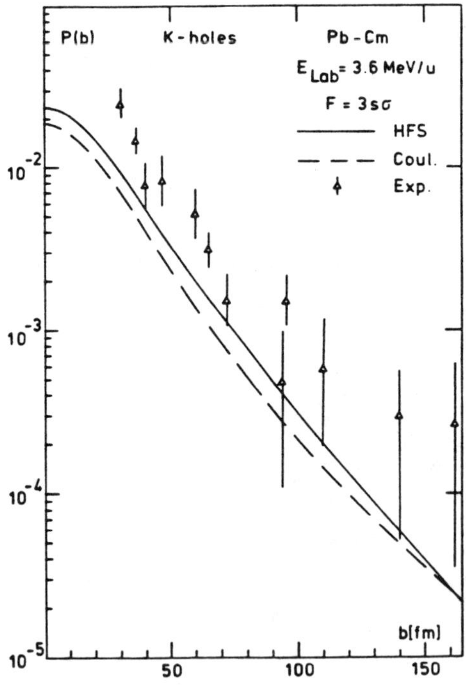

Fig. 8: The same as in fig. 7 for E_{Lab} = 3.6 MeV/u.

For the evaluation of the δ-electron distribution we used a basis of 8 sσ-bound and 18 sσ-continuum states, 6 $p_{\frac{1}{2}}\sigma$-bound and also 18 $p_{\frac{1}{2}}\sigma$-continuum states. The relativistic Hartree-Fock-Slater potential was calculated assuming a loss of 50 electrons in the collision system due to pre-collision stripping techniques. All calculations have been performed twice using a Fermi surface of F = 3 and F = 4 (i.e., all levels up to $4p_{\frac{1}{2}}\sigma$ and $5p_{\frac{1}{2}}\sigma$, respectively, are filled) to get an estimate of the influence of the Fermi level. The differences in the coincident spectra were of the order of 1%. In the total spectra the contribution of the higher Fermi level varies from 35% in the electron energy region of 200 keV down to 15% at 600 keV. The theoretical spectra presented here have been calculated assuming a Fermi level F = 3.

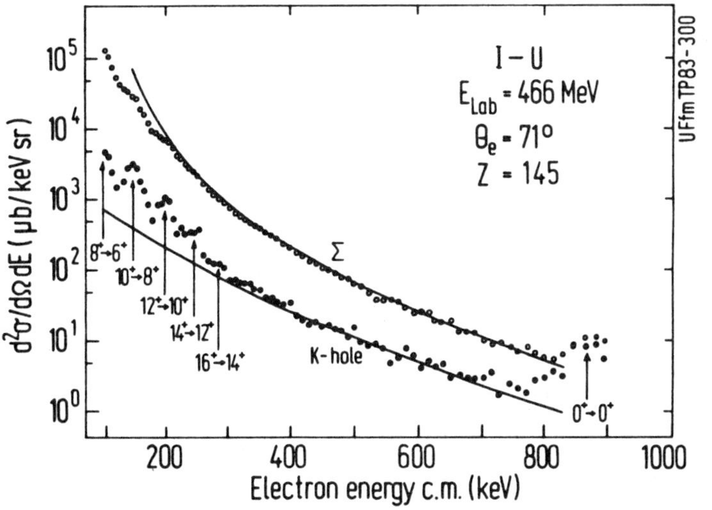

Fig. 9: Experimental double differential cross sections for the δ-electron yield (Σ, open dots) and for the 1s atomic coincident yield (K, full dots) versus kinetic electron energy. The system I - U at 466 MeV bombarding energy is considered. The full lines are the results of our coupled channel calculations[13]. Experimental data by W. Koenig, M.A. Herath-Banda and collaborators[47].

In figure 9 we compare our results[13] for the double differential cross section with the experimental data of Herath-Banda et al.[47] for the system I - U. The double differential cross section is shown versus the kinetic electron energy in the center of mass system. The lower curve represents coincidences between electrons and K-holes. The upper curve indicates our result for the total δ-electron emission rate. Remarkable agreement is achieved for the total and the coincident spectra for the slope as well as for the absolute numbers. Note, that in the calculations no scaling or fitting has been applied. In addition it should be mentioned that there is no influence of vacancy sharing in the asymmetric system I - U. The bumps in the coincidence spectra arise from nuclear Coulomb excitation and subsequent internal conversion in the U-nucleus. The corresponding nuclear transitions are indicated. The electrons were observed at certain angles with respect to the beam axis. No anisotropies in the coincident spectra were detected for these systems, which reflects the validity of the monopole approximation.

In conclusion we may state that our theoretical considerations correctly describe inner-shell excitation and δ-electron production in superheavy systems. The strong relativistic effects are reflected in the rather high production rates of high energy electrons.

POSITRON CREATION IN ELASTIC HEAVY ION COLLISIONS

Several experiments[39-42,44,45,48-53,66-71] have been performed to establish atomic positron production and to study its dependence on the kinetic positron energy, impact parameter, projectile energy and combined nuclear charge. Positron production rates of collision systems with different total charge were measured in coincidence with the scattered ions in an angular window $\theta = 45^{o} \pm 20^{o}$ by a collaboration of GSI, Yale and Frankfurt[45]. The large window ensures that nearly the total positron production rate was detected. In fig. 10 the good agreement between experimental data[45] and theoretical results[4,5,11,36] can be seen. The dashed lines indicate the positrons created by nuclear conversion only. Those yields have been calculated by folding the γ-spectrum with the theoretical conversion coefficients[72]. The sum of nuclear and theoretically deduced atomic positron rates is shown by the full lines. In order to check the method of determining the number of positrons created in nuclear reactions via the γ-spectrum, fig. 10 also shows the positron spectrum of a U + Sm (Z = 154) collision, consisting only of positrons of nuclear origin. The results are in agreement with calculated rates, indicating that the nuclear background in positron emission can be handled correctly by folding the γ-spectrum.

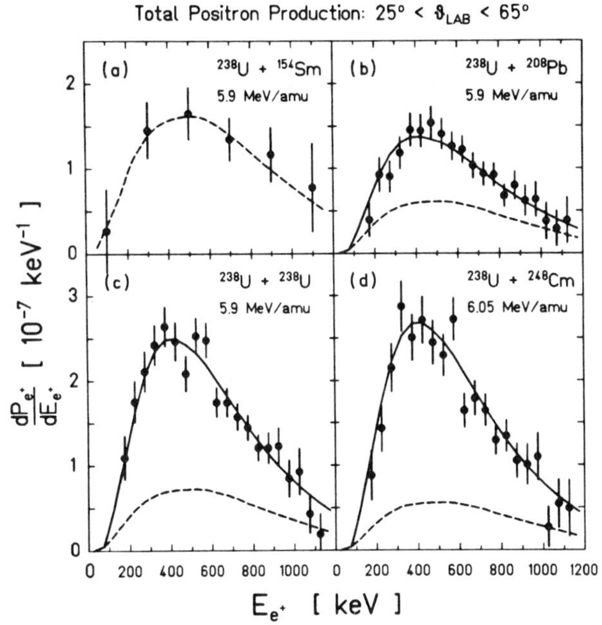

Fig. 10: Kinetic energy distribution of positron production probabilities over the indicated ion lab angular region[45]. Dashed lines: Nuclear background. Solid lines: Quasiatomic theory[4,5] assuming Rutherford trajectories together with nuclear background.

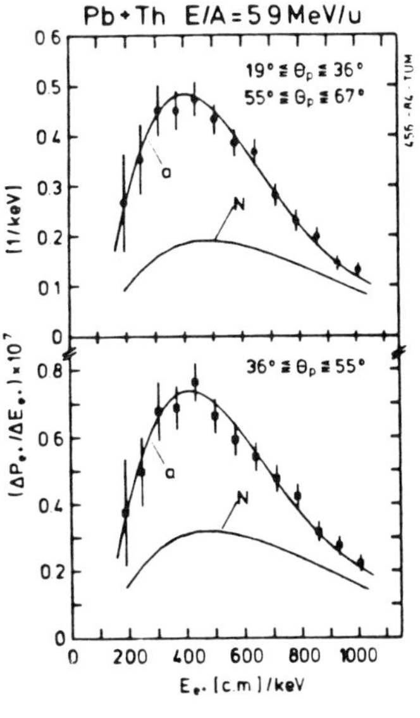

Fig. 11: Positron spectra measured[66] in the subcritical system Pb + Th. The nuclear background (curves N) has been subtracted. The considered lab ion scattering angles are indicated. Solid lines a: Coupled channel calculations[5] taking into account screening corrections.

The positron spectrum for the subcritical system Pb + Th (Z = 172) measured by Tsertos et al.[40,66] is displayed in fig. 11. For ion scattering angles Θ_p between 19° and 67° it is a smooth distribution reflecting the available Fourier frequencies during the collision and the strong Coulomb distortion of the positron wavefunctions. Again the nuclear contribution (curves N) was determined by a simultaneous measurement of the γ-yield. It has been subtracted from the total positron production yield which allows a direct comparison with results of a coupled channel calculation including screening corrections[5]. Perfect agreement is achieved provided the theoretical data are scaled by a common factor of s = 0.9.

For a further discussion of the dependence of the experimentally investigated positron production on kinematical parameters we refer to refs. 41, 45 and 51. Coupled channel calculations and experimental data generally are in fair agreement. Especially the highly non-perturbative nature of the positron creation in the strong electric fields present in heavy ion collisions has been verified. This behaviour is reflected in the unusual strong dependence of the positron production yield on the combined nuclear charge Z as demonstrated in fig. 12. The atomic positron creation probability increases between Z = 164 (Pb + Pb) and Z = 184 (U + U) like the 22nd power of Z. If one would attempt to describe these processes perturbatively by summing Feynman diagrams, the average diagram would imply the exchange of many virtual photons.

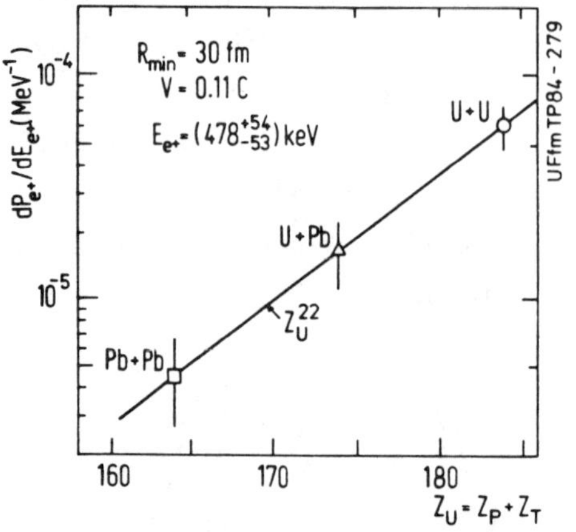

Fig. 12: Probability for positron creation as function of combined nuclear charge Z_u.

POSITRON PRODUCTION IN DEEP-INELASTIC NUCLEAR COLLISIONS

Deep-inelastic reactions have been discussed in terms of many nuclear models[73] with different degrees of refinement. For the description of the experiment to be discussed in this section, it seems to be possible to adopt trajectories calculated from a nuclear model which is consistent with elastic and inelastic scattering data. In our calculation[3,4,36,57] we employed, besides others, the following macroscopic friction model for the nuclear motion in U + U collisions. The model, proposed by R. Schmidt et al.[74], incorporates nuclear intrinsic rotation and handles a set of dynamical variables, i.e., the orientation of the individual nuclei and the internuclear distance as well as their corresponding conjugate momenta. Accounting for neck formation in the separating system in a simple way, it is able to explain the experimentally observed high energy loss, where up to ~30% energy dissipation for b ≃ 0 can be achieved.

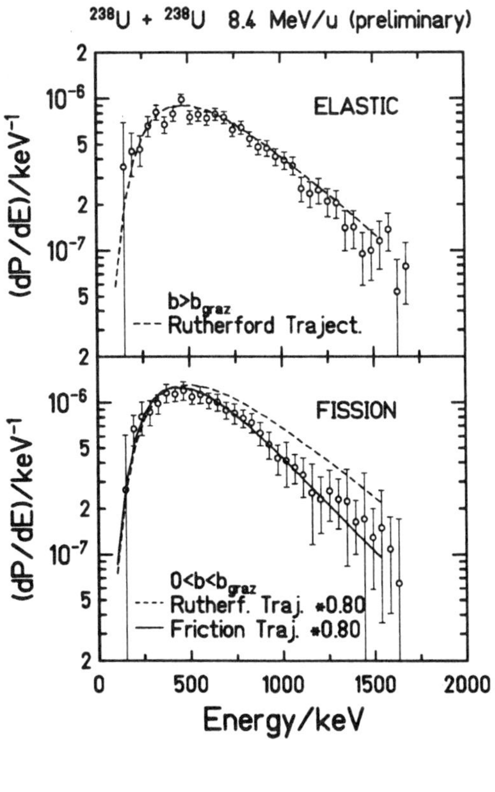

Fig. 13: Experimental positron data[49,50] for elastic and deep inelastic nuclear collisions in comparison with theoretical results[3,36].

In fig. 13 experimental positron data[49,50] for heavy ion collisions of U + U at 8.4 MeV/u are compared with theoretical results. In the upper part elastic scattering has been measured. Thus those results are compared with calculations using Rutherford trajectories only. In the lower part comparison is additionally made with theoretical data calculated for nuclear trajectories derived from the macroscopic friction model mentioned. Here the theoretical values have been reduced by a factor of 0.8. Although this discrepancy in absolute numbers is not yet understood, one may conclude that the slope of the experimental data, as the relevant measure, can be reproduced only if a nuclear time delay in the order of about 10^{-21} s is taken into account for those bombarding energies high above the Coulomb barrier. These conclusions are also well supported by independent results from δ-electron[3,36,44,49,50,70,71,75] and K-hole spectroscopy[36,76].

In recent experiments[49,50] measurements of δ-electron and positron spectra in deep inelastic U + U collisions have been extended up to lab energies of 10 MeV/u. For a comparison with our semiclassical calculations we again adopted certain classical models for the trajectories in the region of nuclear overlap. We employed two models: Model I, of J.R. Birkelund et al.[77], denotes trajectories based on the proximity potential of Blocki et al. using the regular Coulomb potential and the one-body nuclear friction in the proximity formalism of Randrup. Schmidt et al.[74] (model II) additionally incorporated neck formation in the separating system with the advantage of being able to explain the experimentally observed high energy loss. In order to compare the resulting probabilities with experimental data[50] which have been measured in coincidence with subsequent fission products, an integration over those impact parameters leading to nuclear contact is necessary, i.e.

$$dP/dE = \int (dP(b)/dE)\, w_f(b)\, b\, db \,/ \int w_f(b)\, b\, db. \qquad (9)$$

The fission probability $w_f(b)$ can be determined experimentally[50], however, we simply used $w_f(b) = 1$ for $b < b_{grazing}$ (spherical nuclei assumed) and $w_f(b) = 0$ elsewhere. Both trajectory models lead to short nuclear delay times ($\tau \sim 10^{-21}$ s) so that no line-structure in the positron spectra are expected.

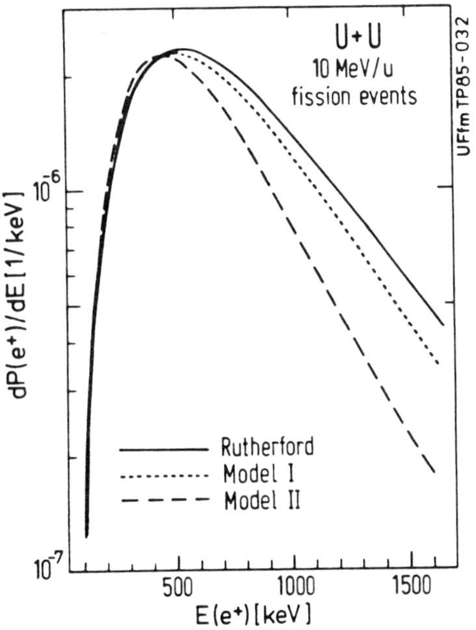

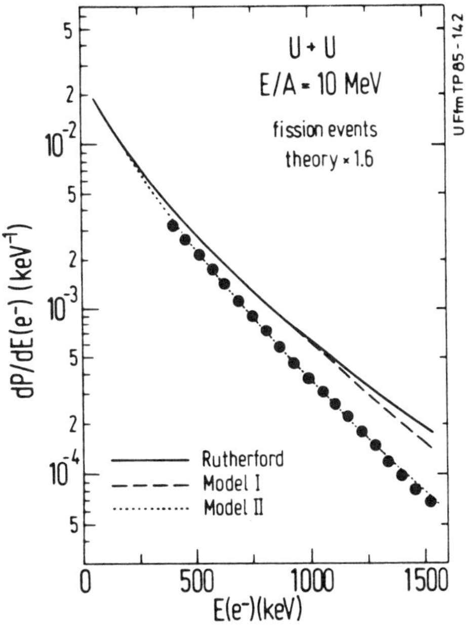

Fig. 14: a) Theoretical positron spectra for U + U collisions at E_{Lab} = 10 MeV/u. Elastic Rutherford collisions as well as deep-inelastic nuclear collisions described by various models are considered. b) Spectrum of δ-electrons in 10 MeV/u U + U collisions measured by Krieg[50] et al. in coincidence with nuclear fission residues. The experimental data are compared with theoretical results assuming different classical trajectories.

In fig. 14a we compare positrons emitted in 10 MeV/u U + U collisions along a Rutherford trajectory (solid line), i.e. assuming transparent nuclear matter, with results along the trajectories of model I (dotted line) and model II (dashed line). A steeper decrease is obtained for the friction- model trajectories being most pronounced for model II with the longest delay time. The positron spectra exhibit a shift of the maximum towards smaller positron energies for inelastic collisions. The logarithmic plot reveals a steeper decrease beyond the kinematic maximum for the friction- model trajectories in close analogy to the calculated δ-electron spectra, being depicted in part b of the figure. The slope of experimentally determined[50] δ-electron emission probabilities is reproduced best by calculations based on the trajectories of model II. Predictions using model I miss both, slope and absolute values. Model II leads to δ-electron spectra having the correct slope, but still a scaling fact of 1.6 is needed to fit the experimental data concerning the absolute values. The latter model favours nuclear delay times around $\tau \sim 10^{-21}$ s.

POSITRON EMISSION FROM GIANT NUCLEAR SYSTEMS?

During the last years two experimental groups have performed experiments with various supercritical collision systems at energies close to the Coulomb barrier, indicating the existence of prominent structures in the positron spectra[39,40]. Looking at the experimental spectra it can be seen that position and width of the peak depend on both, scattering angle and beam energy. It was argued[57] that this fascinating experimental finding signals the formation of long-living ($T > 10^{-20}$ s) giant nuclear systems and the associated spontaneous positron emission[78,79]. Taking this hypothesis at face value it was successfully demonstrated that a ratio of $q \sim 10^{-3}$ of the nuclear reaction cross section relative to the Rutherford cross section in the detection windows considered and delay times $T > 5 \cdot 10^{-20}$ s would reproduce these peaks in the theoretical calculations. Thus, two parameters, the width Γ and the area (e^+-cross section) under the line, were fitted. They correspond to the mean lifetime of the giant system and to the nuclear reaction cross section leading to fusion of the giant system, respectively.

Experimental positron data[39,45,52] for the heaviest colliding system investigated until now, U + Cm, are displayed in fig. 15 for the bombarding energy of E_{Lab} = 6.05 MeV/u and $100° \le \Theta_{cm} < 130°$. The spectra are not corrected with respect to the detector efficiency. To allow for a comparison with theoretical results the calculated probabilities have been folded with the experimental efficiency. The nuclear background contribution has been added to the theoretical values. The lower part of the figure shows the positron emission rates for kinematical conditions ($50^\circ < \Theta_{cm} < 80^\circ$), which suppress close collisions. The dashed lines give the theoretical yield for quasiatomic positron production[4] assuming elastic Rutherford trajectories together with the nuclear background. Fair agreement with the experimental values is achieved. For the angular window $100^\circ \le \Theta_{cm} < 130^\circ$ a distinct peak emerges in the spectrum at E_{e^+} = (316 ±10) keV with a width of about $\Delta E \sim 80$ keV[39]. Within the framework of our hypothesis of spontaneous positron production in giant nuclear molecules which are formed transiently during nuclear contact, we analysed this narrow peak structure. Taking the parameters R ~ 16 fm, which corresponds approximately to the nuclear touching configuration, $T = 10^{-19}$ s and $q \sim 1.3 \cdot 10^{-3}$ we obtain agreement with the experimental facts.

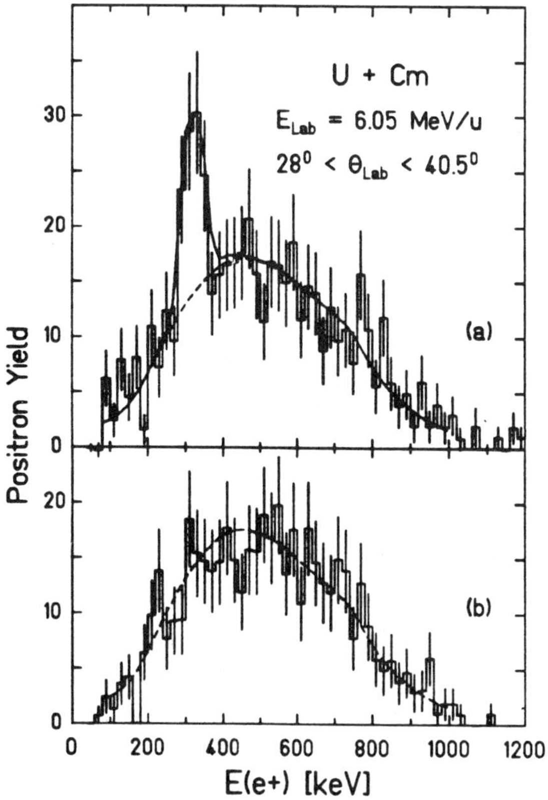

Fig. 15: Positron yield as observed for $^{238}U + ^{248}Cm$ collisions at a beam energy of 6.05 MeV/u[39]. No corrections with respect to the detector efficiency are performed. (a) $100^\circ \le \Theta_{CM} < 130^\circ$. (b) $50^\circ < \Theta_{CM} < 80^\circ$. Dashed lines: Theoretical yield for elastic Rutherford trajectories plus nuclear background including detector efficiency and corresponding folding. Solid line: Calculated result assuming in addition spontaneous positron formation in giant nuclear molecules[4,36].

A deeper physical insight can be achieved by inspection of the Doppler-broadening behaviour of the positron peaks. Fig. 16 shows the measured line width ΔE of the line-like positron distribution detected in the U + Cm and the Th + Cm system at two different angular regions. Assuming a negligible intrinsic width (< 40 keV) of the e^+-line, the dashed and dotted lines display the calculated[48] ion angular dependence for the slow and fast collision partner, respectively, as emitting system. Both constancy and absolute height strongly argue for associating the positron peak being emitted from a system moving with the center of mass velocity[48,52]. According to these data an emission at asymptotic velocities from either the projectile-like or recoil-like frame is inconsistent with the observed data.

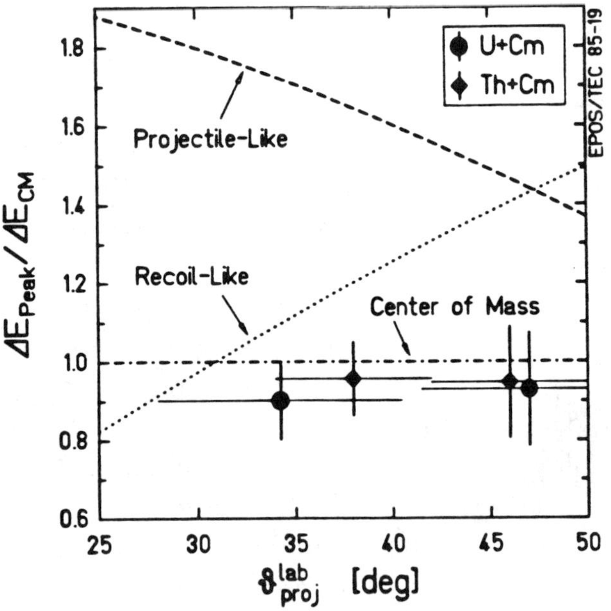

Fig. 16: Measured line width ΔE_{e^+} of the positron line observed[48] in U + Cm and Th + Cm collisions at two different angular regions. The intrinsic line width was assumed to be negligible (< 40 keV).

By a careful analysis of the photon and δ-electron distribution it could be furthermore concluded that the distinct lines at a kinetic energy of E ~ 300 keV do not originate from conversion processes in the separated atoms[39,40]. In this connection also β-decay processes via K-electron capture could be excluded theoretically due to their extremely long decay times[21]. The nuclear physics properties of the giant nuclear systems were studied in refs. 7 and 22. However, it should be mentioned that independent measurements[44,49] of a third experimental group did not yet reveal indications for sharp positron lines.

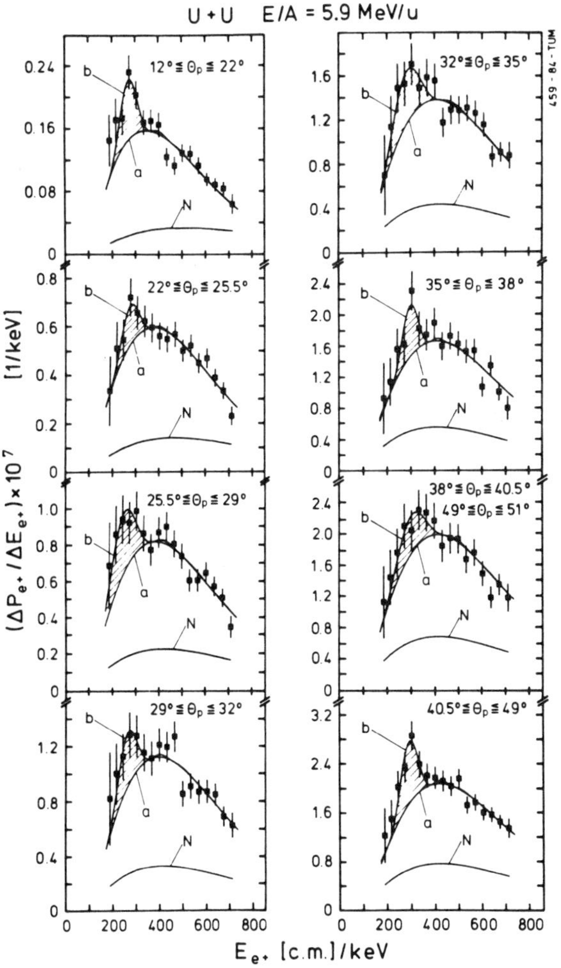

Fig. 17: Positron spectra[53,66,80] for U + U collisions at 5.9 MeV/u for 8 angular ranges of the scattered ions. Solid line N: Nuclear background which has been subtracted from the experimental data. Solid lines a and b: Theoretical fits for the dynamically induced and the spontaneous e^+-emission to the observed data.

Next we turn to the discussion of the dependence of the detected positron line on the ion scattering angle[53,66,80]. Kienle's group measured the positron spectra for U + U collisions at E_{Lab} = 5.9 MeV/u in the angular range between 12° and 49° in the laboratory system. Fig. 17 illustrates the new fact that at all angles the prominent peak structure appears. The curves (N) indicate the nuclear background which was subtracted from the observed data. The solid lines (a) and (b) show theoretical spectra[4,5] for the dynamically induced positron contribution and the spontaneous positron emission, respectively, fitted to the measured e^+-production probability. Here the line energies and widths exhibit no significant dependence on the ion scattering angle Θ. The average energy[53] of the positron line is (280 ±6) keV and (277 ±6) keV for U + U and U + Th collisions, respectively, with corresponding line widths of (86.5 ±6) keV and (74.5 ±5) keV. Especially for the U + Th system the line position deviates considerably from the equivalent result of the EPOS collaboration[48].

The dependence of the prominent line structure on the combined nuclear charge number Z_u was investigated by the EPOS collaboration[48]. Fig. 18 shows their surprising result that for the five different collision systems ranging from ^{232}Th + ^{232}Th (Z_u = 180) to ^{238}U + ^{248}Cm (Z_u = 188) the peak position seems to be independent of the combined nuclear charge. Although the experimental spectra are not corrected with respect to detector efficiency and there is no nuclear background subtraction performed this represents a very exciting experimental outcome. The Z_u-dependence of the lines does not show the theoretically expected behaviour, if the shape of the giant nucleus is the same in all cases. However, a quantitative comparison with calculated results suffers from the fact that the experimental positron yield is presented rather than a production probability or cross section.

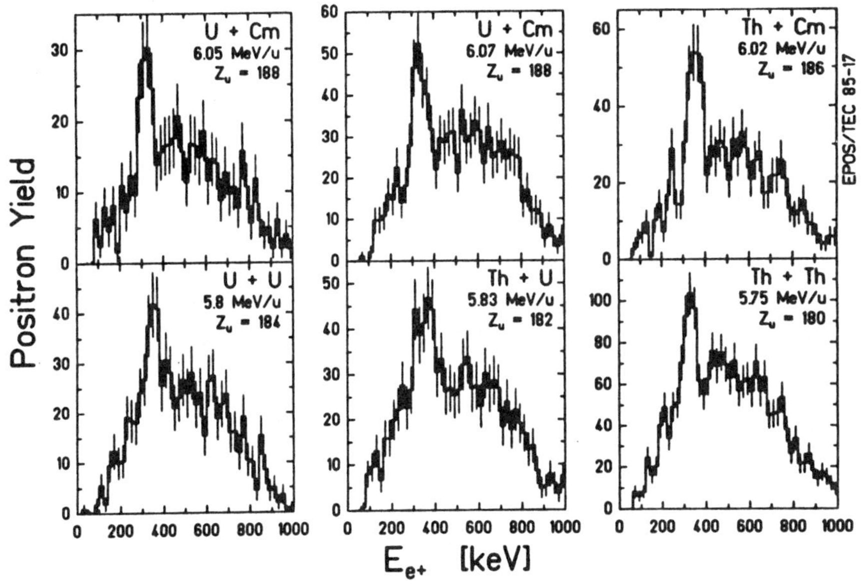

Fig. 18: Positron energy spectra measured by T. Cowan et al.[48] for the five collision systems and bombarding energies indicated.

In our theoretical studies[81] we investigated the interplay between quasiatomic positron emission during the Rutherford scattering of heavy ions and the spontaneous positron creation process due to the existence of long-living giant nuclear molecules in a range of systems from Z_u = 180 (Th + Th) to Z_u = 188 (U + Cm). For the kinematics we choose the bombarding energies[48] E_{Lab} = 5.75 MeV/u for Th + Th up to E_{Lab} = 6.05 MeV/u for U + Cm. No distinction between backward and forward scattering in the angular detection window $28^\circ < \Theta < 40.5^\circ$ is assumed. For the ratio q defining the number of nuclear molecule residues per elastically scattered ion in the angular window considered we take $q = 1.3 \cdot 10^{-3}$ for the U + Cm system. For the other collision systems q is renormalized to yield the same nuclear reaction cross section.

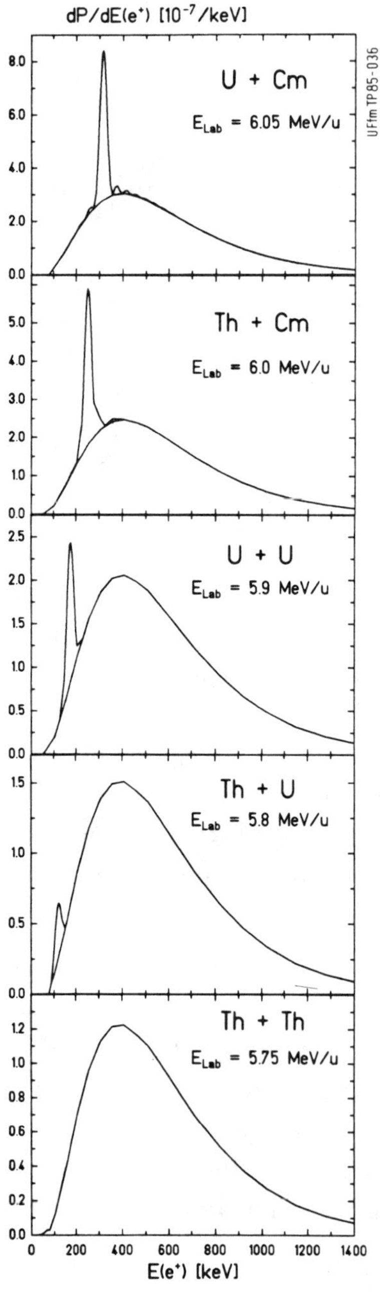

Fig. 19: Theoretical positron spectra for the systems and bombarding energies indicated. This figure demonstrates the theoretically expected dependence of the peak position on the combined nuclear charge.

The shape of the giant nuclear molecules with total charges Z_u is parametrized by the two-center distance R, here the distance of half density overlap for the most favourable elongated orientation. The corresponding energetic positions of the spontaneous lines, calculated[5] within the relativistic Hartree-Fock-Slater (HFS) formalism, are E_{e^+} ~ 60 keV, ~ 115 keV, ~ 170 keV, ~ 250 keV, and ~ 315 keV, respectively. The life time of the giant nuclear molecules is assumed to have the sharp value $T = 10^{-19}$ s.

Fig. 19 shows the superposition of positrons emitted during ordinary Rutherford scattering events and spontaneous positrons from the assumed long-living giant nuclear molecules. Doppler broadening as well as other broadening effects are not taken into account. Thus the width of all peaks (FWHM) approximately is ~ 40 keV. (The satellite finestructure in the U + Cm system appears due to a finer grid in the numerical energy resolution.)

While for the chosen parameters in the spectrum of U + Cm the spontaneous positron line at E_{e^+} ~ 315 keV exceeds the ordinary positron spectrum by a factor of nearly 3, the relative strength of the peak gets reduced step by step for decreasing charge Z_u. Finally, for

Th + Th the remaining structure at $E_{e^+} \sim 60$ keV is strongly suppressed due to the small decay width for spontaneous positron creation. In conclusion, assuming equal conditions for the nuclear process, the model of spontaneous positron production from a giant nuclear molecule predicts a shift of the positron line and a distinct reduction in intensity if Z_u is decreased. The observed constancy[48,53,80] of the peak position certainly necessitates further theoretical as well as experimental examinations. One possible reason might be that the position of the spontaneous positron peak strongly depends on the actual configuration of the di-nuclear system, which could change from system to system. But rather intricate nuclear physics properties must be invoked to enforce the constancy of the e^+-line. In consequence it seems to be legitimate to search for alternative explanations.

CRITICAL ANALYSIS OF ALTERNATIVE EXPLANATIONS OF THE POSITRON LINE STRUCTURE

Up to now no plausible alternative for the positron peak structure has been found. Here we want to argue[32] that the mechanism recently proposed by Lichten and Robatino is not well founded and should be dismissed as an explanation for the observed positron lines.

The argument of ref. 31 is based on the observation that in a many electron system multiple excited states may decay through autoionization. In systems with very deeply bound states this may lead also to positron emission. Even if the K-shell is not yet supercritical, the missing energy may be supplied by transition of an electron from a higher lying state (e.g. $2p_{1/2}\sigma \to 1s\sigma$). Multiple (at least double) inner shell holes are required to support this process. The authors of ref. 31 claim that <i> multiple excitations have not been treated correctly up to now and <ii> that two-electron transitions play an important role when determining positron creation.

Regarding <i> we want to mention that refs. 55 and 61 give an exact description of excitations of the many-particle systems as long as electron correlation interactions are neglected. Coupled channel calculations have been performed solving the system of coupled differential equations

$$d/dt\ a_{ij} = -\Sigma_k \left(\langle\phi_j|\partial/\partial t|\phi_k\rangle - i\langle\phi_j|H_o|\phi_k\rangle\right) a_{ik}\ e^{i\int^t dt'(E_j-E_k)}. \tag{10}$$

Here a_{ij} is the amplitude that an electron initially in state ϕ_i undergoes a transition to state ϕ_j. H_o is the single particle two-center Dirac Hamiltonian. The main effect of the electron-electron interaction may be included using a mean screening potential. Two kinds of couplings are taken into account in eq. (10): the dynamical coupling acting through the time dependence of the internuclear distance R(t) and a "spontaneous coupling" which is present in supercritical systems and accounts for the fact that the K-shell no longer is a discrete eigenstate of the Hamiltonian. The full set of amplitudes a_{ij} contains all the information on excitation rates including many-particle -many-hole configurations[55,61]. The only ingredient missing in this description is the explicit electron- electron two-body interaction H_{ee}.

There is no indication that this interaction plays a substantial role in inner-shell excitations. In fact, the proposed two-electron transition has been found[82] to lead to an extremely small transition width of about 1 eV compared to several keV for the spontaneous decay width of the supercritcal K-shell[78,79]. The claim that the strength of two-electron transitions is determined by the dynamical coupling matrix elements, thus being very large, is unfounded. Dynamical excitations are induced by the one-body operator $\dot{R}\partial/\partial R$ (and possibly rotational coupling). Thus multiple excited configurations can be reached only by the repeated action of this operator. This is fully accounted for in eq. (10).

The explicit calculation presented in ref. 31 is based on the assumption that positron excitation is restricted predominantly to a narrow region around R = 500 fm internuclear distance, where the radial matrix element between the states $2p_{3/2}\sigma$ and $2p_{1/2}\sigma$ has a maximum[83]. This postulate, for which we can see no justification leads to oscillations in the emission spectrum due to interference between transitions on the incoming and outgoing collision path (at least if first order perturbation theory is valid, leading to a symmetric integrand in time, which also may be questioned).

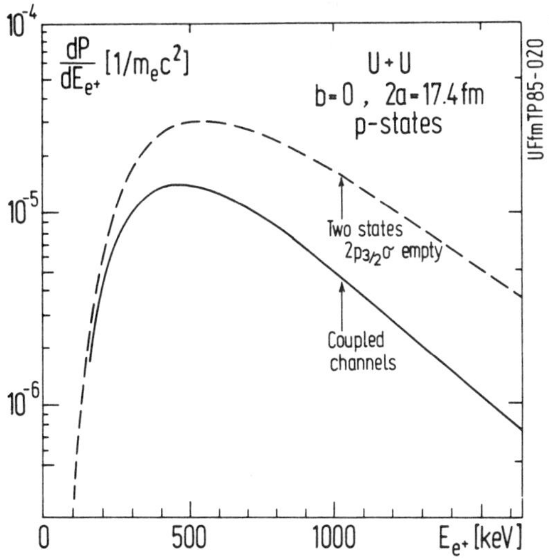

Fig. 20: Positron spectrum[32] for central U + U collisions. Solid line: Coupled state calculations in the monopole approximation. Dashed line: Coupled state calculations employing molecular basis states. In contrast to ref. 31 no peak structures appear.

While we can see no justification for the postulated localization of the positron excitation, it is interesting to study the effect of the $2p_{3/2}\sigma \rightarrow 2p_{1/2}\sigma$ coupling. It might be conceivable, that this coupling rapidly "switches on" the amplitude for holes in the $2p_{1/2}\sigma$ level. This discontinuity might make itself felt in the positron spectrum, one part of which is (roughly speaking) the Fourier transform of the $2p_{1/2}\sigma$ amplitude, multiplied by the coupling matrix element $\langle\phi_{E_{e^+}}|\dot{R}\partial/\partial R|\phi_{2p\frac{1}{2}\sigma}\rangle$. The result of a coupled channel calculation following the method outlined above is shown in Fig. 20. The broken line gives the positron spectrum calculated for a central (b = 0) 5.9 MeV/u U + U collision. Only positrons emitted from the $2p_{1/2}\sigma$ state are included. Holes are fed into this state via radial coupling from the $2p_{3/2}\sigma$ level, which is considered to be empty initially. The spectrum is compared with full coupled channel result including six $np_{1/2}\sigma$ states and the positive energy continuum, all described in monopole approximation presented in ref. 55. Although the shapes of the curves differ somewhat, obviously no oscillatory features are introduced. Comparing the magnitude of the curves it is noticed that the presence of a $2p_{3/2}\sigma$ level would dominate the whole positron spectrum, due to the very large transfer rate into the $2p_{1/2}\sigma$ state, which in turn is quite strongly coupled to the positron continuum.

In summary, we have demonstrated that multi-electron transitions can not be used to explain the observed line structures in the spectrum of positrons emitted in the collision of very heavy ions.

The observed independence of the positron line on Z also leads to the conjecture that the creation of a new, previously undetected particle could provide an unexpected source[48] for monoenergetic positrons. Although this presumption is of highly speculative nature its physical consequences should be discussed[19]. The following possibilities can be considered: (a) A neutral particle X is formed which decays into an electron-positron pair $X \rightarrow e^+ + e^-$. The mass of X is determined by $m_X = 2(m_e + E)$ where E is the kinetic energy of the positron. Assuming $E = 330$ keV[48], we have $m_X = 1680$ keV. (b) The unknown particle carries positive charge and decays into a positron and a neutrino, $X \rightarrow e^+ + \nu$. Its mass is $m_X = m_e + E + [E(E + 2m_e)]^{\frac{1}{2}} = 1510$ keV. The new particle may be produced by two different sources. It may couple to the atomic electrons including the virtual electron-positron pairs excited during the collision or, alternatively, the source may be the nuclei, i.e. the quarks.

However, let us note that there may be a related mechanism which completely avoids the introduction of a new particle[37]. This is based on the fact that the Glashow-Salam-Weinberg (GSW) model of electroweak interaction includes a 'hidden' charged particle, namely the Goldstone boson. As discussed in ref. 37, in sufficiently strong external electromagnetic potentials the Higgs vacuum of the GSW model is expected to become unstable against condensation of negatively charged Higgs particles and formation of a positively charged Higgs vacuum excitation that finally decays into a positron and electron neutrino. This phenomenon is similar to the decay of the neutral vacuum in QED[1,2], since the nuclear charge has to surpass a critical value Z'_{cr} for which, unfortunately, no quantitative predictions exist yet. As the massive Higgs vacuum excitation can only appear if the Higgs vacuum changes locally, no such particles should be observed in other (low-Z) experiments.

While one can hardly disprove the existence of an unknown particle, it is possible to put quite stringent limits on the production cross section. This is required in order to avoid contradictions to the existing body of well established experimental facts. Experimental searches for anomalous decays of elementary particles and excited nuclei constitute an additional important source of information.

Since the mass of the proposed particle is very light, it contributes to virtual processes. This enables us to deduce upper limits for the coupling strength from the high-precision experiments of QED. We will discuss two such arguments. First the anomalous magnetic moment of the electron (or muon) is known and understood to a high degree of accuracy. The presence of a new neutral particle coupling to the electron (muon) will necessarily provide an additional contribution to the g factor. Let us assume an interaction of the form

$$L^{e,X} = g_i \bar{\Psi}_e \Gamma_i \Psi_e \Phi_X \tag{11}$$

where the index i enumerates the Lorentz-invariant couplings of scalar, pseudoscalar, vector, axial vector and tensor type, i.e., $\Gamma_S = 1$, $\Gamma_P = i\gamma_5$, $\Gamma_V = \gamma_\mu$, $\Gamma_A = \gamma_\mu \gamma_5$, $\Gamma_T = \sigma_{\mu\nu}$. The additional contribution to the electron anomaly is calculated from the vertex correction to the electron-photon interaction. The result depends on the ratio between the mass of the fermion and that of the virtual boson and therefore is different for electrons and muons. According to the calculations of A. Schäfer et al.[19] the couplings $\alpha_i = g_i^2/4\pi$ are suppressed by at least seven orders of magnitude compared with the electromagnetic finestructure constant $\alpha \simeq 1/137$ in order to avoid contradictions with the g-2 experiments. For further reference we note that the lifetime of the decay $X \to e^+ + e^-$ is determined by the coupling constant through

$$\tau = \hbar/\tfrac{1}{2}m_X\alpha_i F(m_X/m_e) \sim 7.8 \cdot 10^{-22}\ \mathrm{s}\ /\alpha_i F(m_X/m_e) \tag{12}$$

where the function $F(m_X/m_e)$ is quite independent of the type of coupling (i = S, P, V, A), taking on values between 0.5 and 0.9 at the given particle mass.

If we now assume that the source of the particle X is the electron shell, then the emission intensity can be determined by adopting the formalism for electronic bremsstrahlung or molecular orbital x-ray radiation (MOX). We obtain as an order-of-magnitude estimate[19] $\sigma_X \simeq 8 \cdot 10^{-11}$ b. This number must be compared with the production cross section of the positron line of several μb. In view of the large discrepancy between these two numbers, the hypothesis that an unknown neutral particle coupling to the electrons only is responsible for the positron line can be safely ruled out.

On the other hand if we assume that the particle X is created from the nuclear current during the heavy-ion collision, some conclusions can be drawn from high-precision QED measurements of atomic binding energies. In addition to the electron-X coupling, described by the coupling constant g_i^e, we also have the quark-X interaction

$$L^{q,X} = \Sigma_f \, g_i^f \, \bar{\Psi}_f \, \Gamma_i \, \Psi_f \, \Phi_X \tag{13}$$

where f counts the quark flavours. Taking into account two flavours we obtain for the effective interaction energy for the one-boson exchange process

$$H = - \int d^4x \, d^4y \, g_i^e \, \bar{\Psi}_e \, \Gamma_i \, \Psi_e(x) \, D(x-y) \cdot \\ [g_i^p \, \Sigma_p \, \bar{\Psi}_p \, \Gamma_i \, \Psi_p + g_i^n \, \Sigma_n \, \bar{\Psi}_n \, \Gamma_i \, \Psi_n](y) \tag{14}$$

where the coupling constant for neutron and proton, g^n and g^p, have been introduced. D(x-y) is the propagator for a boson of mass m_X. In the non-relativistic limit for the nucleon current this expression leads to an additional Yukawa-type interaction potential of range $1/m_X$. It couples the atomic electron to the scalar nuclear density (for i = S and V) or to the nuclear spin density (i = A, T). For pseudoscalar coupling (i = P) the interaction vanishes in the non-relativistic limit. We have investigated the consequences of such an additional potential in three cases. (i) The Lamb-shift in hydrogen. (ii) Transitions between high-lying states in muonic atoms. (iii) The electronic K_α transition energy in heavy atoms. For (i) and (ii) there is no significant difference between the cases i = S, V, A, and T due to the non-relativistic behaviour of the electron (muon) wavefunction. From the present agreement between theory and experiment for the high precision atomic physics measurement we deduce $g^e\ g^p < 2\cdot 10^{-8}$.

At energies at or below the nuclear Coulomb barrier we expect that the emission of light (scalar or pseudoscalar) particles is mainly caused by the collective deceleration of the colliding nuclei, i.e. by a mechanism of bremsstrahlung type. To obtain an estimate of the emission probability we have performed a calculation[91] based on the semiclassical approximation.

In accordance with the well established treatment of photon bremsstrahlung and subthreshold pion production we seek a solution of the inhomogeneous wave equation

$$(\Box + m_X^2)\ \phi(\vec{x},t) = \rho(\vec{x},t). \tag{15}$$

The source term $\rho(\vec{x},t)$, which in principle should be determined from the dynamics of the nucleon fields, will be approximated by two localised density distributions travelling on Rutherford trajectories. Solving (1) with the retarded Green's function, the energy flux in the radiation zone can be calculated. Division by the energy carried by a single particle leads to an expression for the number of particles emitted per energy interval dp_o and per solid angle $d\Omega$

$$dn/(d\Omega\ dp_o) = p/(16\ \pi^3)\ |\int d^4x\ e^{i(p_o t - \vec{p}\cdot\vec{x})}\ \rho(\vec{x},t)|^2 \tag{16}$$

with $p \equiv |\vec{p}|$. Note that the emission of electromagnetic bremsstrahlung is determined by a similar formula

$$dn/(d\Omega\ dp_o) = 1/(4\ \pi^2)\ 1/p_o\ |\ \vec{p} \times \int d^4x\ e^{i(p_o t - \vec{p}\cdot\vec{x})}\ \vec{j}(\vec{x},t)|^2 \tag{17}$$

where $\vec{j}(\vec{x},t)$ is the electromagnetic current.

To evaluate the Fourier integral in (16) we have to specify the source term $\rho(\vec{x},t)$. In the case of scalar coupling it will be determined by the nucleon density. Assuming full coherence of the nucleon motion in each nucleus, we will approximate

$$\rho^S(\vec{x},t) = \sum_{i=1}^{2} C_i \, \delta^3(\vec{x}-\vec{R}_i(t)) \tag{18}$$

where the index i enumerates the colliding nuclei. If we further assume for simplicity isospin independence of the interaction, the source strength C_i simply is the product of the scalar coupling constant and the nucleon number

$$C_i = g_N^S \, A_i . \tag{19}$$

The finite extension (formfactor) of the nucleus can easily be incorporated in (18), but this should be unimportant as long as the deBroglie wavelength of the emitted particle is largest compared with the nuclear radius. (Note that this condition is satisfied for low energy emission of particles with mass in the MeV range but not for pions, for which the nuclear extension must be taken into account). Using (18) the emission spectrum (16) is reduced to a one-dimensional integral in time

$$\left(\frac{dn}{d\Omega \, dp_o}\right)_S = p/(16\pi^3) \; \left| \int_{-\infty}^{\infty} dt \, e^{ip_o t} \sum_{i=1}^{2} C_i \, e^{-i \, \vec{p}\cdot\vec{R}_i(t)} \right|^2 \tag{19}$$

The construction of a source term in the case of pseudoscalar coupling is less straightforward. The original expression in terms of the nucleon field Ψ

$$\rho^P = -i \, g_N^P \, \bar{\Psi} \gamma^5 \Psi \tag{20}$$

does not have a classical limit. To obtain an expression that lends itself to interpretation in terms of collective variables, we make use of the long-known approximate equivalence between pseudoscalar coupling and pseudovector derivative coupling. With the help of the Dirac equation

$$(\not{p} - M)\,\Psi = i\, g_N^P\, \gamma^5\, \Psi\, \phi \tag{21}$$

and its adjoint the expression (20) may be transformed to the form

$$\rho^P = g_N^P/(2M)\ \partial_\mu(\bar{\Psi}\,\gamma^5\,\gamma^\mu\,\Psi) - (g_N^P)^2/M\ \bar{\Psi}\,\Psi\,\phi. \tag{22}$$

Here M is the nucleon mass. The last term in (22) is of second order in the coupling constant. It contains a self-interaction of the boson field and will be neglected in the following discussion. The first term contains a four-divergence of a pseudo-vector quantity which is recognized to represent the nucleon spin density. We assume that the spin density is proportional to the number density of nucleons $\rho(\vec{x})$, i.e.

$$\bar{\Psi}\,\gamma^5\,\gamma^\mu\,\Psi = \rho(\vec{x})\ \sigma^\mu(\vec{x}). \tag{23}$$

The spin vector σ^μ is obtained from the spin vector $\sigma^\mu = (0,\vec{s})$ in the particle restframe via a Lorentz boost with velocity $\vec{v}$

$$\sigma^\mu = (-\vec{s}\cdot\vec{v}\ ,\ \vec{s}+(\gamma-1)/v^2\ (\vec{s}\cdot\vec{v})\vec{v}). \tag{24}$$

Inserting (22) and (23) into (16) after an integration by parts and neglect of the surface term leads to

$$\left(\frac{dn}{d\Omega dp_o}\right)_P = p/(16\pi^3)\ \left|\int d^4x\ e^{i(p_o t-\vec{p}\cdot\vec{x})}\ g_N^P/(2M)\ \Sigma_\alpha\ \rho_\alpha(\vec{x})\ \sigma_\alpha{}^\mu p_\mu\right|^2, \tag{25}$$

where the sum runs over all the nucleons. We now again approximate the source by two localized pointlike nuclei moving on classical trajectories (eq.(18))

$$\left(\frac{dn}{d\Omega dp_o}\right)_P = p/(16\pi^3)\ \left|\int_{-\infty}^{\infty} dt\ e^{ip_o t}\ \sum_{i=1}^{2}\ C_i/(2M)\ \vec{s}_i\cdot\vec{p}\ e^{-i\vec{p}\cdot\vec{R}_i(t)}\right|^2. \tag{26}$$

To further simplify the expression here we have replaced $\sigma^\mu p_\mu$ by the nonrelativistic approximation $\vec{s}\cdot\vec{p}$. Note that $\vec{s}$ is the average spin per nucleon.

To solve the Fourier integrals in eqs. (19) or (26) the trajectory R(t) has to be specified. At energies below the Coulomb barrier the nuclei move on Rutherford hyperbolae. We use the convenient parameter representation for the internuclear distance coordinate $\vec{R} = \vec{R}_1 - \vec{R}_2$

$$\begin{aligned} R(w) &= a\ (\varepsilon\ \mathrm{ch}\ w + 1), \\ X(w) &= a\ (\mathrm{ch}\ w + \varepsilon), \\ Y(w) &= a\ [\varepsilon^2-1]^{\frac{1}{2}}\ \mathrm{sh}\ w, \\ Z(w) &= 0, \\ t(w) &= a/v_o\ (\varepsilon\ \mathrm{sh}\ w + w), \end{aligned} \tag{27}$$

where $a = Z_1Z_2/2E_{cm}$ is half the distance of closest approach for head-on collisions, $\varepsilon = [1 + (b/a)^2]^{\frac{1}{2}}$ is the eccentricity of the hyperbola and v_o is the asymptotic relative nuclear velocity. The trajectories of the two colliding nuclei (in the center of mass frame) are given by

$$\vec{R}_1(t) = m_2/(m_1+m_2)\ \vec{R}(t), \qquad \vec{R}_2(t) = -\ m_1/(m_1+m_2)\ \vec{R}(t). \tag{28}$$

As in the case of photon bremsstrahlung it is advantageous to perform a Taylor expansion of the exponential function

$$e^{-i\,\vec{p}\cdot\vec{R}_i(t)} = 1 - i\,\vec{p}\cdot\vec{R}_i(t) - \tfrac{1}{2}(\vec{p}\cdot\vec{R}_i(t))^2 + \ldots \tag{29}$$

This corresponds to a multipole expansion of the radiation field. It is quite sufficient to retain only the lowest order nonvanishing term. Up to second order the integral in (6) for scalar bremsstrahlung reads

$$\begin{aligned} &\int_{-\infty}^{\infty} dt\ e^{ip_ot} \sum_{i=1}^{2} C_i\ e^{-i\,\vec{p}\cdot\vec{R}_i(t)} = \\ &\int_{-\infty}^{\infty} dt\ e^{ip_ot}\ (M - ip\ n_k\ D_k - \tfrac{1}{2}\ p^2\ n_k\ n_l\ Q_{kl} + \ldots) \end{aligned} \tag{30}$$

with the monopole, dipole and quadrupole moments

$$M = C_1 + C_2,$$

$$D_k = \mu\,(C_1/m_1 - C_2/m_2)\,R_k, \qquad (31)$$

$$Q_{kl} = \mu^2\,(C_1/m_1{}^2 + C_2/m_2{}^2)\,R_k\,R_l.$$

Here $\mu = m_1m_2/(m_1+m_2)$ is the nuclear reduced mass and we have introduced the normal vector $\vec{n} = \vec{p}/p$. Since the source strength according to (19) is assumed to be constant in time, no monopole radiation is emitted Furthermore, if the factors C_i are proportional to the nuclear mass, the dipole moment vanishes (Note that in the electromagnetic case this complete cancellation occurs only, if both nuclei have the same charge to mass ratio Z_i/A_i). Therefore the scalar bremsstrahlung will be mainly of quadrupole type. For completeness, however, we will also include the dipole term in the following discussion. Using the trajectory (27) the Fourier integrals entering (30) can be solved explicitly. Here we will content ourselves in deriving the emission probabilities integrated over solid angle. The angular averages of the particle momentum unit vector are

$$\langle n_i n_j \rangle = 1/3\ \delta_{ij},$$

$$\langle n_i n_j n_k n_l \rangle = 1/15\ (\delta_{ij}\delta_{kl} + \delta_{ik}\delta_{jl} + \delta_{il}\delta_{jk}).$$

With (27) this leads to the emission spectrum

$$\left(\frac{dn}{dp_o}\right)_S = \frac{p^3}{4\pi^2}\,[1/3(|D_X|^2 + |D_Y|^2)$$

$$+ p^2/60\ (3|Q_{XX}|^2 + 3|Q_{YY}|^2 + 4|Q_{XY}|^2 + 2Q_{XX}Q_{YY})] \qquad (32)$$

where the Fourier transforms $D_k(p_o)$ and $Q_{kl}(p_o)$ of the multipole moments defined in (31) enter. These integrals can be solved analytically in terms of modified Bessel functions. The final result is

$$(dn/dp_o)_S = (2/\pi)\,(p^3/p_o{}^2)\,a^2\,\{F_1\,h_1 + (p/p_o)^2\,v^2\,F_2\,h_2\} \qquad (33)$$

with the abbreviations

$$F_1 = 1/(4\pi)\ \mu^2\ (C_1/m_1 - C_2/m_2)^2,$$

$$F_2 = 1/(4\pi)\ \mu^4\ (C_1/m_1{}^2 + C_2/m_2{}^2)^2. \quad (34)$$

If the C_i are proportional to the nuclear mass (eq. (19)) these multipole factors reduce to

$$F_1 = 0,$$

$$F_2 = (g^S)^2/(4\pi)\ A_R{}^2.$$

The shape of the emission spectrum is determined by the functions

$$h_1 = 2/3\ e^{-\pi\nu}\ [(\varepsilon^2-1)/\varepsilon^2\ K^2 + K'^2],$$

$$h_2 = 2/15\ e^{-\pi\nu}\ [1/\varepsilon^4\ (3\varepsilon^4 - 4\varepsilon^2 + 4 + 4\nu^2(\varepsilon^2-1)^3)\ K^2$$
$$+ 4\nu\ (\varepsilon^2-1)/\varepsilon^3\ (4-3\varepsilon^2)\ KK' + (\varepsilon^2-1)/\varepsilon^2\ (4\nu^2(\varepsilon^2-1)+4)\ K'^2]. \quad (35)$$

ν denotes the ratio between the particle energy p_o and the characteristic Fourier frequency ω_o of the collision

$$\nu = p_o/\omega_o = p_o a/v_o. \quad (36)$$

Finally, K denotes the MacDonald function of imaginary order

$$K \equiv K_{i\nu}(\varepsilon\nu) = \tfrac{1}{2} \int_{-\infty}^{\infty} d\eta\ e^{-\varepsilon\nu\eta + i\nu\eta} \quad (37)$$

and K' its derivative with respect to the argument. Except for powers of p/p_o which approach unity in the limit of vanishing rest mass, eq. (32) is very similar to (but not identical with) the corresponding result for electromagnetic bremsstrahlung. To obtain an expression for pseudoscalar bremsstrahlung the Fourier integral of eq. (26) has to be solved. Expanding the exponential function to first order gives

$$\int_{-\infty}^{\infty} dt\, e^{ip_o t}/(2M) \sum_{i=1}^{2} C_i\, \vec{s}_i \cdot \vec{p}\, e^{-i\, \vec{p}\cdot\vec{R}_i(t)} =$$

$$\int_{-\infty}^{\infty} dt\, e^{ip_o t}\, p/(2M)\, (n_k \tilde{D}_k - ip\, n_k n_l\, \tilde{Q}_{kl} + \ldots) \tag{38}$$

with the multipole moments

$$\tilde{D}_k = (C_1 \vec{s}_1 + C_2 \vec{s}_2)_k,$$

$$\tilde{Q}_{kl} = \mu\, (C_1/m_1\, \vec{s}_1 - C_2/m_2\, \vec{s}_2)_k\, R_l. \tag{39}$$

Eq. (38) can be evaluated further only if an assumption on the time dependence of the nuclear spin is made. We will make the (perhaps unwarranted) assumption that the vectors s_i are constant in time. Then no dipole radiation arises. If (26) is integrated over solid angle $d\Omega$ and averaged over the spin directions we obtain

$$\left(\frac{dn}{dp_o}\right)_P = (2/\pi)\, (p^3/p_o^2)\, a^2\, (p/2M)^2\, 1/3\, G_1\, h_1, \tag{40}$$

where h_1 is the function defined in (35). The magnitude of the factor G_1 depends on the relative orientation of the nuclear spins. It takes its maximum value

$$G_1 = 1/(4\pi)\, \mu^2\, (C_1/m_1\, \bar{s}_1 + C_2/m_2\, \bar{s}_2) \rightarrow (g^P)^2/(4\pi)\, A_R^2\, 4\, \bar{s}^2 \tag{41}$$

if $\vec{s}_1$ and $\vec{s}_2$ are completely anticorrelated and vanishes in the case of complete correlation (parallel spins).

The differential emission cross section $d\sigma/dE_x$ for scalar or pseudoscalar particles is obtained by integrating eqs. (33) or (40) over impact parameter b. As an example the results are shown in Fig. 21 for a collision of U + U at 6 MeV/u bombarding energy and assuming a particle mass of m_x = 1.68 MeV. The spectra exhibit broad maxima at $E_x \simeq 600$ keV and decay exponentially at high energies. At low kinetic energy E_x the intensity is suppressed due to the powers of p entering eqs.

(33), (40). Note that this would be different in the case of vanishing rest mass of the emitted particle (e.g. photon bremsstrahlung) due to the identity $p = p_0$. In the limit $E_x \to \infty$ the fall-off constant is well described by the characteristic Fourier frequency ω_o of the trajectory divided by 2π, i. e. $d\sigma/dE_x \sim \exp(-E_x/E_o)$ with $E_o = h\omega_o/2\pi = hv_o/(2\pi a) = 420$ keV.

The total emission cross section is found as

$$\sigma_x^S = 6.4\ 10^{-3}\ \text{b} \times (g_S^N)^2/4\pi ,$$

$$\sigma_x^P = 5.3\ 10^{-7}\ \text{b} \times (g_P^N)^2/4\pi \times (\text{Spin/Nucleon})^2 .$$

Note that the numbers are multiplied by the coupling constants $(g_i^N)^2/4\pi$. The pseudoscalar result furthermor has to be multiplied by the square of the average spin per nucleon. The suppression of σ_x^P relative to σ_x^S is mainly due to the factor $(p/2M)^2$ in eq. (40) since the relevant energies are small compared to the nucleon mass M.

Now we will try to answer the question, whether the creation of a new particle through the discussed bremsstrahlung-type mechanism can be invoked to explain the measured positron lines.

First we note that the spectra $d\sigma/dE_x$ displayed in Fig. 21 certainly will not lead, without further assumptions, to a narrow positron line. Due to the suppression of $d\sigma/dE_x$ at small energies the positron energy distribution will be rather broad. Only in the case of particles with a long life-time, i.e. when most of the fast emitted particles decay outside the detector, a narrow structure can emerge. In this case the line intensity is furthermore suppressed, since $d\sigma/dE_x(E_x=0) = 0$.

Taking into account the life-time argument, to explain the positron production cross section we have to identify $\sigma_{e^+}(\exp) \approx \sigma_x\ (1-e^{-T/\tau})$. The experimental cross section is of the order $\sigma_{e^+}(\exp) \geq 50$ µb. The effective 'detector escape time' T is typically of the order 10^{-10} s. If, as required from the linewidth argument, $\tau \gg 10^{-10}$ s we have to require $\sigma_{e^+}(\exp) \approx \sigma_x\ T/\tau$. Let us first discuss the case of scalar particles. Using the approximative expression for the life-time[19], $\tau \approx 10^{-21}\text{s}/\alpha^e$, we arrive at the condition

$$\alpha_S^N \alpha_S^e \geq (50\mu b/6400\mu b)\ (10^{-21}s/10^{-10}s) \simeq 10^{-13}.$$

This conclusion is independent of the assumption of long life-time. In the opposite limiting case ($\tau < 10^{-10}$ s) we would have found separately $\alpha_S{}^e > 10^{-11}$ and $\alpha_S{}^N \geq 10^{-2}$ (from requiring $\sigma_{e^+}(\exp) = \sigma_X{}^S$) which give the same result for $\alpha_S{}^N \alpha_S{}^e$.

For this product of coupling constants, however, an upper limit of $\alpha_S{}^N \alpha_S{}^e \leq 10^{-18}$ has been found from the atomic physics data. Clearly these constraints are not compatible.

If we turn to the case of pseudoscalar particles, we have not deduced an upper limit for the product of coupling constants $\alpha_P{}^N \alpha_P{}^e$. The very small value of $\sigma_X{}^P$, however, is sufficient to rule out this possibility. Using $\sigma_{e^+}(\exp) = \sigma_X{}^P$, i.e. omitting the additional reduction factor due to long life-times, we have $\alpha_P{}^N \geq 10^4$ if the average spin per nucleon is (quite arbitrarily) taken as 0.1. Clearly such a large coupling constant is not meaningful.

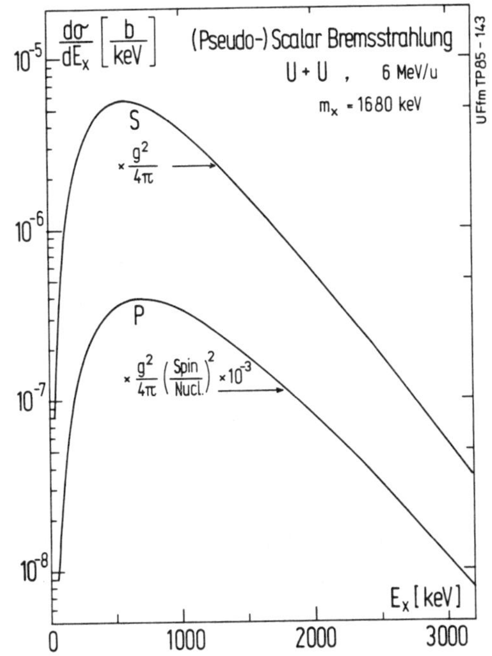

Fig. 21: Differential cross section for the bremsstrahlung production of scalar or pseudoscalar particles with mass m_X = 1.68 MeV in U + U collisions at 6 MeV/u impact energy. To obtain absolute numbers the curves have to be multiplied by the squared coupling constant $g^2/4\pi$. In addition the pseudoscalar result (lower curve) has to be multiplied by the square of the average spin per nucleon and by the numerical factor 10^{-3}.

Let us summarize the result of this discussion. We have studied the hypothesis that the positron lines observed in heavy ion collisions are caused by the decay of a neutral massive boson. Avoiding the need to postulate the existence of a whole family of new particles with adjusted masses and coupling constants, most possible cases were found to lead to conflict with established high-precision data of atomic physics. We are left with one alternative, that cannot be absolutely excluded at present, namely the creation of a pseudoscalar particle from the nuclear current. The particle would have to interact much more strongly with up and down quarks than with leptons and charmed quarks. Its mass would be about 1.68 MeV, the coupling constant to the electron $\alpha^e < 2 \cdot 10^{-9}$, but probably $\alpha^e > 10^{-10}$, to explain the negative outcome of the axion searches, the lifetime in the region $5 \cdot 10^{-13}$ s to 10^{-11} s. However, the bremsstrahlung model indicates that the particle production cross section required to explain the positron line intensity can not be obtained with reasonable values of the coupling constant.

K - HOLE COINCIDENT POSITRON DETECTION IN HEAVY ION COLLISIONS

In ordinary undelayed Rutherford scattering of highly stripped heavy projectiles on targets of heavy atoms the dominant dynamical positron production processes obstruct the search for spontaneous positron creation. Here we want to discuss the consequences of coincidence measurements of positrons with K-shell vacancies. For head-on U + Cm collisions at a bombarding energy of E_{Lab} = 6.2 MeV/u we calculated positron spectra for various delay times T = 0 (pure Rutherford scattering), 1, 2, 4, and $6 \cdot 10^{-20}$ s using the schematic trajectory model[3,4,57]. Considering the contribution of the supercritical κ = -1 channel for one specific spin direction only, for T = 0 the coincidence condition leads to a suppression of probabilities by a factor of more than 500. For various definite delay times this factor is reduced, namely ~ 300 for $T = 4 \cdot 10^{-20}$ s, and ~ 100 for $T = 6 \cdot 10^{-20}$ s. However, a global statement for the behaviour of the suppression factor is not possible, since the corresponding K-hole vacancy probability oscillates very rapidly[36] as a function of T. Thus, the suppression factor, resulting from the interplay between the accidental coincidence term and the diminishing coherence term[55,61], depends very strongly on the delay time considered. The same is true for the subcritical $\kappa = +1$ channel where again the asymptotic K-hole probability plays a dominant role.

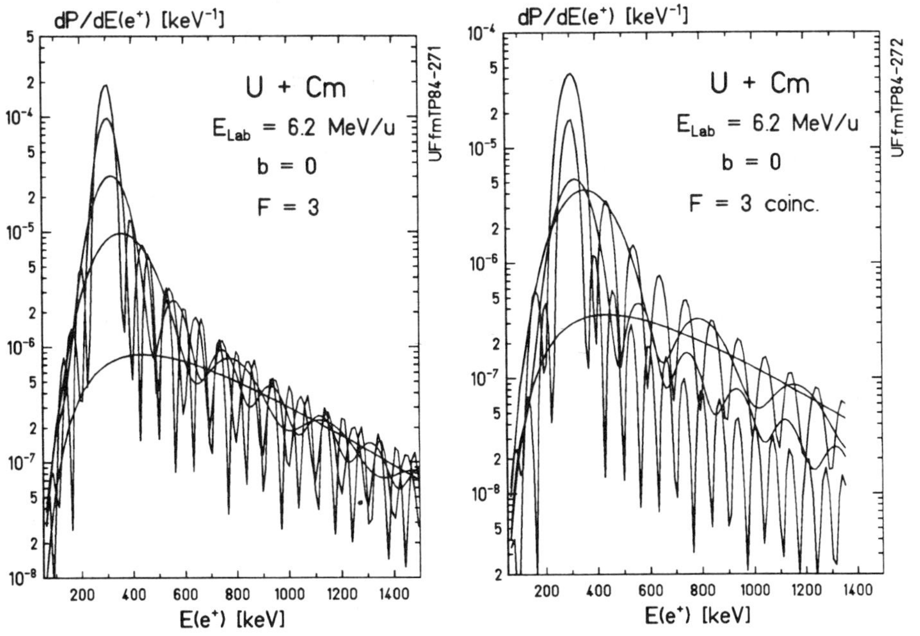

Fig. 22: Comparison between total positron spectra (left) and K-hole coincident spectra (right) in U + Cm collisions for various nuclear delay times. T = 0, 1, 2, 4, and $6 \cdot 10^{-20}$ s.

While the kinematically caused position of the maximum in the energy distribution of emitted positrons is shifted for $\kappa = -1$ by ~ 70 keV and for $\kappa = +1$ by ~ 30 keV, respectively, towards higher kinetic positron energies, the position of the spontaneous positron line at $E_{peak} = E^B_{1s}(2a) - 2mc^2$ and the line shape are not affected.

Since spin direction as well as the angular momentum quantum number κ are not measured in present experimental setups, additional accidental coincidences for all combinations of spin direction and quantum number κ have to be taken into account. Fig. 22 shows the final results, i.e., the total positron spectra (left) and the K-hole coincident spectra (right). It is assumed that the K-holes in uranium and in curium are not distinguished. Since the K-hole probabilities $P_{K(U)}(T)$ and $P_{K(Cm)}(T)$ oscillate with different frequencies, the resulting dominant accidental coincidence term depends on T quite irregularly. This can be seen in fig. 22b, where the usual monotonic increase of the probability in the positron line is partly reversed. The suppression factors are ~ 2, 2.25, 6, 10, and 2.25 for T = 0, 1, 2, 4, and $6 \cdot 10^{-20}$ s, respectively. Thus, the quantitative effect strongly depends on the delay time considered, and for realistic nuclear reactions, where T does not have a sharply defined value, it exhibits no clear trend. Also the shift of the maximum in the energy distribution of emitted positrons cannot be detected due to the overwhelming contribution of accidental coincidences. In summary we conclude that a K-hole coincident detection of positrons using the present experimental setups probably will not yield deeper insight in the spontaneous decay to the charged vacuum.

POSITRON PRODUCTION IN CROSSED BEAMS OF BARE URANIUM NUCLEI

Up to now atomic positron production processes have been studied experimentally by bombarding solid targets of, e.g., Th, U, and Cm with highly stripped heavy ion projectiles of the same region of nuclear charge. In one type of new experiments positron production in collisions of two bare nuclei, e.g., uranium on uranium, might be investigated. This situation can be experimentally realised with crossed beams of high-energy fully stripped heavy ions as discussed in context with the SIS - proposal[84,85].

Obviously, spontaneous positron production will be enhanced by more than a factor of 40 due to the absence of bound electrons occupying the dived 1sσ-state. This is shown in Fig. 23 for head-on U + U collisions at a relative kinetic energy of 680 MeV, calculated within the schematic trajectory model[3-4,36,57] for various delay times T = 0 (pure Rutherford scattering), 1, ..., $5 \cdot 10^{-20}$ s. The present experimental situation is represented by the choice F = 3 for the Fermi level (left part), whereas future experiments with colliding beams of bare U-nuclei are simulated by F = 0 (right part). Since in both calculations (i.e., for F = 0 and for F = 3) the influence of electron-electron interaction on the electronic states has been neglected, the energetic position of the spontaneous positron line is the same ($E_{peak} = E^B_{1s}(2a) - 2mc^2$). In reality, electron screening will shift this position towards lower positron energies as the charge state of the ions is reduced. This dependence may serve to prove the atomic origin of the effect.

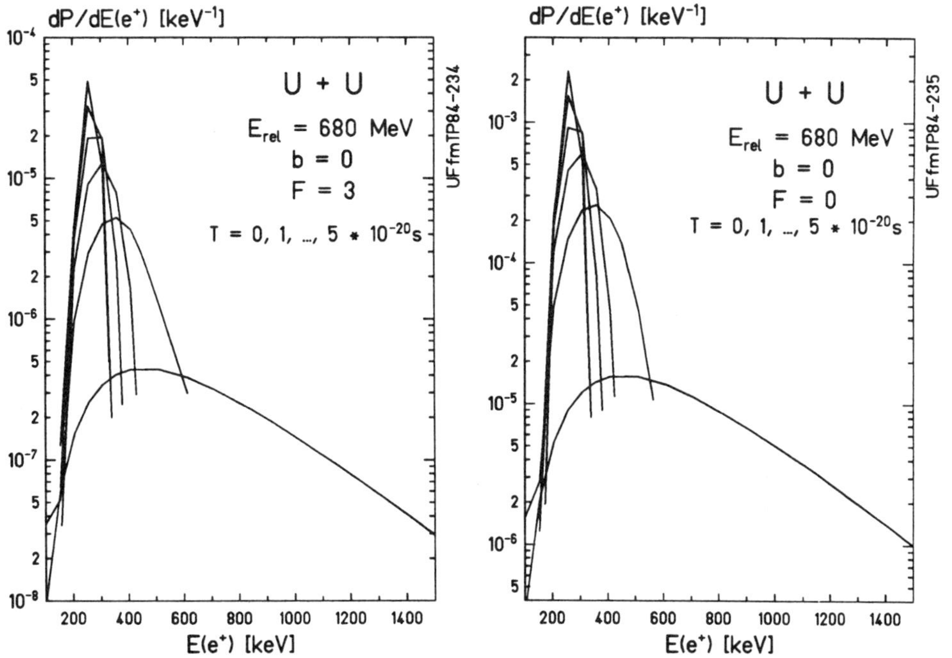

Fig. 23: Comparison of the positron production probability in central U + U collisions for different Fermi surfaces F. The considered nuclear delay times are indicated. A schematic trajectory model is employed.

However, for F = 0 also the induced positron production processes (but not the direct pair creation process) are enhanced. In the discussed system this enhancement amounts to a factor of ~ 35 in central collisions and up to a factor of ~ 175 in peripheral collisions (b ~ 40 fm). While again, due to the neglection of electron screening, the position of the maximum in the energy distribution of induced positrons is not affected, the slope of the high-energy part of the spectra becomes steeper in the bare U + U case for peripheral collisions.

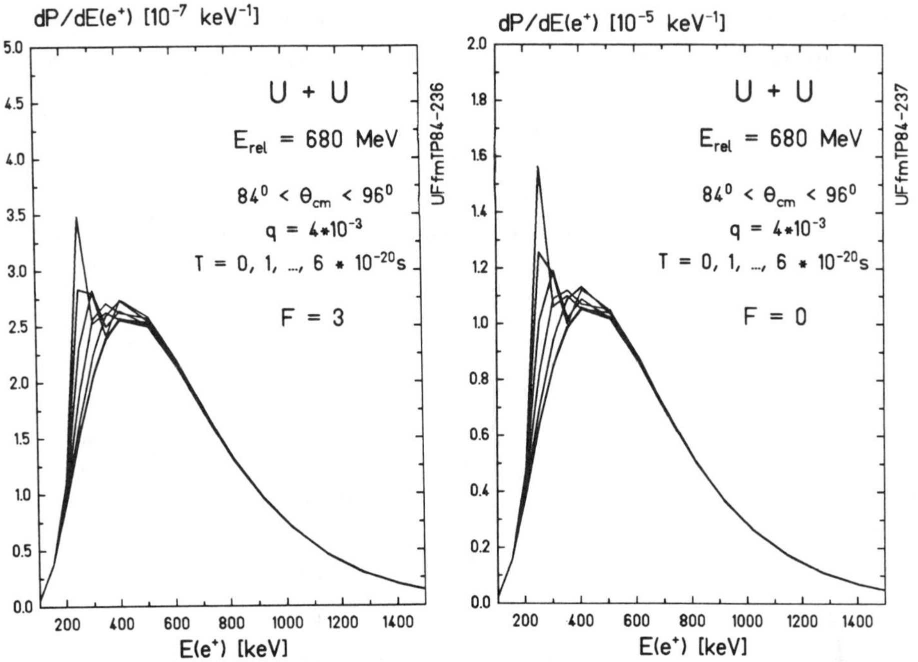

Fig. 24: The same comparison as in fig. 23. Now a superposition of positrons from elastic Rutherford scattering events and, to a fraction q, from the formation of long-lasting nuclear molecules is assumed.

The question, whether spontaneous positron creation can be separated better from dynamical processes when dealing with collisions of bare nuclei, thus is a subject of kinematical conditions. Fig. 24 shows a hypothetical superposition of positrons, originating from usual Rutherford scattering of U + U at E_{rel} = 680 MeV in an angular detection window of $\Theta_{cm} = 90° \pm 6°$, plus spontaneous positrons from long-lasting nuclear molecule formation of lifetime T with a relative cross section $\sigma_N / \sigma_R = q = 4 \cdot 10^{-3}$ in the assumed angular window. Although there is no significant relative increase of the peak structure when comparing Fig. 24a with Fig. 24b, a preference of more central collisions or the choice of higher bombarding energies, i.e., different kinematical conditions, could lead to slightly enhanced detection of spontaneous positrons. However, not more than a relative factor of ~ 1.5 could be gained in our calculations.

In summary the major advantages of the crossed beam technique employing completely stripped uranium nuclei are the following[86]: a) An overall increase of the positron yield by a factor of 40 can be achieved. In consequence background processes caused by nuclear internal conversion are suppressed by the same factor. In addition the absence of δ-electrons reduces possible background processes. However, monoenergetic positron production due to conversion processes could be enhanced. c) The forward geometry should allow for technical improvements. With a relatively small detector one could identify all reaction partners. d) Coincidences with specific ionic charges after the collision could help to investigate distinguished lepton production processes, e.g., the spontaneous positron emission.

CONVERSION PROCESSES IN GIANT ATOMS

Due to their distinct Z-dependence conversion processes in principle may also serve as detection method to identify superheavy elements. Beside the traditional nuclear physics methods and the direct detection of characteristic X-rays precise measurements of the δ-electron and positron yield thus could provide additional indication for the formation of a superheavy element.

In this section therefore theoretical results on the conversion of bound state electrons and on the internal e^+,e^- - pair creation are presented. We consider various nuclear transitions of multipolarities E0, EL, and ML ($L \geq 1$) in superheavy atoms with $Z > 109$. The pecularities of conversion processes in overcritical systems are illuminated. These theoretical studies are confronted with the prominent peak structures observed in positron spectra of very heavy collision systems. In particular we want to examine more closely the possible role of internal pair formation in the combined system with $Z = Z_1 + Z_2$ with respect to the measured positron production rate.

We will present theoretical results on the conversion coefficients of bound state electrons $\alpha = P_{e^-}/P_\gamma$. P_γ denotes the photon emission probability in a nuclear transition of multipolarity EL or ML ($L \geq 1$) and transition energy ω. P_{e^-} signifies the complementary probability for ionizing a bound state electron. Likewise we consider the differential conversion coefficient[72] $d\beta/dE$ for internal e^+,e^--pair creation with respect to the kinetic positron energy E, which is defined by $d\beta/dE = (dP_{e^+,e^-}/dE)/P_\gamma$. P_{e^+,e^-} is the probability for electron-positron pair formation. The corresponding quantity exclusively related to nuclear E0 transitions will be denoted by $d\eta/dE = (dP_{e^+,e^-}/dE)/P_{e^-}$.

The underlying theoretical formalism has been worked out in detail in refs. 72, 87 and 88. Since there is no principal modification it will not be repeated here. For the computation of the various conversion coefficients we employed electron wavefunctions that are solutions of the Dirac equation for finite-size nuclei. Electron screening effects as well as radiative corrections are neglected.

The competing modes of a nuclear transition are predominantly determined by the transition energy ω, the angular momentum selection rules, and the electron density inside the nuclear interior. After investigating the Z- and ω-dependence of conversion coefficients in the domain $110 \leq Z \leq 170$ some characteristic features for supercritical systems accessible to experimental observations ($170 < Z \leq 188$) are examined. Finally it should be mentioned that also extraordinary Auger processes may lead to e^+,e^--pair formation[89].

First we restrict our considerations to conversion processes in subcritical systems. In fig. 25 the K-shell conversion coefficient is displayed for nuclear transitions above the pair production threshold. To survey the dependence of $\alpha(1s)$ on the nuclear charge number Z we regard as example the nucleus $Z = 114$, which is already outside the known periodic table but which is assumed to be stable again. Furthermore we envisage a medium system ($Z = 140$) and a heavy system ($Z = 169$) which almost reaches criticality. Nuclear transitions of the most probable multipolarities E1, E2, M1, and M2 are studied. Obviously in the very heavy systems transitions of magnetic type preferably would occur via ionization of 1s-electrons whereas γ-emission is always dominant for transitions of electric type in the considered energy range. The most frequent collective modes are electric quadrupole

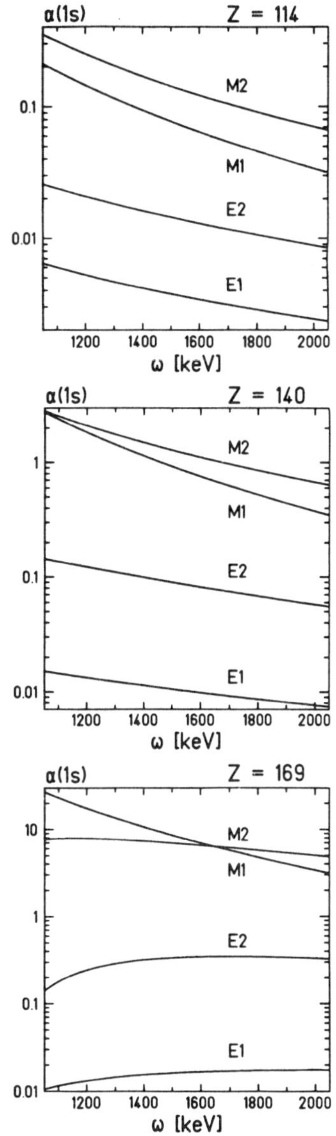

Fig. 25: K-shell conversion coefficient versus nuclear transition energy ω for the superheavy systems Z = 114, Z = 140, and Z = 169. The multipolarities E1, E2, M1, and M2 are considered.

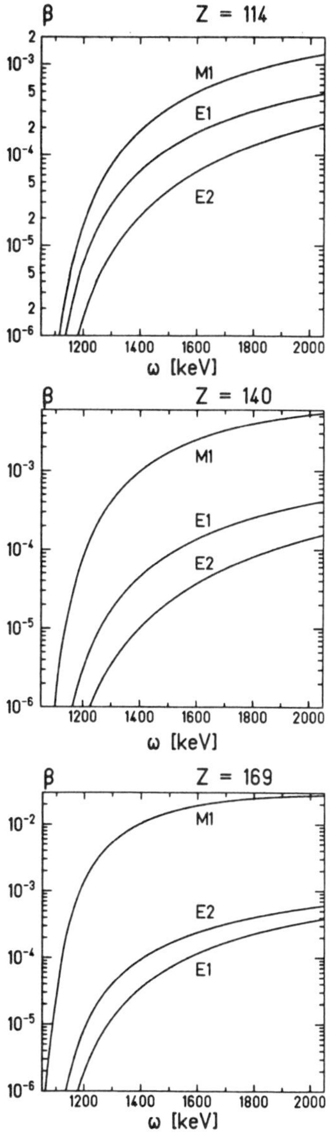

Fig. 26: Total conversion coefficient for e^+,e^--pair creation versus nuclear transition energy ω for the superheavy systems Z = 114, Z = 140, and Z = 169. The multipolarities E1, E2, and M1 are considered.

transitions. Here the conversion coefficient $\alpha(1s)$ displays a slight increase if one enlarges the central charge.

The total pair conversion coefficient β is presented in fig. 26 for the same parameters. With increasing nuclear charge number no variation of β is obtained for transitions of electric type. As a typical feature for very heavy systems transitions with magnetic multipolarity exhibit a relatively large tendency to produce e^+,e^--pairs. For comparison fig. 27 shows the pair conversion coefficient η for electric monopole transitions where the single photon emission is a strictly forbidden channel. Again we find no striking Z- or ω-dependence. Even for rather high nuclear transition energies pair production is suppressed compared with K-shell ionization.

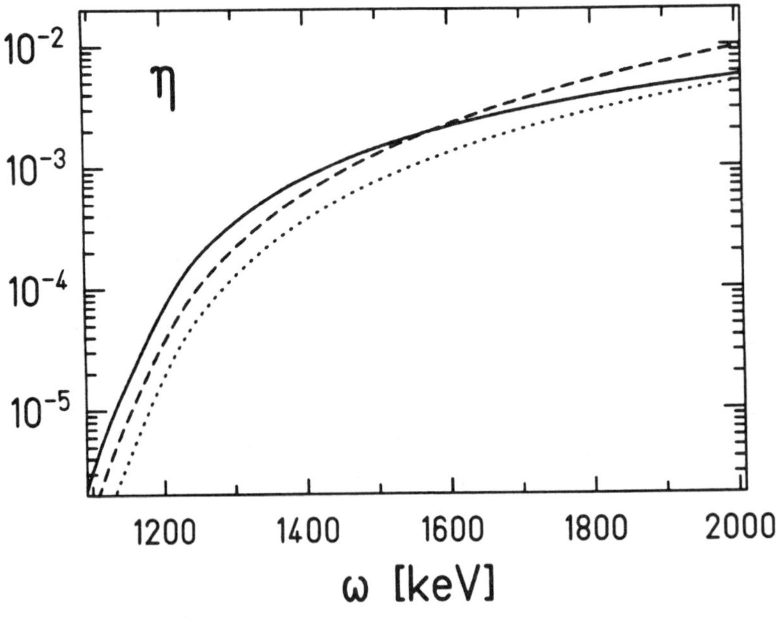

Fig. 27: Total conversion coefficient η for electric monopole transitions versus nuclear transition energy ω. The total e^+,e^--pair production probability is compared with the K-shell ionization probability. The superheavy systems Z = 114 (dashed line), Z = 140 (dotted line) and Z = 169 (solid line) are considered.

The spectral distribution of emitted positrons is illustrated in figs. 28 and 29 for nuclear transitions in the atom Z = 169 with ω = 1323,

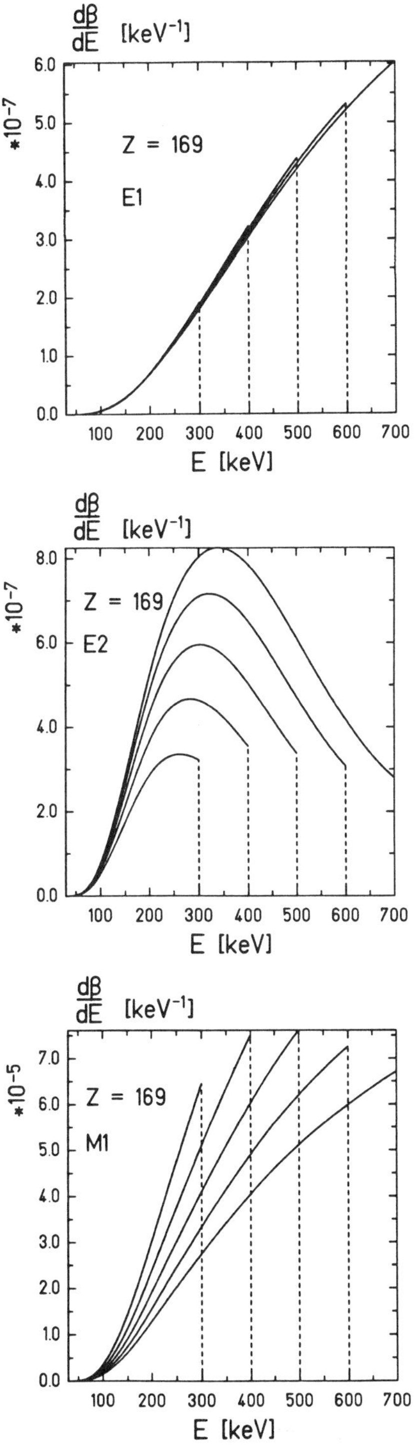

Fig. 28: Differential conversion coefficient for e^+,e^--pair creation versus kinetic positron energy E. dβ/dE is depicted for nuclear transition energies ω = 1323, 1423, 1523, 1623, and 1723 keV in the superheavy system Z = 169. The multipolarities E1, E2, and M1 are considered.

1423, 1523, 1623, and 1723 keV, respectively. The dashed lines indicate the end points of the corresponding positron spectra. In each case the width of the energy distribution is much to large to be responsible for the narrow lines observed in the differential e^+-production cross sections[39,40,48].

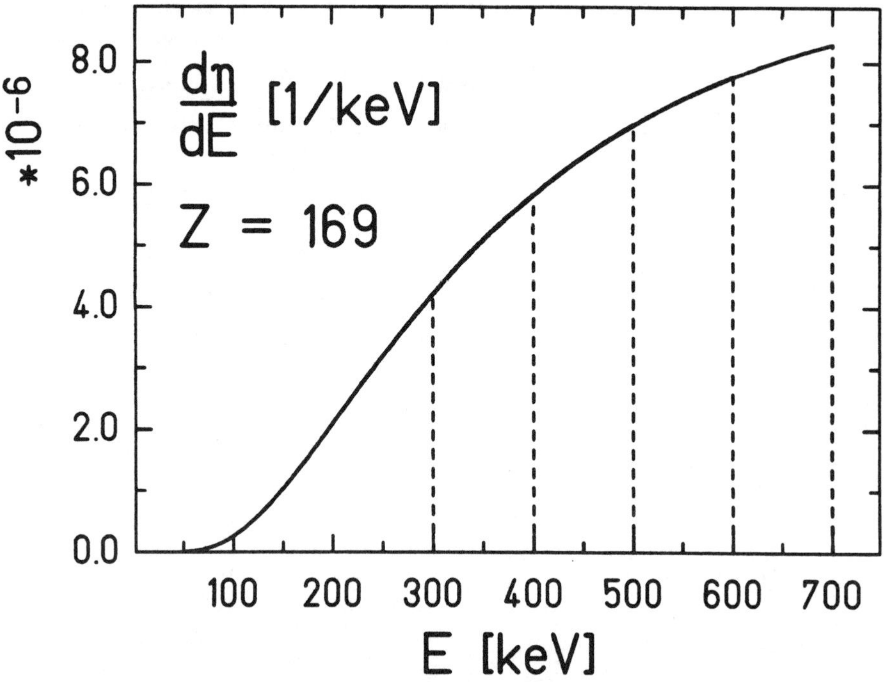

Fig. 29: The differential conversion coefficient dη/dE for electric monopole transitions in the superheavy system Z = 169 as function of the kinetic positron energy E. The differential e^+,e^--pair production probability is compared with the ionization probability of 1s-electrons for the nuclear transition energies ω = 1323, 1423, 1523, 1623, and 1723 keV. The dashed lines indicate the end points of the distinct positron spectra.

We now turn to the discussion of electron-positron pair conversion in supercritical compound systems. We consider a collision of very heavy ions such as $_{92}U + _{96}Cm$ with a combined charge $Z = Z_1 + Z_2 = 188$. According to the ideas of Reinhardt et al.[57] the projectile and target nucleus may come into nuclear contact even for bombarding energies slightly below the nuclear Coulomb barrier. Related to these theoretical investigations we considered the question whether positrons may be created due to conversion processes during the sticking period T. In this case pair conversion is not distinguishable from ordinary K-shell conversion[10]. This situation is visualized in fig. 30. The energy spectrum of the Dirac equation is shown with the upper continuum and the lower continuum, which includes the supercritical K-shell denoted by its energetical location E_{res}.

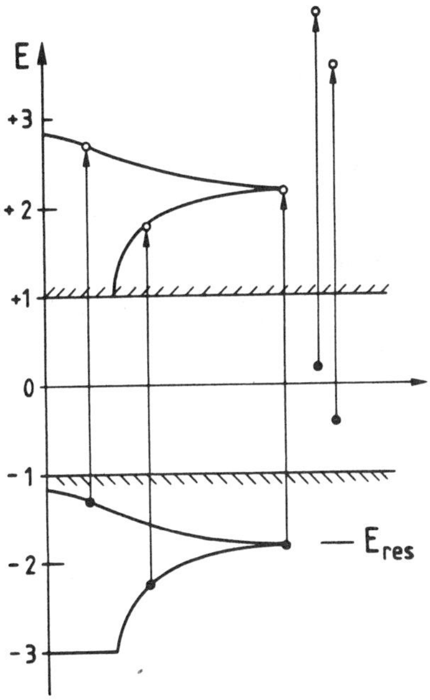

Fig. 30: Conversion processes indicated by arrows in supercritical systems. The energy spectrum of the Dirac equation is shown with the upper continuum and the lower continuum, which includes the supercritical K-shell denoted by the energetical location E_{res}. The curves in the lower and in the upper continuum display the spectral shape of the positron and of the electron distribution, respectively. The two arrows on the right-hand side denote the conversion of bound state electrons into high-lying states.

A supercritical nucleus Z = 184 (U + U) which undergoes a transition with $\omega > 2$ during the sticking period T may transfer this excitation energy to one of the electrons in the negative energy continuum. The remaining hole is defined as positron. But also the K-shell electron can be lifted to the upper continuum. If T is longer than the spontaneous decay width of the K-shell resonance, the K-vacancy will be filled again leading to spontaneous positron emission. The conversion processes are indicated by the arrows in fig. 30. The curves in the lower and in the upper continuum display the spectral shape of the positron and of the electron distribution, respectively. In the positron spectra a sharp peak appears at $E = E_{res}$. The two arrows on the right-hand side denote the conversion of bound state electrons into high-lying states. The conversion coefficients β and the differential distribution dβ/dE are calculated in precisely the same manner as described extensively in ref. 72. The differential ratio dβ/dE for the electron-positron pair creation compared with the photon emission probability as function of the positron energy E is presented in ref. 10. Most striking is the appearance of a pronounced peak in the spectral distribution at $E = E_{res}$. However, one has to emphasize that the contribution of the discussed conversion processes to the positron yield in a realistic heavy ion collision strongly depends on the specific nuclear structure of the compound system. The observed positron production probability per collision is about 10^{-5}. This requires transition times shorter than 10^{-14} s in the compound nucleus in order that internal pair conversion can contribute considerably. It should be stressed that internal pair creation connected with the bound state resonance in the negative energy continuum cannot be distinguished from spontaneous positron formation. The only difference is provided by the production mechanism of the initial K-hole. But electron excitations as consequence of the Fourier frequencies along a Rutherford trajectory or as result of a nuclear transition are not separable.

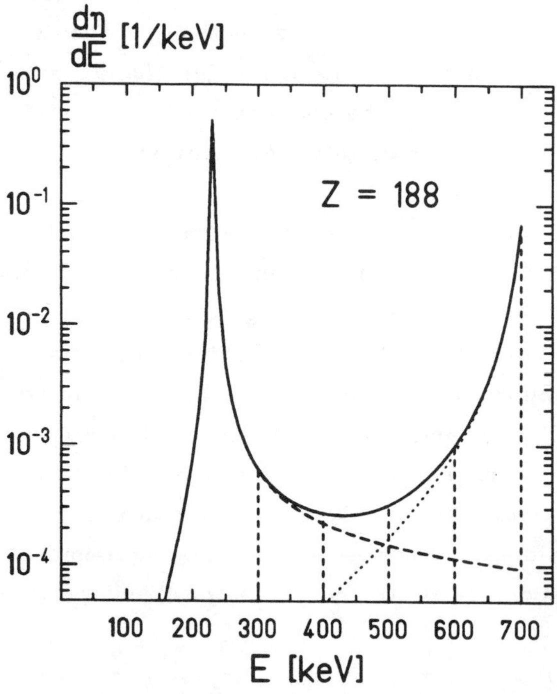

Fig. 31: The same as in fig. 29 for Z = 188. The differential e^+,e^--pair production probability is compared with the ionization probability of 2s-electrons. The dashed and dotted lines give the contributions of the κ = 1 and κ = -1 states, respectively. The cusp at E ≃ 230 keV signals the $2p_{1/2}$-resonance imbedded into the negative energy continuum.

The bound state resonance imbedded in the negative energy continuum is also reflected in the differential conversion coefficient $d\eta/dE$ for nuclear E0-transitions which is presented in fig. 31 for the supercritical system Z = 188 (U + Cm). The cusp in the spectrum at E ≈ 230 keV signals the $2p_{1/2}$-resonance while the peak at the endpoint of the positron distribution for ω = 1723 keV is caused by the $1s_{1/2}$-resonance. The possible relationship of monoenergetic conversion processes in supercritical systems and the prominent peak structures observed in the positron spectra will be reported elsewhere[90].

CONCLUSIONS

Electron configurations of superheavy atoms are transiently produced in collisions of very heavy ions. The measured K-vacancy probabilities and δ-electron cross sections generally are in fair agreement with our coupled channel calculations employing a quasimolecular basis set. It was demonstrated that relativistic effects completely dominate the electron excitation processes. Our calculations were extended to describe collisions incorporating nuclear contact. The possibility[75] to employ the spectrum of emitted leptons as an atomic clock for deep-inelastic nuclear reactions was verified experimentally[44,49,50,70]. First measurements of the δ-electron and positron spectrum clearly revealed a nuclear reaction time of about $\tau \sim 10^{-21}$ s. Recently T. de Reus et al.[96] proposed to deduce also nuclear decelleration times in high-energy nucleus-nucleus collisions from high-energetic δ-electron spectra.

The observed peak structures[39,40] in positron distributions of U + Cm systems could be understood as spontaneous pair creation after formation of giant nuclear systems with A ~ 500. However, this interpretation does not in a natural way explain the new experimental finding[48] that the energetical location of this peak structure seems to be independent of the combined nuclear charge Z. We have argued that an alternative explanation of the narrow positron lines based on a purely molecular mechanism is not well founded. Furthermore we investigated the hypothesis that a new previously undetected elementary particle is created in the strong electric field of a superheavy quasiatomic system which decays into an electron-positron pair. Here we could impose some stringent limits on the formation of such an exotic particle. Within the framework of a bremsstrahlung's model for the creation of a neutral pseudoscalar particle with mass m = 1.68 MeV we totally failed[91] to reproduce the measured production yield and energy distribution of positrons. Finally we investigated the possibility of conversion processes in the combined nuclear system and its relationship to the mysterious peak structures. But also this, more conservative explanation[90] is accompanied with many questionable features.

Also the experimental data do not provide an absolute consistent framework. The precise position of the positron peaks is uncertain. For instance, the various observations[48,53] for the U + Th system differ by about 70 keV. In addition there is conflicting information concerning the excitation function and the dependence on the ion scattering angle available. Restricting the informations to the published data the intensities in the positron spectra of the different groups are not comparable. It is still an open question whether there is any corresponding peak in subcritical systems or whether there are additional peaks at different positron energies.

For future experimental investigations it would be highly desirable to perform a careful analysis of the kinetic energy distribution of positrons in the subcritical Pb + Pb system where spontaneous positron emission is strictly prohibited. In addition due to the specific nuclear structure of ^{208}Pb no low-energy conversion lines should be visible. The hypothesis of a new elementary particle which decays into an electron-positron pair can be tested by a coincidence measurement of positrons and electrons at $E_{e^\pm} \approx 300$ keV. A remeasurement of the excitation function of the positron spectra in a small range around the nuclear Coulomb barrier would provide support for the existence of a pocket in the internuclear potential leading to a giant nuclear system and ultimately to the spontaneous positron creation.

In conclusion we note that due to the present experimental uncertainties there can be no final theoretical explanation of the positron line structures which explains all observed features.

Acknowledgement

We are grateful for many fruitful cooperations with H. Backe, K. Bethge, H. Bokemeyer, F. Bosch, M. Clemente, T. Cowan, J.S. Greenberg, P. Kienle, C. Kozhuharov, R. Krieg, D. Schwalm , J. Schweppe, P. Senger, K. Stiebing, and H. Tsertos, who have made their results available to us prior to publication. G.S. would like to thank the members of the Institut für Theoretische Physik at the Justus-Liebig-Universität for the hospitality which he received during his stay in Gießen.

REFERENCES

1. W. Greiner, ed., Quantum Electrodynamics of Strong Fields, NASI series B80, (Plenum, New York, 1983)
2. W. Greiner, B. Müller, J. Rafelski, Quantum Electrodynamics of Strong Fields, (Springer, Berlin, 1985)
3. U. Müller, G. Soff, J. Reinhardt, T. de Reus, B. Müller, W. Greiner, Phys. Rev. C30, 1199 (1984)
4. U. Müller, G. Soff, T. de Reus, J. Reinhardt, B. Müller, W. Greiner, Z. Physik A313, 263 (1983)
5. T. de Reus, J. Reinhardt, B. Müller, W. Greiner, G. Soff, and U. Müller, J. Phys B17, 615 (1984)
6. U. Heinz, B. Müller, and W. Greiner, Ann. Phys. 151, 227 (1983)
7. U. Heinz, J. Reinhardt, B. Müller, W. Greiner, and U. Müller, Z. Physik A314, 125 (1983)
8. U. Heinz, J. Reinhardt, B. Müller, W. Greiner, W.T. Pinkston, Z. Physik A316, 341 (1984)
9. U. Heinz, U. Müller, J. Reinhardt, B. Müller, W. Greiner, Ann. Phys. 158, 476 (1984)
10. G. Soff, U. Müller, T. de Reus, P. Schlüter, A. Schäfer, J. Reinhardt, B. Müller, W. Greiner, in: Atomic Physics of Highly Ionized Atoms, ed.: R. Marrus, (Plenum, New York, 1983), p. 177.
11. B. Müller, Decay of the vacuum in heavy ion collisions, Preprint GSI-84-62
12. J. Reinhardt, W. Greiner, Phys. Bl. 41, 38 and 93 (1985)
13. G. Mehler, T. de Reus, J. Reinhardt, G. Soff, U. Müller, Z. Physik A320, 355 (1985)
14. G. Soff, P. Schlüter, K.-H. Wietschorke, U. Müller, J. Reinhardt, T. de Reus, U. Heinz, S. Schramm, B. Müller, W. Greiner, Ionization and vacuum decay in the field of giant nuclear molecules, X84, X-ray and inner-shell processes in atoms, molecules and solids, Conference Proceedings, Ed.: A. Meisel, J. Finster, Karl-Marx-Universität Leipzig, DDR, 1984, p. 101.
15. E. Friedman, G. Soff, J. Phys. G11, L37 (1985)

16. P. Schlüter, G. Soff, K.-H. Wietschorke, W. Greiner, J. Phys. B18, 1685 (1985)
17. M.J. Rhoades-Brown, V.E. Oberacker, M. Seiwert, and W. Greiner, Z. Physik A310, 287 (1983)
18. J. Reinhardt, B. Müller, W. Greiner and U. Müller, Phys. Rev. A28, 2558 (1983)
19. A. Schäfer, J. Reinhardt, B. Müller, W. Greiner, G. Soff, J. Phys. G11, L69 (1985)
20. H.-J. Bär, G. Soff, Physica 128C, 225 (1985)
21. S. Schramm, U. Müller, J. Reinhardt, T. de Reus, A. Schäfer, U. Heinz, G. Soff, W. Greiner, B. Müller, in: Fundamental Problems In Heavy-Ion Collisions, Eds.: N. Cindro, W. Greiner, R. Caplar, (World Scientific, Singapore, 1984), p. 453.
22. P.O. Hess, W. Greiner, W.T. Pinkston, Phys. Rev. Lett. 53, 1535 (1984)
23. R.H. Lemmer, W. Greiner, Collective enhancement of positron photo-production in nearly critical strong fields, preprint 1984
24. T. Tomoda, H.A. Weidenmüller, Phys. Rev. A26, 162 (1982)
25. T. Tomoda, Phys. Rev. A26, 174 (1982)
26. T. Tomoda, H.A. Weidenmüller, Phys. Rev. C28, 739 (1983)
27. T. Tomoda, Phys. Rev. A29, 536 (1984)
28. P. Manakos, H.A. Weidenmüller, Z. Phys. A320, 173 (1985)
29. C. Bottcher, M.R. Strayer, Phys. Rev. Lett. 54, 669 (1985)
30 J. Krause, M. Kleber, Phys. Rev. A31, 113 (1985)
31. W. Lichten, A. Robatino, Phys. Rev. Lett. 54, 781 (1985)
32. J. Reinhardt, B. Müller, W. Greiner, Phys. Rev. Lett. 55, 134 (1985)
33. W. Lichten, A. Robatino, Phys. Rev. Lett. 55, 135 (1985)
34. A.B. Balantekin, C. Bottcher, M.R. Strayer, S.J. Lee, Axion production in heavy-ion collisions, preprint 1985
35. W. Fleischer, G. Soff, Z. Naturforsch. 39a, 703 (1984)
36. U. Müller, Atomare Anregungen in Schwerionenstößen mit nuklearem Kontakt, Report GSI-84-7
37. A. Schäfer, B. Müller, W. Greiner, Phys. Lett. 149B, 455 (1984)
38. P. Schlüter, T. de Reus, J. Reinhardt, B. Müller, G. Soff, Z. Physik A314, 297 (1983)

39. J. Schweppe, A. Gruppe, K. Bethge, H. Bokemeyer, T. Cowan, H. Folger, J.S. Greenberg, H. Grein, S. Ito, R. Schule, D. Schwalm, K.E. Stiebing, N. Trautmann, P. Vincent, M. Waldschmidt, Phys. Rev. Lett. 51, 2261 (1983)
40. M. Clemente, E. Berdermann, P. Kienle, H. Tsertos, W. Wagner, C. Kozhuharov, F. Bosch, and W. Koenig, Phys. Lett. 137B, 41 (1984)
41. H. Backe and C. Kozhuharov, in: H.J. Beyer and H. Kleinpoppen, eds., Progress in Atomic Spectroscopy, Part C, (Plenum, New York, 1984), p. 459.
42. C. Kozhuharov, in: S. Datz, ed., Physics of Electronic and Atomic Collisions, (North Holland, Amsterdam, 1982), p. 179.
43. P.H. Mokler, D. Liesen, X rays from superheavy collision systems, in: Progress in atomic spectroscopy, Part C, eds.: H.J. Beyer and H. Kleinpoppen, (Plenum, New York, 1984), p. 321.
44. H. Backe, P. Senger, W. Bonin, E. Kankeleit, M. Krämer, R. Krieg, V. Metag, N. Trautmann, and J.B. Wilhelmy, Phys. Rev. Lett. 50, 1838 (1983)
45. D. Schwalm, in: Electronic and Atomic Collisions, eds.: J. Eichler, I.V. Hertel, N. Stolterfoht, (North-Holland, Amsterdam, 1984), p. 295.
46. M.A. Herath Banda, A.V. Ramayya, W. Koenig, B. Martin, H. Skapa, J. Soltani, Phys. Rev. Lett. 53, 1646 (1984)
47. M.A. Herath-Banda, A.V. Ramayya, C.F. Maguire, F. Güttner, W. Koenig, B. Martin, B. Povh, H. Skapa and J. Soltani, Phys. Rev. A29, 2429 (1984) and private communication
48. T. Cowan, H. Backe, M. Begemann, K. Bethge, H. Bokemeyer, H. Folger, J.S. Greenberg, H. Grein, A. Gruppe, Y. Kido, M. Klüver, D. Schwalm, J. Schweppe, K.E. Stiebing, N. Trautmann, P. Vincent, Phys. Rev. Lett. 54, 1761 (1985)
49. R. Krieg, E. Bozek, U. Gollerthan, E. Kankeleit, G. Klotz, M. Krämer, U. Meyer, H. Oeschler, P. Senger, Positron spectroscopy in elastic and dissipative heavy ion collisions, preprint IKDA 84/19
50. R. Krieg, Positronen und δ-Elektronen als Sonden der Stoßdynamik schwerer Ionen, thesis, Darmstadt, 1985.
51. H. Backe, Nucl. Phys. A400, 451c (1983)

52. H. Bokemeyer, Positron spectroscopy of supercritical heavy ion collision systems, Proc. of the XIX Winter School on Physics, Zakopane, Poland, Preprint GSI-84-43
53. P. Kienle, in: Fundamental Problems In Heavy-Ion Collisions, Eds.: N. Cindro, W. Greiner, R. Caplar, (World Scientific, Singapore, 1984), p. 429.
54. G. Soff, J. Reinhardt, B. Müller, W. Greiner, Phys. Rev. Lett. 38, 592 (1977)
55. J. Reinhardt, B. Müller, W. Greiner, Phys. Rev. A24, 103 (1981)
56. J. Rafelski, B. Müller, and W. Greiner, Z. Phys. A285, 49 (1978)
57. J. Reinhardt, U. Müller, B. Müller, and W. Greiner, Z. Physik A303, 173 (1981)
58. B. Müller and W. Greiner: Z. Naturforsch. 31a, 1 (1976)
59 P. Schlüter, K.-H. Wietschorke and W. Greiner: J. Phys. A16, 1999 (1983)
60. K.-H. Wietschorke, P. Schlüter and W. Greiner: J. Phys. A16, 2017 (1983)
61. G. Soff, J. Reinhardt, B. Müller, W. Greiner, Z. Physik A294, 137 (1980)
62. R.L. Becker, A.L. Ford, J.F. Reading, Phys. Rev. A29, 3111 (1984)
63. D. Liesen, P. Armbruster, F. Bosch, S. Hagmann, P. Mokler, H.J. Wollersheim, H. Schmidt-Böcking, R. Schuch, J.B. Wilhelmy, Phys. Rev. Lett. 44, 983 (1980)
64. S. Ito, P. Armbruster, H. Bokemeyer, F. Bosch, H. Emling, H. Folger, E. Grosse, R. Kulessa, D. Liesen, D. Maor, D. Schwalm, J.S. Greenberg, R. Schule, N. Trautmann, GSI Scientific Report 1982, GSI 82-1, p. 141.
65. G. Soff, B. Müller, W. Greiner, Phys. Rev. Lett. 40, 540 (1978)
66. H. Tsertos, Positronenerzeugung in superschweren Stoßsystemen - Eine empfindliche Sonde für neue atom- sowie kernphysikalische Phänomene, thesis, Report GSI-85-13.
67. H. Backe, L. Handschug, F. Hessberger, E. Kankeleit, L. Richter, F. Weik, R. Willwater, H. Bokemeyer, P. Vincent, Y. Nakayama, J.S. Greenberg, Phys. Rev. Lett. 40, 1443 (1978)

68. C. Kozhuharov, P. Kienle, E. Berdermann, H. Bokemeyer, J.S. Greenberg, Y. Nakayama, P. Vincent, H. Backe, L. Handschug, E. Kankeleit, Phys. Rev. Lett. 42, 376 (1979)
69. A. Gruppe, Messungen zur Positronenerzeugung im Schwerionenstoß, thesis, Report GSI-85-4.
70. P. Senger, Spektroskopie von Positronen und δ-Elektronen nach dissipativen Schwerionenstößen, thesis, Darmstadt, 1983.
71. U. Gollerthan, δ-Elektronenspektroskopie nach ^{238}U + ^{238}U Stößen - Experiment und Monte Carlo Simulationen zu 3- und 4-Körper Ereignissen, diploma thesis, Darmstadt, 1984.
72. P. Schlüter, G. Soff, W. Greiner, Phys. Rep. 75, 327 (1981)
73. S. Yoshida, Ann. Rev. Nucl. Sci. 24, 1 (1974)
74. R. Schmidt, V.D. Toneev, and G. Wolschin, Nucl. Phys. A311, 247 (1978)
75. G. Soff, J. Reinhardt, B. Müller, W. Greiner, Phys. Rev. Lett. 43, 1981 (1979)
76. Ch. Stoller, M. Nessi, E. Morenzoni, W. Wölfli, W.E. Meyerhof, J.D. Molitoris, E. Grosse, Ch. Michel, Phys. Rev. Lett. 53, 1329 (1984)
77. J.R. Birkelund, L.E. Tubbs, J.R. Huizenga, J.N. De, and D. Sperber, Phys. Rep. 56, 107 (1979)
78. B. Müller, J. Rafelski, and W. Greiner, Z. Physik 257, 62 and 183 (1972)
79. Ya.B. Zeldovich and V. S. Popov, Sov. Phys. Usp. 14, 673 (1972)
80. H. Tsertos, E. Berdermann, F. Bosch, M. Clemente, P. Kienle, W. Koenig, C. Kozhuharov, W. Wagner, submitted to Phys. Lett. B
81. U. Müller, T. de Reus, J. Reinhardt, B. Müller, W. Greiner, to be published.
82. T. de Reus, U. Müller, B. Müller, W. Greiner and G. Soff, GSI Scientific Report 1981, GSI 82-1, 1982, p. 174.
83. G. Soff, W. Greiner, W. Betz, and B. Müller, Phys. Rev. A20, 169 (1979)
84. P. Kienle, Die Ausbaupläne der GSI, März 1984.
85. P. Kienle, The SIS-ESR Project at GSI, Report GSI 85-16.
86. P. Kienle, H. Backe, private communication

87. G. Soff, P. Schlüter, W. Greiner, Z. Physik A303, 189 (1981)
88. P. Schlüter, Die Diracgleichung in der lokalen Darstellung - Beiträge zur Quantenlektrodynamik starker Felder, Report GSI-85-15.
89. G. Soff, W. Greiner, J. Phys. B15, L681 (1981)
90. P. Schlüter, G. Soff, T. de Reus, W. Greiner, to be published.
91. J. Reinhardt, A. Schäfer, B. Müller, W. Greiner, Phenomenological consequences of a hypothetical light neutral particle in heavy ion collisions, preprint 1985
92. G. Hardekopf, J. Sucher, Phys. Rev. A31, 2020 (1985)
93. H. Backe, B. Müller, Positron production in heavy-ion collisions, preprint UFfmTP-139-84.
94. J. Rafelski, B. Müller, Die Struktur des Vakuums - Ein Dialog über das Nichts, (Harri Deutsch, Thun, 1985).
95. G. Soff, U. Müller, T. de Reus, J. Reinhardt, B. Müller, W. Greiner, Nucl. Instr. Meth. B10/11, 214 (1985)
96. T. de Reus, J. Reinhardt, B. Müller, W. Greiner, U. Müller, G. Soff, Z. Physik A321, (1985)
97. J. Hoppe, J. Reinhardt, Complex dilatations in relativistic quantum mechanics and the Lamb shift for resonances, UFTP preprint 151/1985
98. J. Kirsch, B. Müller, W. Greiner, MO-radiation interference phenomena as a "clock" for nuclear reaction times, preprint 1985

PAIR PRODUCTION AT GeV/u ENERGIES*

C. Bottcher and M. R. Strayer
Physics Division, Oak Ridge National Laboratory
Oak Ridge, Tennessee 37831

ABSTRACT

Electron and positron production in relativistic ion-atom collisions is discussed within the context of the time-dependent Dirac-Hartree approximation to a fully relativistic field theory of the collision. The time-dependent fields are treated classically, and the numerical methods employing basis splines are discussed in detail and contrasted with results obtained from the case of non-relativistic velocities. The results of a one-dimensional model are presented and show a moderately large probability for pair production followed by electron capture.

INTRODUCTION

We propose to discuss electron and positron production in collisions between heavy ions moving at relative velocities exceeding half the speed of light. Using the notation A,B for bare nuclei [e.g. U,Cm] and A(n) for a nucleus associated with a single electron in state n, delta-electron production may be written,

$$A + B(1s) \rightarrow A + B + e^{-}(E > mc^{2}). \qquad (1)$$

Positron production may be pictured as the removal of an electron from the Dirac sea,

$$A + B + e^{-}(E < -mc^{2}) \rightarrow A + B(1s). \qquad (2)$$

If we time reverse (2), we see that both processes are in essence no different from the ionization phenomena familiar in low energy atomic physics. Unless otherwise stated, m is the electron mass.

While the systems experimentally studied so far should strictly be treated as many electron problems, we believe that the independent particle model is a good first approximation to the relativistic phenomena considered, and we shall use this language throughout in the interest of clarity.

Heretofore theoretical and experimental studies of pair production[1,2] have focused on relative collision velocities around 0.1c. At these velocities two heavy nuclei just touch at the classical turning point of the collision, thus maximizing the time during which

*Research sponsored in part by the U.S. Department of Energy under contract DE-AC05-840R21400 with Martin Marietta Energy Systems, Inc.

a combined "supercritical" system can be said to exist. The possibility of forming a long-lived nuclear molecule is also maximized, since at higher velocities binary and ternary fission results in a rapid decay of the supercritical system.

The present work has two motivations. We are interested in relativistic many-body problems in nuclear and particle physics, e.g. dynamics of hadronic matter and the confinement of quarks in hadrons. It would seem appropriate to begin by understanding the relativistic many-electron problem for which we believe we can write down a correct formalism. From other articles in this volume, it is clear that the relativistic many-electron system is very much an open problem, even as regards bound-state calculations. In either atomic or nuclear physics, the dynamical problem is almost untouched. A more practical motivation is to calculate possible charge changing mechanisms which might affect the design of a relativistic heavy ion collider, as discussed in this volume by H. Gould. Other applications may be found in the realm of heavy-ion-induced fusion, in which interest has recently been revived.

Based on preliminary calculations (described below), we reach the conclusion that atomic phenomena do not "run out" at GeV energies, but become more rich and interesting, and continue so to energies as high as we are likely to reach experimentally in the near future.

The processes (1) and (2) can be better understood from Fig. 1 which illustrates the spectrum of an electron in the field of a heavy nucleus. Our approach is conceptually very simple: we integrate the single-particle, time-dependent Dirac equation

$$H_D\psi = i\,\frac{\partial\psi}{\partial t} \tag{3}$$

where H_D contains the external vector potential of the moving nuclei. Initially ψ corresponds to a 1s state; at the end of the collision, we project onto the discretized spectrum of the final state Hamiltonian to obtain the probabilities for (1) or (2),

$$p_n = \left|\langle\phi_n|\psi\rangle\right|^2 \tag{4}$$

This process is explained in detail in Section 5 below. We must stress that the time integration is achieved by representing ψ as a superposition of space localized finite elements: until the last step (4), we do not deal with a basis of eigenfunctions. It is also worth repeating that the positron production (2) is computed by time reversing the physical process.

We conclude this section by noting that we shall use a system of units natural to the Dirac equation:

energy	mc^2	511 keV
length	αa_0	386 fm
time	$\alpha^2 t_0$	1.21×10^{-21} s

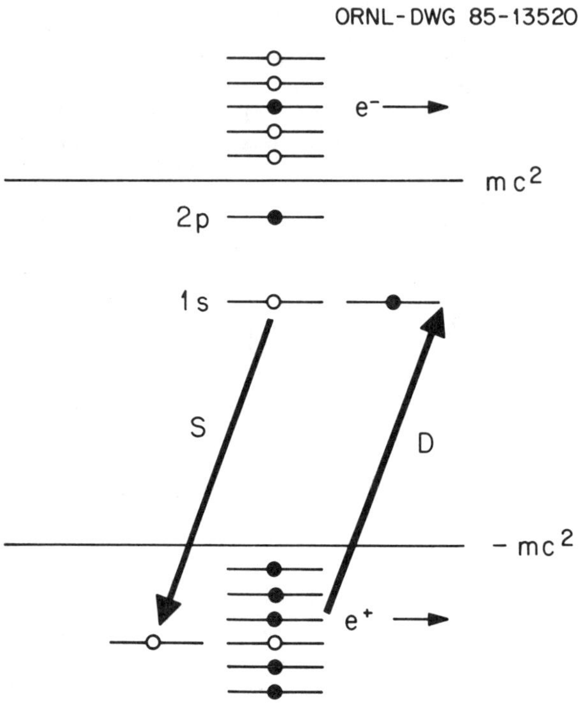

Fig. 1. Energy spectrum of particle and hole states for an electron in the field of a heavy nucleus. The arrows S and D distinguish schematically the spontaneous and dynamical mechanisms of positron production (Ref. 1).

Then the equations can be written with m=c=h=1; however, we sometimes display m in the interests of clarity. The above length is the Compton wavelength of the electron, and the time unit is also natural for nuclear heavy-ion processes. This reflects in part the relativistic collapse of the electronic orbit to nuclear dimensions in the presence of very strong fields. Figure 2 shows the three length scales which enter pair production problems for a U atom: the radius of the U nucleus, the Compton wavelength of the electron, and the radius of the U K-shell. When an electron approaches within αa_0 of a heavy nucleus, its potential energy becomes comparable with mc^2. As two heavy nuclei are brought together slowly, the K-shell radius collapses from a few αa_0 to ~0.2 αa_0, nearly the size of the nucleus.

PAIR PRODUCTION RESULTS AT A FEW MeV/u

We shall not labor the results of our earlier calculations at collision velocities ~0.1c since they are adequately described in

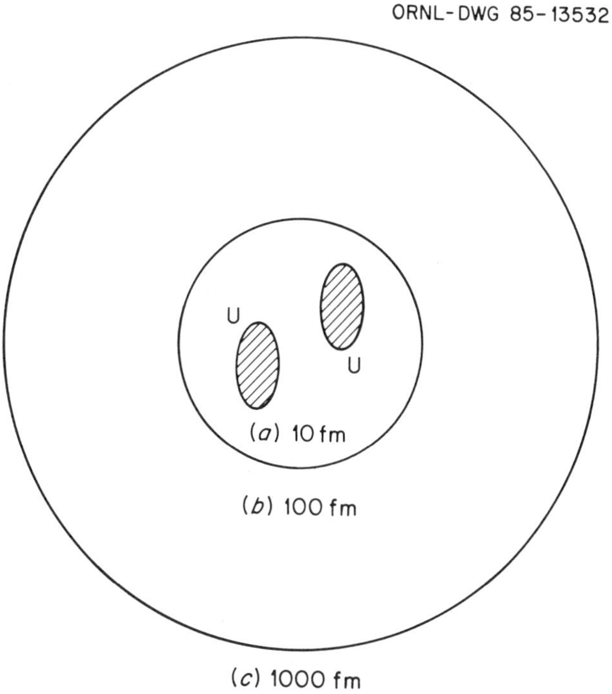

Fig. 2. Length scales in a collision between two heavy ions: (a) nuclear radius; (b) electron Compton wavelength; (c) radius of uranium K-shell.

Ref. 2 and in a more extensive forthcoming article. However, we feel it is worthwhile to highlight the undoubted insights derived from our numerical approach.

Our Hamiltonian is based on the monopole approximation which is well established for relative velocities up to ~0.3c. Since electron and positron production are dominated by the rapid collapse of the K-shell when the nuclei approach within 0.3 αa_0, it is a good approximation to expand the projectile and target interactions about some point on the internuclear axis and retain only the monopole terms. At higher velocities the retarded electrostatic and magnetic interactions break the monopole symmetry and indeed all the symmetries of three-space dimensions. We have been able to confirm many of the earlier predictions of the Frankfurt group at the 20% level of accuracy, a significant achievement in view of the vast differences in the two approaches.

The numerical method has enabled us to unify the apparently physically distinct mechanisms of spontaneous and dynamical positron production. In the spontaneous mechanism we imagine the 1s hole energy adiabatically depressed to an energy $E < -mc^2$: an electron in the Dirac sea then spontaneously decays into the depressed state creating a positron of energy $|E|$ (see Fig. 1). This decay has a natural width of about one KeV. In the dynamical mechanism the time Fourier components of the perturbation induce transitions from the Dirac sea to the 1s hole. However, if we imagine that the collision time is tuned from a small value to values which are comparable to or larger than the spontaneous decay time of about 10^{-20} sec, then the dynamical process must slide continuously into the spontaneous. In the adiabatic basis method[1] the 1s hole must be represented as a decaying state whose rate of decay is separately calculated. By contrast both limits emerge natually from the numerical solution of the time-dependent Dirac equation. We can demonstrate how this works by considering a head-on U + Cm collision at 6.05 MeV/u.

In Fig. 3c is plotted the perturbation in the wavefunction (initially a Cm 1s state) at the distance of closest approach as a function of the distance from the Cm nucleus. Figure 3d shows the result of introducing a time delay by simply holding the nuclei fixed for a period $T = 50\ \alpha^2 t_0$: the localized state has decayed into an outgoing Coulomb wave at large distances. In Fig. 3a,b are plotted the amounts of probability in the continua $E > mc^2$, $< -mc^2$ as a function of time: these are calculated by projecting on eigenstates of the final-state Hamiltonian. We see that while the nuclei are fixed, electron probability is converted into positron probability which is observed as a sharp spike in the positron emission spectrum. Figure 4 shows the positron spectrum as a function of our "ad hoc" time delay T. The broad hump at T = 0 evolves continuously into an oscillatory structure and ultimately a single sharp peak. The merits and shortcomings of the long-lived nuclear state hypothesis are discussed elsewhere in these proceedings. We merely note here that the peak position is sensitive to the impact parameter and target charge (Fig. 4d).

Figures 3 and 4 make a further point which probably applies to all energetic ion-atom collisions. The wavefunctions are highly localized in space and present a rather simple picture which is easy to interpret. As a corollary, the Fourier components of the wavefunction are spread over a wide range of momenta and present a complicated picture. For these reasons, we believe that numerical methods which represent the wavefunction in space offer the best hope for future progress.

RELATIVISTIC ION-ATOM COLLISIONS

Consider a system of electrons moving under the influence of an external current provided by the motion of one or more heavy nuclei. The electron field operator satisfies the Dirac equation

$$[\vec{\alpha}\cdot(\vec{p}+\vec{A}) + \beta - A_0]\psi = i\,\frac{\partial\psi}{\partial t} \qquad (5)$$

while the electromagnetic four-potential satisfies

ORNL-DWG 84-13898

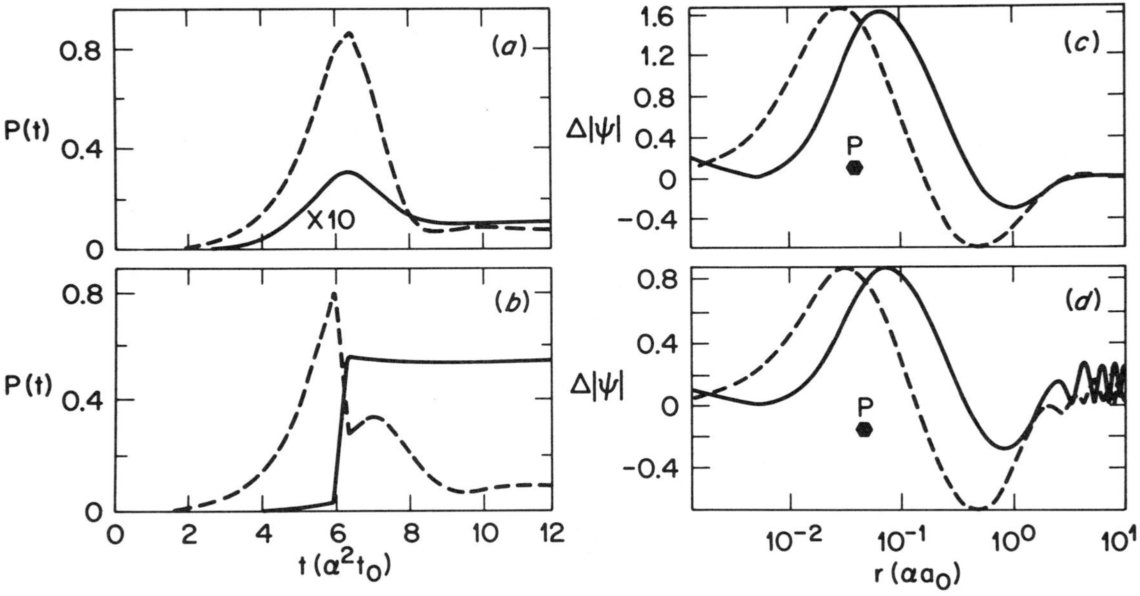

Fig. 3. U + Cm head-on collision at a bombarding energy of 6.05 MeV per particle. (a),(b): time evolution of the positron (full line) and electron (dotted line) probabilities in the finite element basis, respectively with and without, a time delay of T = 50 $\alpha^2 t_o$ occurring at t = 6 $\alpha^2 t_o$. (c),(d) shows the change from the initial state in the modulus of the 1s wavefunction vs. the distance from the target nucleus before and after a time delay of T = 50 $\alpha^2 t_o$; full and dotted lines refer to the large and small wavefunction components, G and F. P denotes the position of the projectile nucleus.

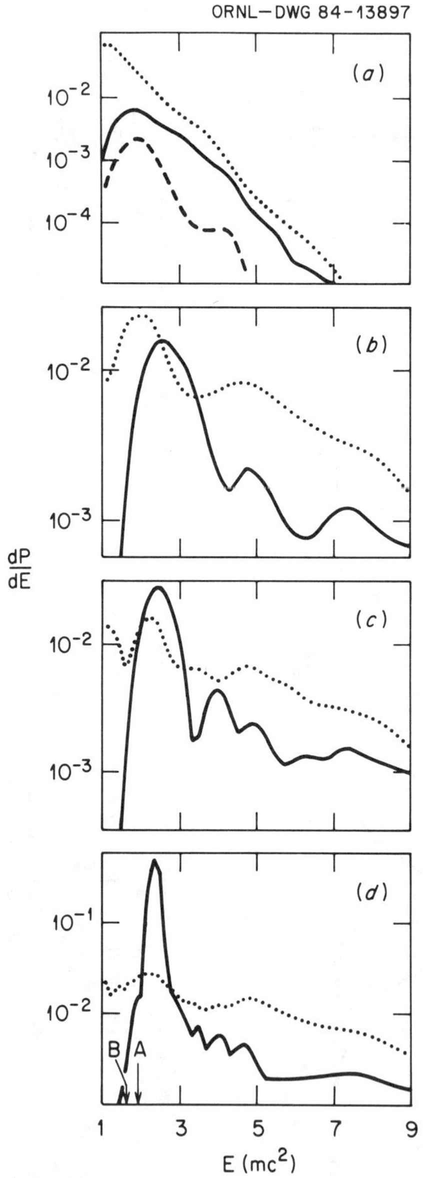

Fig. 4. U + Cm head-on collision at a bombarding energy of 6.05 MeV per particle. Positron (full line) and electron (dotted line) spectra in the target frame, in units of $1/mc^2$ per collision, for (a): no time delay; (b), (c), (d) respectively T = 2.5, 5, 20 $\alpha^2 t_o$. The positron spectrum for b=19 fm is shown by the dashed line in (a). In (d) the arrows A,B indicate the peak position in a head-on U+Pu collision, and a U+Cm collision at b = 19 fm (both at the same bombarding energy.

$$\Box\, A_\mu = \vec{J}^{ext} + \psi^\dagger \alpha_\mu \psi \qquad (6)$$

where ψ and A are field operators. For the appropriate choice of gauge, (5) and (6) contain most of quantum electrodynamics. For electrons moving independently in strong external fields, we are justified in treating A as a classical field, using a wavefunction expansion for ψ, and the expectation of the current operator on the righthand side of (6).

The first step in solving (5) and (6) is to retain only the external source as the righthand side of (6). Then we have only to solve the one-particle Dirac equation with a prescribed retarded potential. In a subsequent section we shall present the results of such calculations for a one-dimensional (1D) model. The next step would be to restore the mean lepton current on the righthand side of (6) and propagate A self-consistently with ψ, thus obtaining a time-dependent Dirac Hartree (TDDH) theory. We shall outline the algorithms needed for such a program, though calculations are not yet complete.

Of necessity, (5) and (6) must be solved numerically. For the reasons given in Section 2, the numerical procedure will represent the wavefunctions and the potentials on a spatial mesh. We shall then demand that the algorithms satisfy two criteria which we shall term fidelity and plausibility. Fidelity means that the finite numerical representation of (5) and (6) goes uniformly to the continuous limit as the density of mesh points becomes infinite. A less formal statement is that the discrete equations must have an intuitive correspondence with their continuous counterparts. Plausibility means that the discrete equations imply well-behaved finite analogues of the correspondence principle for the Dirac fields

$$(p+A)^2 \ \psi = \psi \tag{7}$$

with current conservation and the gauge condition,

$$\partial^\mu J^{tot}_\mu = 0 \quad , \quad \partial^\mu A_\mu = 0. \tag{8}$$

Our view is that (7) and (8), as well as (5) and (6), form the set of discrete equations which must all be satisfied in order to obtain a plausible physical interpretation.

In order to test our numerical methods and obtain some feeling for the physics of relativistic heavy-ion collisions, we have introduced a 1D simplification of (5)-(8). The target T is at rest while the projectile P moves at a velocity v. The target and projectile potentials in their rest frames are cut off Coulomb fields

$$V_C = - \frac{Q_C}{\left(x^2 + a_C{}^2\right)^{1/2}} \ (C = T,P). \tag{9}$$

The wavefunction reduces to a two-component spinor associated with the Hamiltonian

$$H_D = \begin{bmatrix} m - A_o & -\frac{\partial}{\partial x} - iA_x \\ \\ \frac{\partial}{\partial x} + iA_x & -m - A_o \end{bmatrix} \tag{10}$$

where the potentials are given by

$$A_o = -V_T(x) - \gamma \ V_p[\gamma(x-vt)] \tag{11}$$

$$A_x = \gamma \ v \ V_p[\gamma(x-vt)] \tag{12}$$

To save writing indices, we shall write the components of the two potentials as $\left(A_o,A_x\right) = (\phi,a)$ and the components of the two-current $\left(J_o,J_x\right) = (\rho,j)$ where ρ is the normal electron density.

$$\rho = \psi^{\dagger}\psi. \tag{13}$$

From (10) we deduce the conservation of the current

$$\frac{\partial\rho}{\partial t} = -\frac{\partial j}{\partial x} \tag{14}$$

where

$$j = \psi^{\dagger}\begin{pmatrix} 0 & -i \\ i & 0 \end{pmatrix}\psi = -i(j_{+} - j_{-}) \tag{15}$$

We have indicated that the current is naturally written as the sum of two complex entities. Then the vector potential satisfies,

$$\left(\frac{\partial^2}{\partial t^2} - \frac{\partial^2}{\partial x^2}\right)a = j^{ext} + j. \tag{16}$$

A similar equation relates ϕ and ρ, whence the gauge condition follows,

$$\frac{\partial\phi}{\partial t} = -\frac{\partial a}{\partial x}. \tag{17}$$

Our next task is to derive a faithful and plausible numerical representation of (10) and (16).

BASIS SPLINE COLLOCATION METHODS

We shall expand the unknown functions (e.g. ψ) in basis splines (often called B-splines) and determine the expansion coefficients by collocation. This combination of techniques represents the state of the art for solving partial differential equations in other fields.[4,5] At this time, we shall only outline the method: details will be given elsewhere.

As an exercise, consider the 1D non-relativistic Schroedinger equation,

$$[T + V(x)]\psi = H\psi = i\,\frac{\partial\psi}{\partial t} \tag{18}$$

We choose a set of points $\{x_j\}$ $\{1 \leq j \leq n\}$ known as "knots" and on the subset $(k \leq j \leq k+N-1)$ construct a piecewise continuous and (N-1) differentiable polynomal, which is zero outside the interval. This function $u_k^{(N)}(x)$ is the N-th order basis spline: Fig. 5 shows a quadratic (N=3) function. The linear superposition

ORNL-DWG 85-13531

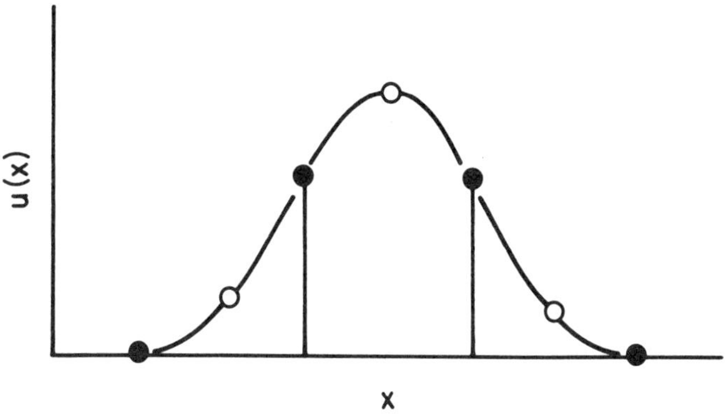

Fig. 5. Quadratic (third order) basis spline. The knots are indicated by the filled circles and the collocation points by the open circles.

$$\psi(x) = \sum_{k=1}^{n} \psi^k u_k(x) \tag{19}$$

provides a spline approximation to ψ on an interval spanned by the subset of the knots. In the collocation method we demand that (18) be satisfied at a set of points $\{\xi_\alpha\}$ ($i \leq \alpha \leq n$) known as "data" or "collocation" points. The choice

$$\xi_\alpha = \frac{1}{2}\left(x_{\alpha+\mu} + x_{\alpha+\mu+1}\right),\ \mu = [N/2] \tag{20}$$

is adequate for N odd. The number of collocation points must equal the number of functions in (19), so that the coefficients may be eliminated in favor of the values of ψ at the collocation points,

$$\psi_\alpha = \psi(\xi_\alpha) = \sum_k \psi^k B_{\alpha k}. \tag{21}$$

The matrix B acts like a metric,

$$B_{\alpha k} = u_k(\xi_\alpha) \quad , \quad B^{k\alpha} = [B^{-1}]_{k\alpha} \tag{22}$$

$$\psi^k = \sum \psi_\alpha B^{k\alpha}. \tag{23}$$

Inserting (19) in (18) and eliminating ψ^k, we find the collocation representation of the Schroedinger equation,

$$\sum_\beta H_\alpha{}^\beta \psi_\beta = i \frac{\partial \psi_\alpha}{\partial t} \tag{24}$$

$$H_\alpha{}^\beta = T_\alpha{}^\beta + V_\alpha{}^\beta \quad , \quad V_\alpha{}^\beta = \delta_{\alpha\beta} V(\xi_\alpha) \tag{25}$$

$$T_\alpha{}^\beta = -\frac{1}{2} \sum_k B''_{\alpha k} B^{k\beta} \quad , \quad B''_{\alpha k} = u''_k(\xi_\alpha). \tag{26}$$

We immediately note that local operators, e.g. V, are represented by diagonal matrices of their values at the collocation points: this apparently trivial result coexists with the more sophisticated functional representation (19). The complexities arise in the kinetic energy matrix, though the non-sparse form of (26) is actually an advantage, in that the metric B spreads the large elements of B over a wide band on either side of the diagonal. As a consequence, the largest eigenvalue of H is much smaller than in other expansion methods. We should also remark that no overlap matrix appears on the righthand side of (24). However, the matrix H is, in general, unsymmetric, requiring the introduction of both direct and adjoint solutions. Most theorems used for symmetric self-adjoint methods have analogs for collocation methods; however, they are somewhat more difficult to state and prove.

We shall jump ahead a little and write down the adjoint problem before considering the Dirac equation in detail. Using matrix notation (24) becomes

$$i \frac{\partial}{\partial t} \underset{\sim}{\psi} = \underset{\sim}{H} \underset{\sim}{\psi} \tag{27}$$

The adjoint eigenvector is defined as the solution of

$$i \frac{\partial}{\partial t} \underset{\sim}{\phi} = \underset{\sim}{H}^\dagger \underset{\sim}{\phi}. \tag{28}$$

It readily follows that the norm

$$Q = \underset{\sim}{\phi}^\dagger \underset{\sim}{\psi} = \sum \phi^*_\alpha \psi_\alpha \tag{29}$$

is conserved in time: conventionally $Q = 1$. The expectation of any operator is given by

$$\langle\Omega\rangle = \underset{\sim}{\phi}^\dagger \underset{\sim}{\Omega} \underset{\sim}{\psi} = \sum \phi^*_\alpha \Omega_\alpha{}^\beta \psi_\beta \tag{30}$$

where $\Omega_\alpha{}^\beta$ is constructed according to the prescription

$$\Omega_\alpha{}^\beta = \sum \Omega_{\alpha k} B^{k\beta} \quad , \quad \Omega_{\alpha k} = (\Omega u_k)_{\xi_\alpha} \tag{31}$$

of which (24) is a special case. In particular, if Ω is a local function of position,

$$\Omega_{\alpha}{}^{\beta} = \delta_{\alpha\beta}\Omega(\xi_{\alpha}) \tag{32}$$

A general feature of collocation methods is illustrated by (29), that regardless of the basis or knot sequence, the metric for constructing the norm is just a Kronecker delta function. It is a further corollary that the particle density associated with each collocation point is given by

$$\rho_{\alpha} = \phi^{*}_{\alpha}\psi_{\alpha} \tag{33}$$

To propagate (27) over a small time interval, we use an exponential propagator constructed by means of a power series,

$$\underset{\sim}{\psi}(t+\tau) = \exp\left[-i\tau\underset{\sim}{H}\left(t+\frac{1}{2}\tau\right)\right]\underset{\sim}{\psi}(t). \tag{34}$$

With a reasonable choice of τ, there is no difficulty in satisfying (29) to one part in 10^{12}.

Finally we point out that it is a simple matter to extend the foregoing formalism to more than one dimension. The expansion (19) now uses products, e.g. in 2D

$$\psi(x,y) = \sum \psi^{k_1k_2} u_{k_1}(x)u_{k_2}(y) \tag{35}$$

these are sometimes called "tensor" splines.[4] The Hamiltonian (25) and (26) is now generalized to

$$\begin{aligned} H_{\alpha_1\alpha_2}{}^{\beta_1\beta_2} &= \left(T_x\right)_{\alpha_1}{}^{\beta_1}\delta_{\alpha_2\beta_2} + \left(T_y\right)_{\alpha_2}{}^{\beta_2}\delta_{\alpha_1\beta_1} \\ &+ V\left(\xi_{\alpha_1},\xi_{\alpha_2}\right)\delta_{\alpha_1\beta_1}\delta_{\alpha_2\beta_2} \end{aligned} \tag{36}$$

Again the potential energy is completely diagonal while the kinetic energy breaks into small blocks. The structure of (36) is exceedingly simple when compared to other methods of comparable accuracy.

APPLICATION TO THE DIRAC EQUATION

Unfortunately it is not possible to apply directly to (10) the prescription which led to (26). This would lead head on to the "Fermion Doubling" pathology,[6] which we will now briefly explain. Suppose we write the Dirac equation corresponding to (10) for a free particle in terms of the large and small components, $\psi = (G,F)$,

$$G' = (m+E)F \quad , \quad F' = (m-E)G. \tag{37}$$

If the derivatives are replaced by the simplest finite difference formulae on a mesh $x = jh$ $(1 \leqslant j \leqslant n)$

$$(G')_j = \left(G_{j+1} - G_{j-1}\right)/2h \tag{38}$$

we have to solve the difference equation

$$G_{j+2} - 2G_j + G_{j-2} = 4h^2 (m^2 - E^2)G_j \tag{39}$$

A solution of (39) is given by

$$G_j = \exp\left(ik_\lambda x_j\right) \quad , \quad k_\lambda = \frac{\lambda\pi}{nh} \tag{40}$$

where we have imposed periodic boundary conditions. However, the eigenvalue

$$E_\lambda^2 = m^2 + \frac{1}{h^2} \sin^2 \left(\frac{\lambda\pi}{n}\right) \tag{41}$$

is also associated with another solution

$$\tilde{G}_j = (-1)^j G_j \tag{42}$$

of unphysical momentum $k_\lambda + (\pi/h)$.

The doubled spectrum (42) appears in all numerical formulations. For time integrations its existence has serious consequences: high momentum components appear more or less at random and grow exponentially. In a stationary problem the spurious solutions can be avoided by eliminating the small component and discretizing an equivalent Schroedinger equation. This is usually not practical for time evolution problems, and in any case, does not extirpate the problem at its root. Our analysis is that (39) is unacceptable as a mass-energy dispersion relation, since it has unphysical solutions. The discrete Dirac equation should be a factorization of a correct dispersion relation,

$$G_{j+1} - 2G_j + G_{j-1} = h^2(m^2 - E^2)G_j \tag{43}$$

This is one part of our requirement of plausibility introduced in Section 3.

Our first attempt to address this problem[2] consisted in using forward and backward differences to discretize the two parts of (37),

$$F_{j+1} - F_j = h(m-E)G_j \quad , \quad G_j - G_{j-1} = h(m+E)F_j . \tag{44}$$

Thus hermiticity is preserved and (43) follows. It might be thought that a generalization of (44) to an arbitrarily complicated discretization scheme, e.g. that described in Section 4, would be difficult to find. We have been fortunate to arrive at a prescription which is entirely satisfactory in practice. Given a representation of the non-relativistic kinetic energy, e.g. (26), we make an LU decomposition (lower-upper triangular),

$$T = -\frac{1}{2} \underset{\sim}{D}_- \underset{\sim}{D}_+ \tag{45}$$

The decomposition is uniquely specified if

$$D^-_{\alpha\alpha} = \left| D^+_{\alpha\alpha} \right| \tag{46}$$

The matrices $D_\pm$ are those used to represent the first derivative in (10). A special case is provided by (44), since it is easily verified (using an obvious notation for tridiagonal matrices) that

$$(1,-2,1) = (-1,1,0)(0,-1,1). \tag{47}$$

Among the possible decompositions, (45) is favored because it always gives a faithful representation of the first derivatives (as in (47)), in the sense of Section 4. We shall return to this point below. It is useful to remark that the elements of $D_\pm$ in the first and last rows define boundary conditions, though we do not have space to go into more detail here.

The basis spline collocation representation of (10) is now constructed from (25) and (26). The potentials are diagonal matrices of their values at the collocation points. The kinetic energy (26) is factorized according to (45) to get the derivative matrices. Then

$$\underset{\sim}{H}_D = \begin{bmatrix} m-\underset{\sim}{A}_o & -\underset{\sim}{D}_+ - i\underset{\sim}{A}_x \\ \underset{\sim}{D}_- + i\underset{\sim}{A}_x & -m-\underset{\sim}{A}_o \end{bmatrix}. \tag{48}$$

Propagation in time is accomplished by applying (34) to (27) and (28).

We suggested in Section 3 that conservation of flux (14) is an important part of solving the complete problem. This requires an appropriate definition of the current, which in turn requires further discussion of the $D_\pm$ matrices. If these are faithful representations of the first derivative, each column sum must be approximately zero (except for the column containing only one non-zero element): we can show that if the diagonal elements are slightly modified to make the column sum exactly zero, these matrices acquire unusual properties. An algorithm, such as (34), which employs ascending powers of $D_\pm$ is hardly affected by the modification: inverse powers are a different matter. Consider the sequence of vectors defined by

$$\underset{\sim}{D}^T_+ \, \underset{\sim}{\omega}^+_{\mu+1} = \underset{\sim}{\omega}^+_\mu \;, \quad \left(\omega^+_o\right)_\alpha = \delta_{\alpha 1} \tag{49}$$

with the new definition, it is identically true that

$$\left(\omega^+_1\right)_\alpha = 1 \tag{50}$$

while, to a good approximation,

$$(\omega^{+}_{\mu+1})_{\alpha} \simeq \xi^{\mu}_{\alpha}/\mu! . \tag{51}$$

Analogous relations may be derived for D_-. The content of (50) is that the inverse of differentiation is integration with the metric (29); (51) generalizes this statement to iterated integrals (Dirichlet's formula). For physics, the significant consequence of (49)-(51) is that the manipulations leading to conservation laws (Green's theorem, etc.) now have a rigorous analog on a finite mesh. The key lemma deduced from (49) is that

$$(\tilde{\omega}^{+}_{\mu+1})^{T}\tilde{f} = (\tilde{\omega}^{+}_{\mu})^{T}\tilde{g} \quad \text{if } \tilde{f} = \tilde{D}_{+}\tilde{g}. \tag{52}$$

Some examples will be given below.

To derive a current operator, we begin with the density (33), which is a vector in collocation space whose components may be written

$$\rho_{\lambda} = \tilde{\phi}^{\dagger}\tilde{\Pi}_{\lambda}\tilde{\psi} \quad , \quad (\Pi_{\lambda})_{\alpha\beta} = \delta_{\lambda\alpha}\delta_{\lambda\beta} \tag{53}$$

Then from (27), (28), (48), and (53) we find that

$$\frac{\partial\rho_{\lambda}}{\partial t} = i(f^{+}_{\lambda} - f^{-}_{\lambda}) \tag{54}$$

where

$$f^{\pm}_{\lambda} = \tilde{\phi}^{\dagger}_{\substack{G\\F}} [\tilde{\Pi}_{\lambda},\tilde{D}_{\pm}]\tilde{\psi}_{\substack{F\\G}} \tag{55}$$

(G, F again refer to large, small components; upper, lower subscripts to +, -.) If we define current vectors (in general complex) by

$$\tilde{j}_{\pm} = \tilde{D}^{-1}_{\pm}\,\tilde{f}_{\pm} \tag{56}$$

(14) and (15) are replaced by the discrete flux conservation theorem,

$$\dot{\tilde{\rho}} = i(\tilde{D}_{+}\tilde{j}_{+} - \tilde{D}_{-}\tilde{j}_{-}) \tag{57}$$

From (52) and (57) it follows that

$$\frac{\partial}{\partial t}(\tilde{\omega}^{T}_{1}\tilde{\rho}) = i(j^{+}_{1} - j^{-}_{n}). \tag{58}$$

We have examined the case of a bound system moving with a uniform velocity in the observer's frame, for which the form of the current operator is the density times the velocity. The prescription (55)-(57) reproduces the correct result to better than 1% for a reasonable mesh spacing providing (50) is satisfied.

The final mark of plausibility is the gauge condition (17). Suppose (16) has the finite representation

$$\left(\frac{\partial^2}{\partial t^2} - \underset{\sim}{\Delta}^2\right)\underset{\sim}{a} = \underset{\sim}{j}(t). \tag{59}$$

In a 2D or 3D model problem Δ^2 is a sum of matrices: the notation does not imply that an explicit form of Δ is available. To propagate (59) in time, we invoke the solution of a driven harmonic oscillator, so that

$$\begin{aligned}\underset{\sim}{a}(\tau) &= \underset{\sim}{\Delta}^{-2}\,[\cosh(\underset{\sim}{\Delta}\tau)-1]\underset{\sim}{j}(\tau/2)\\ &\quad + \cosh(\underset{\sim}{\Delta}\tau)\underset{\sim}{a}(0) + \underset{\sim}{\Delta}^{-1}\,\sinh(\underset{\sim}{\Delta}\tau)\dot{\underset{\sim}{a}}(0)\end{aligned} \tag{60}$$

with error $O(\tau^3)$ over a small interval τ. The operators in (60) are all power series in Δ^2, which may thus be applied in 2D or 3D.

Let us define

$$\underset{\sim}{a} = -i\left(\underset{\sim}{a}_+ - \underset{\sim}{a}_-\right) \tag{61}$$

where $a_\pm$ are generated by the sources $j_\pm$ using representations $\Delta_\pm^2$ of the Laplacian; ϕ is generated from ρ. Then by analogy with (57),

$$\underset{\sim}{\phi} = i\left(\underset{\sim}{D}_+\underset{\sim}{a}_+ - \underset{\sim}{D}_-a-\right) \tag{62}$$

if and only if

$$\left[\underset{\sim}{D}_\pm, \underset{\sim}{\Delta}_\pm^2\right] = 0. \tag{63}$$

The simplest way of satisfying (63) is just to make Δ identical with D, i.e. to make $a_\pm$ satisfy

$$\left(\frac{\partial^2}{\partial t^2} - \underset{\sim}{D}_\pm^2\right)\underset{\sim}{a}_\pm = \underset{\sim}{j}. \tag{64}$$

Applying (52), we obtain a discrete form of Gauss's theorem: in the static limit

$$\omega_1^T j_+ = -\left(D_+a_+\right)_1 \quad , \quad \omega_1^T j_- = -\left(D_-a_-\right)_n . \tag{65}$$

In 2D or 3D, $\Delta_\pm^2$ would have a block form similar to (36), but still lower or upper triangular. This LU property makes the static analog of (64) vastly more attractive than Possion's equation discretized in a conventional manner.

ANALYSIS OF FINAL STATE PROBABILITIES

We have made calculations on the 1D model explained in Section 3. Our procedure is to solve (3) numerically and then project on a set of final eigenstates to obtain the probabilities (4). The geometry of a collision is illustrated in Fig. 6. The target is fixed at T, while the projectile starts at P_o and ends at P. The mesh extends from x=-L to x=L. In the final configuration we project on eigenstates of the target and moving projectile separately. Though everyone has a clear intuitive picture of bound and continuum states associated with either the target or projectile, the mathematical and computational description of these states has always been problematic. We translate our intuition directly by introducing a spatial projection of the final-state wavefunction into parts associated with the target and projectile,

$$\psi_T = \Omega(-x)\psi \quad , \quad \psi_P = \Omega(x)\psi \tag{66}$$

where

$$\begin{aligned} \Omega(x) &= 0 & x < 0 \\ &= 1 & x > L\theta \\ &= \frac{1}{2}\left[1-\cos\left(\frac{\pi x}{L\theta}\right)\right] & 0 < x < L\theta \end{aligned} \tag{67}$$

ORNL-DWG 85-13530

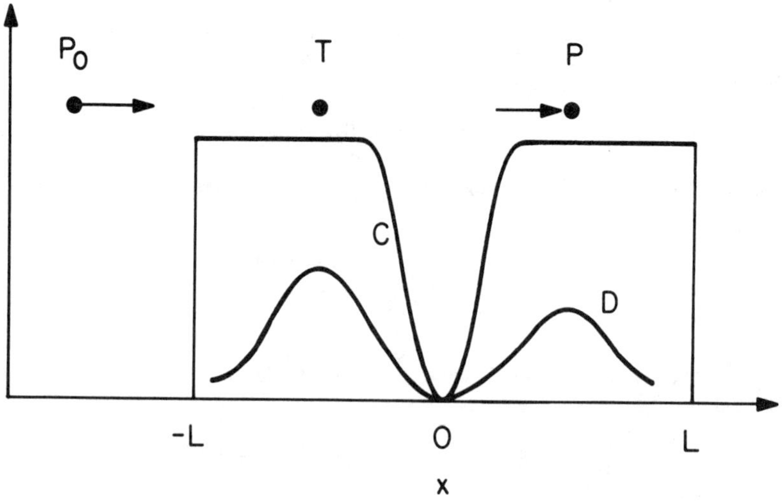

Fig. 6. Geometry of a 1D collision. We show the position of the target (T) and the initial and final positions of the projectile (P_o,P) relative to the box -L < x < L. D is a schematic final state which has been folded into the cosine bell C.

The "cosine bell" shape of Ω minimizes spurious Fourier components due to discontinuties at $x = 0$; we normally take $\theta = 0.2$.

The target probabilities are calculated by projecting ψ_T on the discrete eigenstates of the target alone (spanning both bound and continuum states) calculated on the part of the mesh spanning $-L < x < 0$. Projectile probabilities are calculated by projecting on the discrete eigenstates of the moving projectile alone, calculated on the part of the mesh spanning $0 < x < L$.

The discretization techniques explained in Sections 4 and 5 reduce the stationary Dirac equation in the observer's (fixed) frame to a matrix eigenvalue problem. To obtain the projectile eigenstates in a frame moving with the projectile, requires that we introduce the Lorentz transformation of the Dirac spinor.[7] Let (x',t') denote the frame moving with a velocity v and (x,t) the observer's frame: the Lorentz transformation is given by

$$x' = \gamma(x-vt) \quad , \quad t' = \gamma(t-vx) \tag{68}$$

where

$$\gamma = (1-v^2)^{-1/2} = \cosh\omega. \tag{69}$$

A spinor $\psi(x,t)$ in the moving frame appears to the observer as

$$\psi' = S(-v)\psi(x't') \tag{70}$$

where

$$S(v) = \exp\ (1/2\ \omega\alpha) = \cosh\ \omega/2 + \alpha\ \sinh\ \omega/2 \tag{71}$$

and in the representation (10)

$$\alpha = \begin{pmatrix} 0 & -i \\ i & 0 \end{pmatrix} \tag{72}$$

Thus consider a stationary state in the moving frame which satisfies

$$H_D' \phi_o = \varepsilon_o \phi_o \tag{73}$$

This appears as

$$\psi' = \exp\left[-i\varepsilon_o\gamma(t-vx)\right]S(-v)\phi_o\left[\gamma(x-vt)\right]. \tag{74}$$

In the non-relativistic limit, $\gamma \to 1$ and $\varepsilon_o \to 1$, to $O(v\ )$, so we recover the familiar result

$$\psi_{NR}' = \exp(ivx)\phi_o(x-vt)\exp(-it). \tag{75}$$

To obtain a consistent finite representation of (74), we introduce the "boosted" Dirac equation

$$[H_D(x',p) - vp]\psi' = [\varepsilon_o/\gamma]\psi'. \quad (76)$$

The projectile eigenstates are obtained by setting up the usual discrete representation of (76) in the fixed frame and solving the resulting matrix eigenvalue problem.

Before leaving this section, we should clear up some technicalities associated with the projection (4). If the wavefunction is propagated in the direct representation (27), we must solve the eigenvalue problem in the adjoint representation,

$$\underset{\sim}{H}_D^{fT} \underset{\sim}{\phi}_n^f = \varepsilon_n \underset{\sim}{\phi}_n^f \quad (77)$$

where f specifies that the final state of either the projectile or target is being considered; we will omit the superscript henceforth in the interests of clarity. The final state probabilities are given by

$$p_n = \left|\phi_n^\dagger \psi\right|^2 \quad (78)$$

if the norm (29) is unity.

If ε_n corresponds to a bound state, the interpretation of (78) is clear. For ε_n in the "box" continuum, we construct a differential probability by interpolating

$$\frac{dp}{d\varepsilon}\left[\varepsilon = \frac{1}{2}\left(\varepsilon_{n+1} + \varepsilon_n\right)\right] = \frac{\frac{1}{2}\left(p_{n+1} + p_n\right)}{\left(\varepsilon_{n+1} - \varepsilon_n\right)}. \quad (79)$$

We further want to distinguish between electrons moving to the right or left. This can be done by observing that the box continuum states tend to be alternately of approximately even or odd character with respect to reflection about the center of the box. By combining these pairs, we can make up eigenfunctions which satisfy

$$\hat{R}\,\psi^{\pm} = \left(\psi^{\pm}\right)^* \quad (80)$$

to a good approximation, $\hat{R}$ being the space reflection operator. Thus, if n = 2m, we diagonalize R in the basis $\left(\phi_n, \phi_{n+1}\right)$ to obtain

$$\hat{R}\,\phi_m^c \simeq \phi_m^c \;, \quad \hat{R}\,\phi_m^s \simeq -\phi_m^s \quad (81)$$

whence

$$\psi_m^{\pm} = \frac{1}{\sqrt{2}}\left(\phi_m^c \pm i\phi_m^s\right) \quad (82)$$

are approximately eigenstates of positive and negative momentum, respectively.

NUMERICAL RESULTS FROM THE ONE-DIMENSIONAL MODEL

We now describe calculations on the model defined by (9)-(12) with the specific choice of parameters:

$$Q_T = 1.0,\ Q_P = 0.9;\ a_T = a_P = 0.75. \qquad (83)$$

The Dirac equation was discretized using the methods explained in Sections 4 and 5, particularly (45)-(48). Propagation in time was accomplished using (34). The interval $-15 < x < 15$ was covered by 150 equally spaced knots. Most of the calculations employed quadratic splines; some checks with quartic splines did not show any serious change in the numbers. The time step τ was chosen so that $0.01 > v\tau > 0.003$.

Accuracy was investigated by studying an isolated moving atom and by considering various meshes. The norm (29) was always conserved to better than one in 10^{10}, from which we conclude that propagation in time is not a source of error for practical purposes. Sensitivity of probabilities to the mesh spacing was 1-5%. However, sensitivity to the size of the box was 10-30%. The larger estimates apply to smaller probabilities. We conclude that efforts to improve accuracy should focus on the location of the finite boundary. It is worth remarking that, by comparison to the earlier studies using finite elements, the solutions exhibited very little noise outside the physical interaction region. It appears that this sensitivity to boundary conditions is an artifact of the primitive discretization algorithms.

The results will be more readily understood if we consider the momentum distribution in the initial (ground) state of the target. Figure 7 shows this distribution for the positive and negative energy components of the 1s state of Cm; our model ground state has very similar features. For the positive components, the mean momentum $\overline{p} = 0.76$ corrsponding to $\beta = 0.6$ (130 KeV, 240 MeV/u beam energy); for negative components $\overline{p} = 1.27$ whence $\beta = 0.8$ (320 KeV, 580 MeV/u beam energy). Thus, symmetric charge capture by a bare nucleus of similar charge will take place where the relative velocities of the electron components match, i.e. about $\beta = 0.6$. Figure 7 shows why the spontaneous emission of positrons is dominant for $\beta < 0.4$: here no matching momentum components are available. However, the dynamical emission, particularly in the projectile frame is increasingly probable above $\beta = 0.4$ and persists to very high energies. The energy spectrum of the spontaneous positron emission (Fig. 4) is essentially the negative energy distribution from united nuclei, so that the resemblence to Fig. 7 is not coincidental. If the "box" is defined by $-L < x < L$, where $L \sim 15$ (the units are those explained in Section 2), we conventionally take the target nucleus to be fixed at $-L/2$, while the projectile moves from $x = -3L/2$ to $+L/2$ in the course of a collision (Fig. 6). Figures 8, 9, and 10 show the squared components $|G|^2$ and $|F|^2$ of the spinor ψ as functions of x for the projectile velocities $v = 0.4, 0.8, 0.98$ ($\psi = 1.09, 1.67, 5.03$; bombarding energy = 0.08, 0.63, 3.8 GeV/u). The evolution of the wavefunction as the collision proceeds is indicated in three frames, which

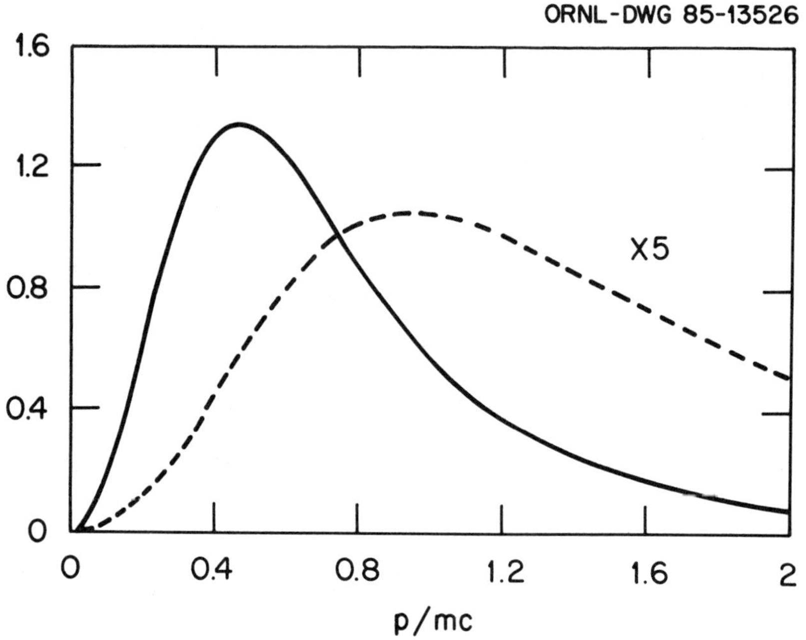

Fig. 7. Momentum distributions for the positive (full line) and negative (dotted line) energy components of the 1s orbital of Cm.

also show the corresponding position of the projectile. Figure 11a-11c shows the time evolution of probabilities for the collisions illustrated in the preceeding figures: specifically we have plotted the quantities (78) summed over the target positive and negative energy continua. At the lowest velocity, the system is responding nearly adiabatically, as suggested by the oscillations as the probabilities relax to their final values. At the highest velocity, the system behaves impulsively, the probabilities relaxing to small final values without oscillating: in the range v = 0.5–0.9 the projectile velocity matches the main components in the initial state wavefunction (Fig. 7) so that the inelastic probabilities do not relax but resonate in the final state at large values. Though G and F are mixed in electron and positron eigenstates, perusal of Figs. 8-10 does suggest qualitative phenomena. Thus, in Fig. 9 we notice that G is large between the nuclei, suggesting a high probability of finding electrons near the projectile. By contrast, the prominence of F ahead of the projectile in Fig. 10 suggests positron formation in this frame of reference.

The processes considered are the formation of bound, positive energy continuum and negative energy continuum states, described in (1), (2), and Fig. 1. In summary,

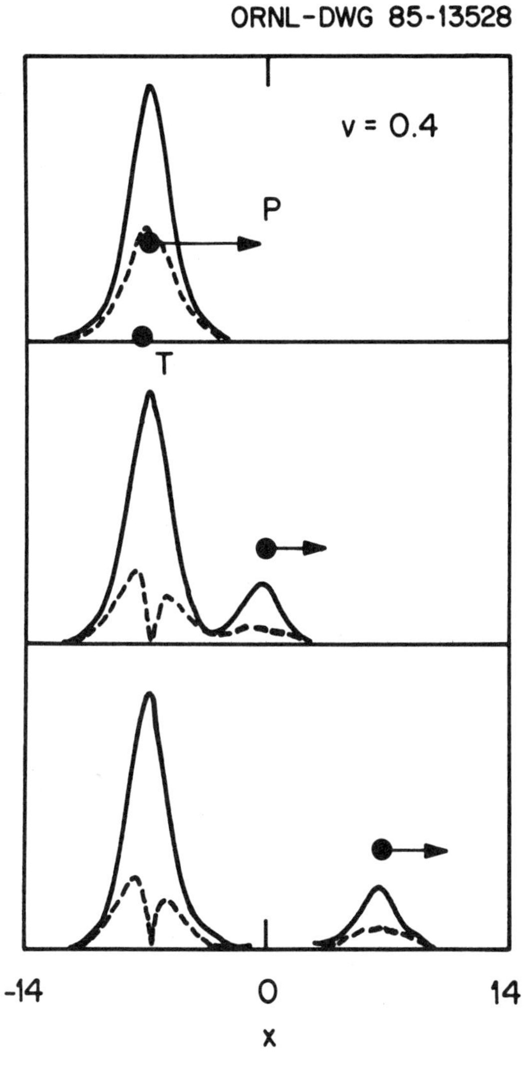

Fig. 8. Time evolution of the particle density in space for a collision at v = 0.4. $|G|^2$ and $|F|^2$ are shown by full and dotted lines. P and T denote the projectile and target nuclei: the passage of time in the three frames is indicated by the placement of P.

ORNL-DWG 85-13527

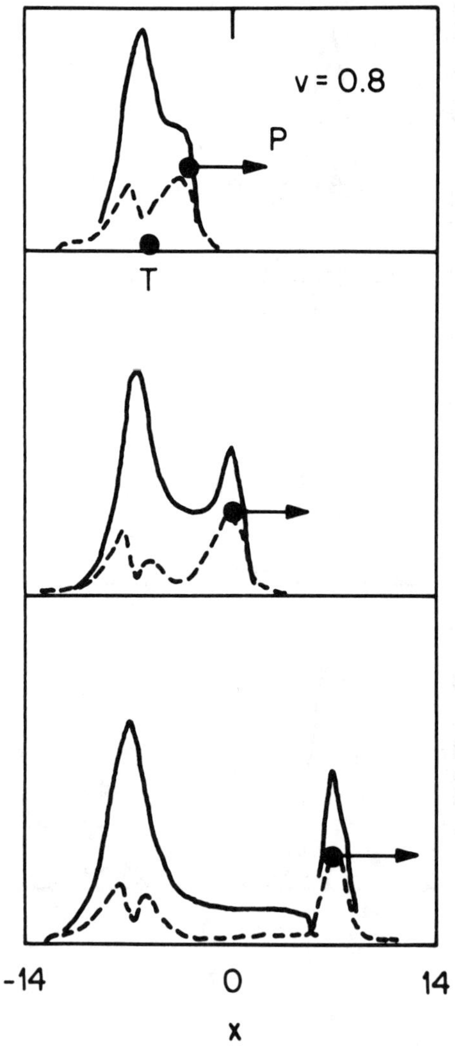

Fig. 9. Same as Fig. 8 with v = 0.8.

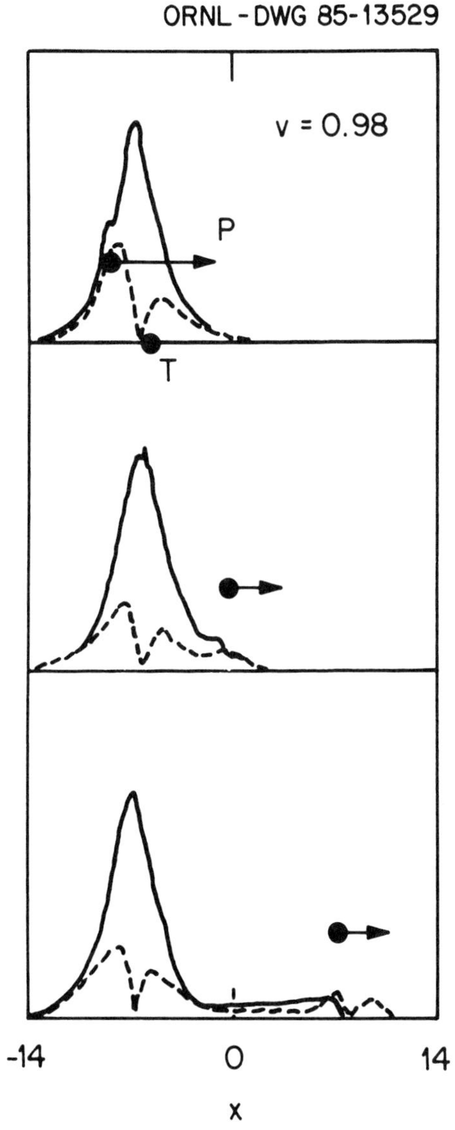

Fig. 10. Same as Fig. 8 with v = 0.98.

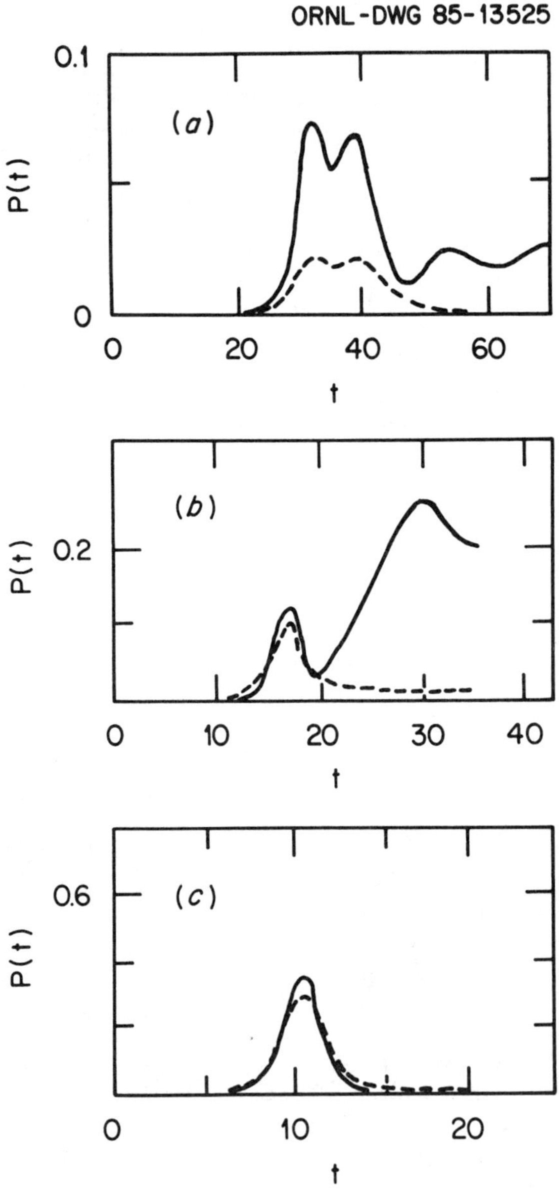

Fig. 11. Probabilities associated with positive (full lines) and negative (dotted lines) energy continua of the target are plotted as a function of time for the collisions considered in Figs. 8–10: (a) v = 0.4, (b) v = 0.8, and (c) v = 0.98.

$$
\begin{aligned}
P + T(1) \rightarrow\ & P + T(n > 1) \\
& P + [T + e^{\pm}] \\
& P(n \geqslant 1) + T \\
& [P + e^{\pm}] + T
\end{aligned}
\tag{84}
$$

Figures 12 and 13 show a variety of probabilities in the projectile and target frames, as functions of projectile velocity. The most striking results are the extent of capture into the ground state of the projectile and the extent of positron production in both frames of reference. Figure 14 shows the combined positron and delta-electron production in both frames, and the total inelastic probability which is dominated by capture.

Finally, in Figs. 15 and 16 we show the energy spectra of electron and positrons in the target and projectile frames for v = 0.8. Right- and left-moving particles are distinguished using the algorithms (78)-(82). Considerable left-right asymmetry is observed in all cases. The oscillations seen in the left and right components separately are due to rescattering by the nuclei: the flux removed from one component reappears in the other component, so the total shows little effect. An attenuated form of the same phenomena[8] is responsible for the cusp asymmetry in real 3D systems. The striking

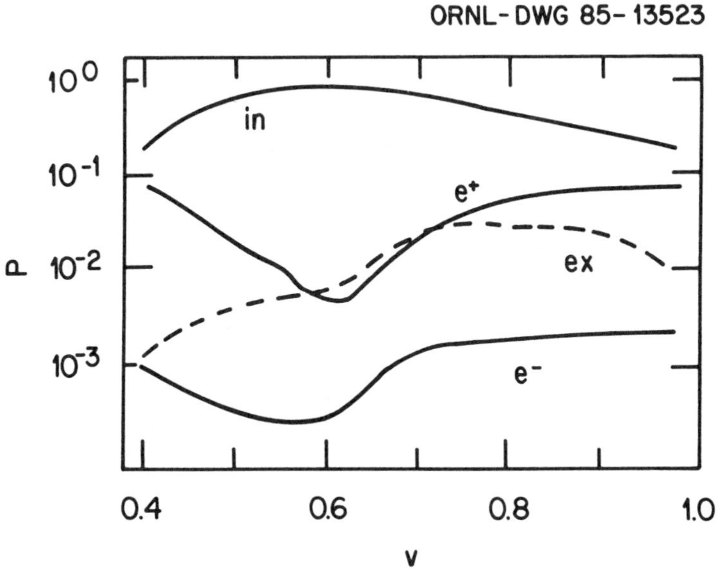

Fig. 12. Final-state probabilities in the target frame as functions of the projectile velocity: (in) total inelastic; (ex) excited bound states of the target; (e^-) positive energy continuum (electrons); (e^+) negative energy continuum (positrons).

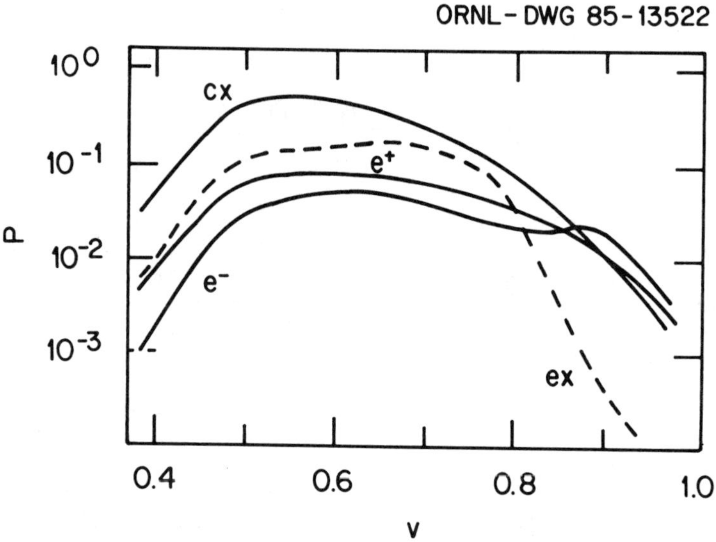

Fig. 13. Final-state probabilities in the projectile frame as functions of the projectile velocity: (cx) capture into the projectile ground state; (ex) excited bound states of the projectile: (e^-) positive energy continuum (electrons); (e^+) negative energy continuum (positrons).

feature here is the enhancement of the rightward and the suppression of the leftward positrons in the projectile frame, Fig. 16b. While electrons are captured, it appears that positrons are repelled into the continuum. Furthermore, they are repelled into the forward direction. More realistic calculations are needed, but it is plausible to speculate that such directional emission of positrons may be observed experimentally forward of the projectile.

We conclude by listing some processes of interest in heavy ion colliders together with estimates of their probabilities in the range 0.1 to 3.0 GeV/u, based on our model. P denotes a bare projectile, T(lab) denotes a fixed target experiment and K(KK) one (two) filled K orbits.

(a) P + T(lab) → T X-ray + anything:

dominated by capture, probability approaches unity.

(b) P + T(lab) → δ-electron + anything:

probability small, usually < 0.05.

ORNL-DWG 85-13524

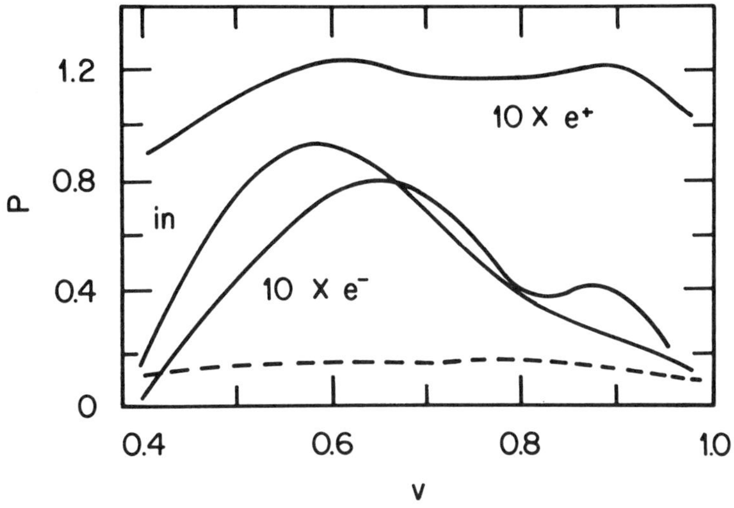

Fig. 14. Combined final-state probabilities for the target and projectile frames as functions of the projectile velocity: (in) total inelastic; (e^-) positive energy continuum (electrons); (e^+) negative energy continuum (positrons). The flux lost by folding with the cosine bell is shown by the dashed line.

(c) $P + T(\text{lab}) \rightarrow P_K + e^+ + \text{anything}, \ P_{KK} + e^+ + \text{anything}.$

This process can take place with or without capture from T; in either case, probability < 0.05.

(d) $P + T \rightarrow P_K + T + e^+, \ P + T_K + e^+$

probability < 0.1 either way.

Clearly more realistic calculations are needed. However, we can convert the above head-on probabilities to cross sections by conservatively estimating that relativistic processes take place within nuclear separations[1,2] of about 100 fm, corresponding to a perfectly absorbing cross section of 100 barns. The resulting estimates are not far out of line with those discussed by H. Gould elsewhere in this volume for higher energies. We note that at these energies (~100 GeV) pair production will occur predominantly by direct excitation out of the negative continuum.

We can also say something about heavy ion inertial confinement. Optimal beam energies are those for which the energy deposition is as large as possible, while the target is not yet transparent. From Figs. 12-14 this is probably about 1 GeV/u.

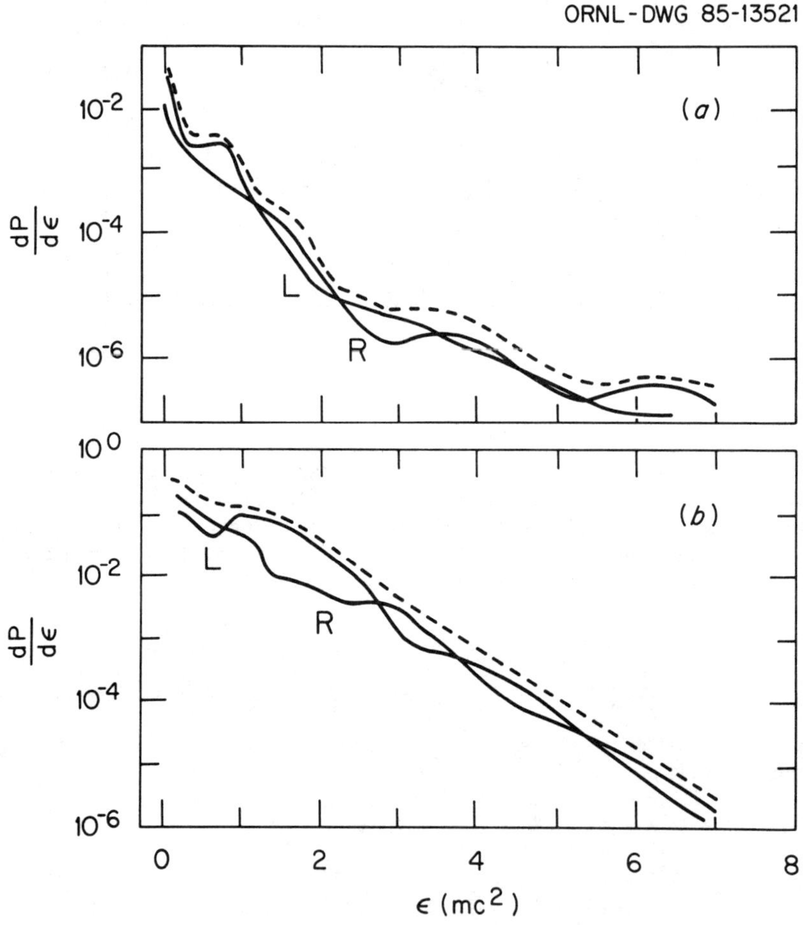

Fig. 15. Energy spectra as functions of kinetic energy in the target frame for v = 0.8: (A) electrons; (B) positrons. The right and leftwards momentum components are labelled R,L and their sum is shown as a dashed line.

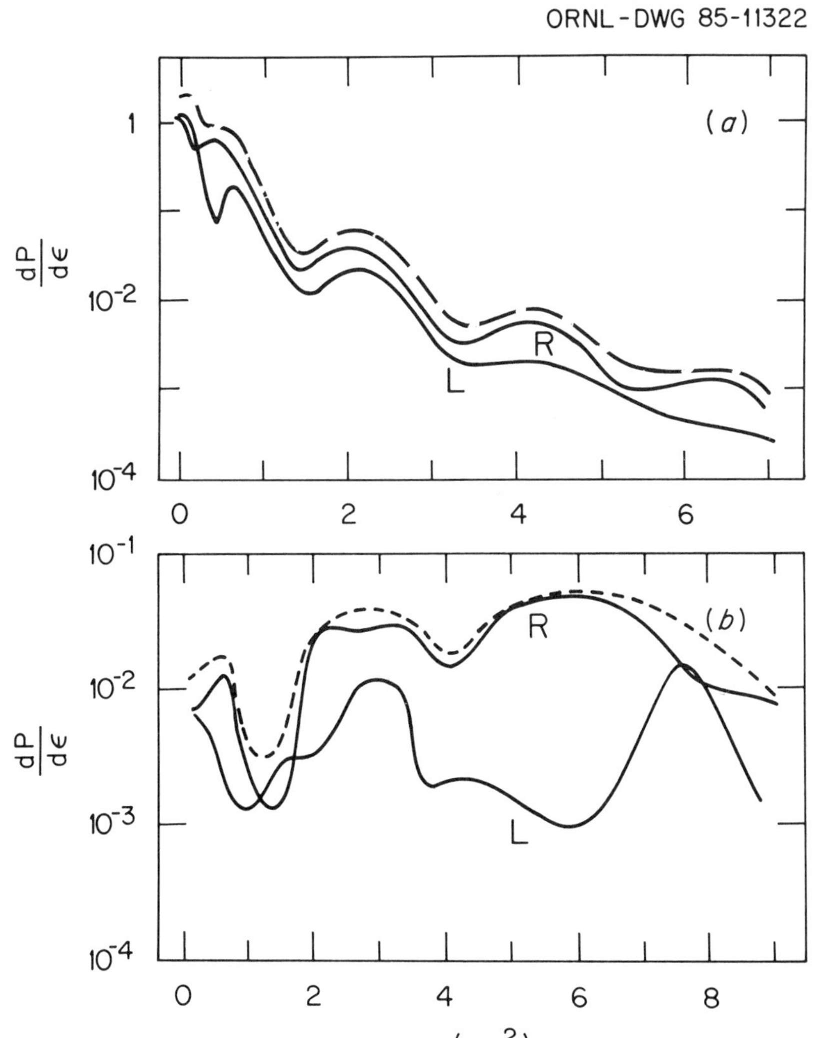

Fig. 16. Same as in Fig. 15 except for the projectile frame.

REFERENCES

1. "Quantum Electrodynamics of Strong Fields", ed. W. Greiner, Plenum Press, New York, 1983.

2. C. Bottcher and M. R. Strayer, Phys. Rev. Letts. 54, 669 (1985).

3. P.A.M. Dirac, "The Principles of Quantum Mechanics" (4th edn.), Claraendon Press, Oxford, 1958.

4. Carl de Boor, "A Practical Guide To Splines", Springer-Verlag, New York, 1978.

5. C.A.J. Fletcher, "Computational Galerkin Methods", Springer-Verlag, New York, 1984.

6. C. M. Bender and D. H. Sharp, Phys. Rev. Letts. 50, 1535 (1983).

7. Claude Itzykson and Jean-Bernard Zuber, "Quantum Field Theory", McGraw-Hill, New York, 1980.

8. R. Shakeshaft and L. Spruch, Phys. Rev. A20, 376 (1979); C. Bottcher, Phys. Rev. Lett. 48, 85 (1982).

COMMENTS ON BOTTCHER'S TALK

I. P. Grant
Department of Theoretical Chemistry, Oxford University
1 South Parks Road, Oxford OX1 3TG, England

We have initiated a project to study relativistically the electronic structure of molecules using finite basis set variational methods of the type pioneered in atoms by Yong-Ki Kim[1] and Kagawa.[2] Attempts to carry out similar calculations for molecules have had mixed results, the most common problem being gross overestimation of the relativistic energy correction (often termed, inaccurately, "variational collapse") and the appearance of spurious, unphysical energy levels in the computed spectrum, making physical interpretation a matter of guesswork. Kutzelnigg[3] lists a large number of attempts to overcome these problems.

From our work,[4] it appears that the trouble lies in the choice of basis sets, which must be such as to reflect the usual boundary conditions implied when we seek bound state solutions. To explain this, consider the Dirac one-electron problem for a single electron in a Coulomb field. The radial large and small components, P(r) and Q(r) respectively, satisfy the equation

$$h_\kappa u(r) \equiv \begin{bmatrix} \frac{-Z}{r} & c\left[\frac{-d}{dr} + \frac{\kappa}{r}\right] \\ c\left[\frac{d}{dr} + \frac{\kappa}{r}\right] & -2c^2 - \frac{Z}{r} \end{bmatrix} \begin{bmatrix} P(r) \\ Q(r) \end{bmatrix} = \varepsilon \begin{bmatrix} P(r) \\ Q(r) \end{bmatrix} \equiv \varepsilon u(r).$$

Approximate u(r) by the expression

$$u_N(r) = \begin{bmatrix} P_N(r) \\ Q_N(r) \end{bmatrix}$$

where

$$P_N(r) = \sum_i^N \pi_i(r)\, P_i \equiv \pi^\dagger p, \quad Q_N(r) = \sum_i^N \rho_j(r)\, q_j \equiv \rho^\dagger q,$$

and then require the functional

$$W[u_N] = \langle u_N | h_\kappa | u_N \rangle / \langle u_N | u_N \rangle$$

to be stationary with respect to variations in p and q. We obtain the matrix Dirac equation

0094-243X/85/1360299-3$3.00

$$\begin{bmatrix} V & c\Pi \\ c\Pi^{\dagger} & -2c^2S'-V' \end{bmatrix} \begin{bmatrix} p \\ q \end{bmatrix} = \varepsilon \begin{bmatrix} Sp \\ S'q \end{bmatrix} \qquad (1)$$

where V, Π, $\Pi^{\dagger}$, S and S' are N × N matrices with elements

$$V_{ij} = \langle \pi_i | -Z/r | \pi_j \rangle, \qquad V'_{ij} = \langle \rho_i | -Z/r | \rho_j \rangle$$

$$\Pi_{ij} = \langle \pi_i | -d/dr + \kappa/r | \rho_j \rangle \qquad \Pi^{\dagger}_{ij} = \langle \rho_i | d/dr + \kappa/r | \pi_j \rangle$$

$$S_{ij} = \langle \pi_i | \pi_j \rangle \qquad S'_{ij} = \langle \rho_i | \rho_j \rangle .$$

By eliminating q between the pair of (N-vector) equations embodied in (1), we get the matrix version of the second-order Dirac equation

$$\{\Pi(S')^{-1}\Pi^{\dagger}/2 + V - C(\varepsilon)\}\, p = \varepsilon Sp. \qquad (2)$$

The matrix $C(\varepsilon)$ contains all the relativistic effects,

$$C(\varepsilon) = \Pi(S')^{-1}W'(1 + W')^{-1}\Pi^{\dagger}/2,$$

with

$$W' = (\varepsilon S' - V')/2c^2;$$

$C(\varepsilon) \to 0$ as $c \to \infty$, and we need to ensure that what is left,

$$\{\Pi(S')^{-1}\Pi^{\dagger}/2 + V\}\; p = \varepsilon Sp, \qquad (3)$$

corresponds to the nonrelativistic matrix Schrödinger equation. Clearly, this will only be true if[4]

$$\Pi(S')^{-1}\Pi^{\dagger}/2 = T_{nr}, \qquad (4)$$

where T_{nr} is the matrix of the nonrelativistic radial kinetic energy operator in the $\{\pi_i\}$ basis. [This is where we make contact with Bottcher's requirement that the matrix product of D^+ and D^- shall yield the matrix representation of the Laplacian.] If we write (4) in terms of matrix elements, and (purely for ease of comprehension) assume that the set $\{\rho_i\}$ is orthonormal, we see that we require

$$\Sigma_k \; \Pi_{ik}\Pi^{\dagger}_{kj}/2 = (T_{nr})_{ij} \qquad (5)$$

The k-summation must run over a complete set of intermediate states if (5) is to be true in general, and it is apparent that in practice a finite set $\{\rho_i, \ldots, \rho_N\}$ can never be complete. Because of this, any expectation, $p^{\dagger}T_{nr}p$, must be greater than the expectation of $p^{\dagger}\Pi(S')^{-1}\Pi^{\dagger}p/2$--unless, that is, we can choose the span of the set $\{\rho_i\}$ to be the image of the set $\{\pi_i\}$ under the action of the differential operator $(d/dr + \kappa/r)$ represented by $\Pi^{\dagger}$. It can be shown that[5]

a) this ensures that (4) and (5) are always true;
b) $P_N(r)$ and $Q_N(r)$ are correctly related near $r = 0$;
c) solutions of (1) approximating bound state eigenfunctions belong to eigenvalues which are rigorous upper bounds to exact eigenvalues of the Dirac equation;
d) solutions corresponding to continuum eigenvalues do not converge to any particular limit.

It seems likely that this can be generalized to Dirac-Fock atomic self-consistent field equations in a rigorous fashion, and numerical tests, so far only in closed shell atoms, support this conjecture. Spurious eigensolutions are completely eliminated, and any "variational collapse" encountered has always proved to be due to a program error. Nothing in the mathematical development restricts the theory to atoms, and tests on systems such as H_2^+, H_2 have behaved as expected. The results will be published elsewhere.[5]

REFERENCES

1. Y.-K. Kim, Phys. Rev. 154, 17 (1967); Phys. Rev. 159, 190(E) (1967).
2. T. Kagawa, Phys. Rev. A12, 2245 (1975); Phys. Rev. A22, 2340 (1980).
3. W. Kutzelnigg, Int. J. Quantum, Chem. 25, 107 (1984).
4. K. G. Dyall, I. P. Grant, S. Wilson, J. Phys. B 17, 493 (1984); J. Phys. B 17 L45 (1984); and other work in course of publication.
5. I. P. Grant, Phys. Rev. A 25, 1220 (1982) gives an early version; in collaboration with J. Wood, H. M. Quiney and S. Wilson, a more complete analysis, with numerical evidence, is being prepared.

PHENOMENOLOGY OF NEW PARTICLE PRODUCTION IN HEAVY-ION COLLISIONS

A. B. Balantekin, C. Bottcher, and M. R. Strayer
Physics Division, Oak Ridge National Laboratory
Oak Ridge, Tennessee 37831

and

S. J. Lee
Wright Nuclear Structure Laboratory, Yale University
New Haven, Connecticut 06511

ABSTRACT

It is shown that the existence of sharp lines in the positron emission spectra observed in low-energy heavy-ion collisions is consistent with the production and subsequent decay of a neutral, pseudoscalar particle with a mass of about 1.6 MeV and a lifetime of about 10^{-13} sec.

INTRODUCTION

Binding energies of the electronic states in atoms are known to increase with the nuclear charge Z. At a critical charge $Z_c \approx 173$, the 1s state dives into the negative energy continuum. Consequently, it was suggested[1] some time ago that positrons would be emitted nonperturbatively from the vacuum in low-energy collisions of heavy ions with a combined charge greater than 173. The spectra of positrons emitted from the center of mass of the heavy-ion system have been experimentally studied; in particular, some sharp lines were identified.[2]

Numerical methods have been developed to solve the time-dependent Dirac equation in the presence of strong electromagnetic fields,[1,3] and it has been shown that if a small fraction of the collisions leads to a nuclear reaction with a very long time delay, then one would expect to see such peaks. A possible mechanism to obtain such a delay time has been advanced by Greiner and co-workers.[4] They suggest that the potential between heavy ions has a pocket, and when the system hits this pocket, quasimolecular nuclear configurations with a lifetime of 10^{-20} sec or more are formed. However, our present understanding of such systems requires that the energy of the peak and the cross section associated with the peak change considerably from system to system. For the five systems U+Cm, U+U, Th+U, Th+Cm, and Th-Th experimentally studied so far, the sharp lines in the positron emission spectra occur at approximately the same positron kinetic energy of 300 KeV with integrated cross sections of about 200 μb and with widths less than approximately 70 KeV.[5] In the following, another possibility for the existence of these sharp peaks will be elucidated. It will be shown that production and subsequent decay of a new neutral, pseudoscalar particle[6] is a viable alternative explanation to the formation of a long-lived superheavy nuclear complex.

DECAY MECHANISM

A standard pseudoscalar interaction Lagrangian will be assumed for the coupling of the new particle to the positrons and electrons

$$\mathcal{L}_{int} = g\, \overline{\psi}_e\, \gamma_5\, \psi_e\, \phi_\alpha . \tag{1}$$

The interaction strength g and the mass of the new particle, m_a, are taken to be free parameters to be fixed by data. If m_a is greater than twice the electron mass, then the two-photon decays of this new particle are suppressed and the dominant decay mode is into electron-positron pairs. Such positrons coming from the decay of the new particle, together with those produced directly by the strong electromagnetic field, give the experimental spectrum.

Let us assume that the lifetime of this particle is much longer than the heavy-ion reaction time which is about 10^{-21} sec. Under this assumption, which will be justified by the consistency of the final results, the positron differential cross section resulting from the decay of such a particle can be written as

$$\frac{d\sigma}{dE} = \sigma_{cl} \frac{dN}{dE} \tag{2}$$

where E is the total positron energy and dN/dE is the energy distribution of the emitted positrons. The characteristic heavy-ion cross section for positron production, σ_{cl}, sets the scale for the yield in Eq. (2). Since strong fields occur mostly in the central collisions, the characteristic interaction length is taken to be $R_D \approx 20$ fm which gives $\sigma_{cl} \approx \pi R_D^2 \approx 12.6$ b. Let $F(\vec{k})$ be the normalized probability for such a particle coming out of the interaction region with momentum between $|\vec{k}|$ and $|\vec{k}| + d|\vec{k}|$. Assuming the new particle decays isotropically into electron-positron pairs, $F(\vec{k}) = F(|\vec{k}|)$, one gets[6]

$$\frac{dN}{dE} = \frac{4\pi\, m_a \mathcal{M}_a}{\sqrt{m_a^2 - 4m_e^2}} \int_{|\vec{k}_-|}^{|\vec{k}_+|} d|\vec{k}|\, |\vec{k}|\, F(|\vec{k}|)\, \theta(E_a - E), \tag{3a}$$

with

$$|\vec{k}_\pm| = \frac{m_a^2}{2m_e^2} \left| E \sqrt{1 - \frac{4m_e^2}{m_a^2}} \pm \sqrt{E^2 - m_e^2} \right| , \tag{3b}$$

where E_a and $\mathcal{M}_a$ are the total energy and the multiplicity of the produced new particles. Since there essentially are three data points (position, cross section, and the width of the peak), one can fit only three parameters in Eq. (3). Choosing two of these parameters to be the mass m_a and the multiplicity $\mathcal{M}_a$, one is forced to take a one-parameter function for $F(|\vec{k}|)$. We have made the analytically convenient choice

$$F(|\vec{k}|) \propto |\vec{k}|^2 \exp[-|\vec{k}|^2/\kappa^2], \tag{4}$$

where the free parameter κ characterizes the momentum of this particle. The mass m_a and the multiplicity $\mathcal{M}_a$ are adjusted to fit the experimental peak energy (300 KeV) and the integrated cross section (200 μb) respectively giving $m_a \approx 3.2\ m_e$ and $\mathcal{M}_a \sim 10^{-5}$ new particles per heavy-ion reaction.[6] The width of the peak (70 KeV) is the order of experimental resolution. When adjusted to fit this width, the momentum κ is found to be $0.1\ m_e c$. The calculated cross section is shown in Fig. (1a) and the function $F(|\vec{k}|)$ is shown by a solid line in Fig. (1b).

In principle, the coupling strengths of this new particle to electrons, g, to up quarks, g_u, and to down quarks, g_d, need not be the same. Consequently, even if we make a model for the production mechanism of such particles during heavy-ion collisions, we would not be able to deduce g uniquely from the phenomenologically determined multiplicity, $\mathcal{M}_a \sim 10^{-5}$. However, the assumptions made in the previous paragraphs would fail unless the lifetime of this particle is longer than the heavy-ion reaction time $(10^{-21}$ sec$)$ but is shorter than the time of flight to the positron detectors $(10^{-9}$ sec$)$. Assuming electron-positron decay is the dominant mode, the lifetime of this particle is

$$\tau = \frac{\hbar}{\Gamma} \tag{5a}$$

where

$$\Gamma = \frac{g^2}{8\pi} \sqrt{m_a^2 - 4m_e^2}. \tag{5b}$$

Consequently, the above restrictions on the lifetime restrict the coupling strength to electrons and positrons to be $10^{-7} < g < 10^{-1}$.

Such particles interacting with electrons and positrons via the Lagrangian in Eq. (1) would contribute to the anomalous magnetic moment of electron. For a mass $m_a \approx 3.2\ m_e \approx 1.6$ MeV, the contribution to the electron g-factor is[6] $\sim 5 \times 10^{-3}\ g^2$. Hence, the current state of the art g-2 measurements further restrict the value of the coupling constant g. Combining these data with the Q.E.D. calculations,[7] the maximum allowed values of g^2 versus m_a are plotted in Fig. 2. The shaded area denotes the allowed values. Based on this figure, in the rest of this analysis we will take $g \sim 10^{-4}$ which yields a lifetime of $\tau \sim 10^{-13}$ sec.

Sometime ago, it was observed[8] that a global U(1) symmetry would suppress CP-violating instanton effects in quantum chromodynamics. As a consequence of breaking this U(1) invariance, a new neutral

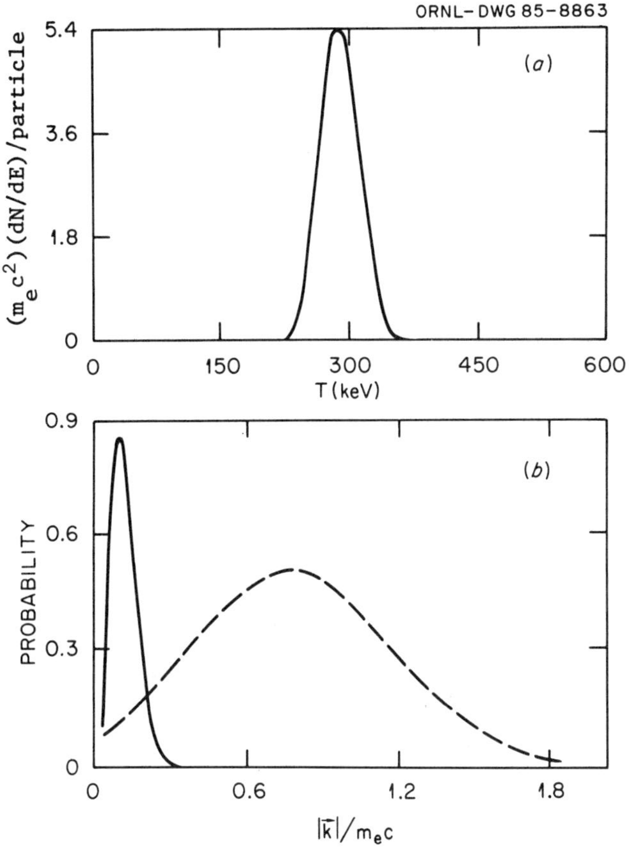

Fig. 1. (a) dN/dE per particle as calculated in Eq. (3a). (b) $F(|\vec{k}|)$ os Eq. (4) (solid line) and 250 $D(|\vec{k}|)$ of Eq. (6) (dashed line).

pseudoscalar boson, called the axion,[9] should exist. So far, no such particles have been observed. In the next section, we will demonstrate that the particle we introduce is not the standard axion. However at this point, one might ask if the existence of such a particle can be ruled out using the existing negative results from the axion search experiments. Since most of those experiments looking for the electron-positron decay mode would be sensitive not only to axions, but also to other pseudoscalar particles in this mass range, one would expect that our new particle, introduced to explain the heavy-ion data, could already be ruled out. Two such experiments have been performed. Calaprice et al. searched[10] for the decay of the 12.7-MeV state in ^{12}C by the pseudoscalar particle emission to ^{8}Be. A model-independent analysis of their result does not rule out

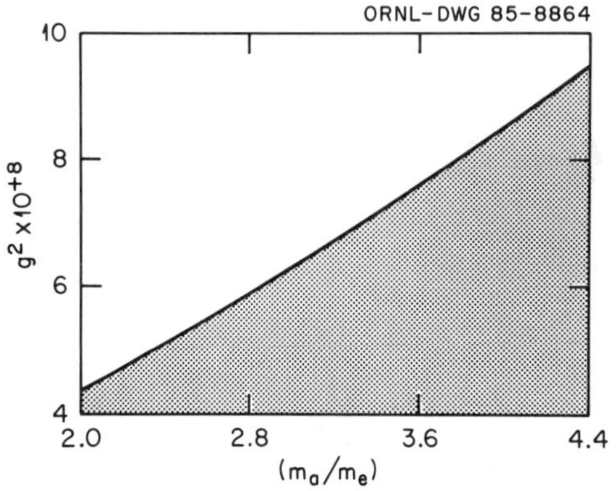

Fig. 2. The values of g^2 and m_a allowed by the anomalous magnetic moment of electrons (shaded area).

the possibility $g \sim 10^{-4}$. In another experiment, Faissner et al.[11] conclude that such particles with a lifetime less than 10^{-9} sec are not ruled out.

The existence of such particles would also have astrophysical implications, particularly for our understanding of stellar evolution. Emission of neutral particles can cause substantial energy loss from red giant stars. The values m_a = 1.6 MeV and $g \sim 10^{-4}$ are consistent with the limits on m_a and g required for the stability of those stars.[12]

PRODUCTION MECHANISMS

So far, we have only studied the decay of this particle. Using the numerical techniques developed in Ref. 3 to calculate pair production during slow heavy-ion collisions, we can try estimating the distribution function $F(\vec{k})$. In the first approximation, let us neglect the nuclear contributions and estimate the production of this particle using first-order perturbation theory. The resulting inclusive singles distribution for the neutral particles is

$$D(|\vec{k}|) = \sum_{\vec{p},\vec{q}} \delta_{\vec{k},\vec{p}+\vec{q}} \left(\frac{dN_+}{d\vec{p}}\right)\left(\frac{dN_-}{d\vec{q}}\right), \tag{6}$$

where $\left(dN_+/d\vec{p}\right)$ and $\left(dN_-/d\vec{q}\right)$ are the differential number of positrons and electrons respectively computed using the techniques of Ref. 3. Since we are performing a first-order calculation, multiplicity is proportional to the mean number of pairs given in Eq. (6) with a proportionality factor g^2. We find the multiplicity to be about 10^{-10} neutral particles per collision. The function $D(|\vec{k}|)$ is given in Fig. 1b as the dashed curve, where it is multiplied by a factor of 250 for plotting purposes. We conclude that nuclear processes and/or non-perturbative electromagnetic processes must contribute very significantly to the production of such particles.

Up to this point, very little has been said about the coupling of this new, neutral pseudoscalar particle to quarks. Let us assume that its coupling to a particular quark q has the same form as in the Lagrangian of Eq. (1), with g replaced by a different parameter g_q. The vector meson V, composed of the quark q and its antiquark $\bar{q}$, would decay into this new particle and a photon. For a vector meson, the ratio of the radiative decay width to that of decay width into $\mu^+\mu^-$ pairs is given by[13]

$$\frac{\Gamma(V \to \alpha + \gamma)}{\Gamma(V \to \mu^+ + \mu^-)} = \frac{g_q{}^2}{2\pi\alpha}\left(1 - \frac{m_a{}^2}{m_V{}^2}\right) \tag{7}$$

where m_V is the mass of the vector meson. For the decay of J/ψ the bound on this ratio is[14] $BR(J/\psi \to a+\gamma) < 10^{-6}$, and for the decay of Υ it is[15] $BR(\Upsilon \to a+\gamma) < 10^{-4}$. These numbers limit the coupling strength of this particle to the charmed and bottom quarks as $g_c < 10^{-4}$ and $g_b < 10^{-3}$.

Using g_c and g_b determined above, one can easily demonstrate that this particle is not the standard axion. For the standard axion, one has[9]

$$\frac{g_e}{g_b} = \frac{m_e}{m_b}, \tag{8}$$

where m_b is the mass of the bottom quark (~ 4.5 GeV). For $g_e \sim 10^{-4}$, Eq. (1) predicts $g_b \sim 0.9$, which clearly violates the experimental limit in the previous paragraph. Since there is no other evidence for the standard axion model, it would be more surprising than otherwise if it explained the heavy-ion data.

CONCLUSIONS

We have shown that the experimentally observed positron emission spectra are consistent with the production and subsequent decay of a neutral, pseudoscalar particle whose mass and the coupling strength are consistent with all presently known experimental and theoretical limits. Moreover, we have demonstrated that if this particle exists, it is not the standard axion, and it must be produced by nuclear and/or non-perturbative electromagnetic processes.

It is, however, clear that those peaks observed in the positron emission spectra are not well understood theoretically. We would like to briefly summarize some of the other proposed explanations and suggest possible experimental tests.

A. Atomic Many-Electron Processes:[16]

The main problem in assessing this explanation is the lack of detailed many-body calculations addressing the question of positron emission. Such calculations should be available in the near future.

B. Long-Lived Superheavy Nuclear Systems:[4]

Nuclear molecules are well established in light nuclear systems.[17] In principle, most detailed theoretical studies predict the existence of metastable, superheavy nuclei; a definitive experimental confirmation, however, is still lacking. The stability of such nuclei is intimately connected to the detailed nature of nuclear shell structure in the superheavy region and is presently not well known. Schematic calculations indicate that the position of a positron peak varies[1] between 200 KeV and 400 KeV for the systems considered, which is not experimentally observed. On the other hand, the experimental situation is somewhat ambiguous since two groups report peak energies which differ by many standard deviations for the same system.

C. Nuclear Pair Conversion:

The longitudinal part of the electromagnetic field can strongly induce transitions between two nuclear 0^+ states, producing electron-positron pairs. However, a detailed study of such processes has not been carried out to understand the relationship with the positron peak.

It is then very important to do further experiments to assess the validity of these different explanations. Several such investigations are possible:

a) Looking for the universal spike in a clearly subcritical system (one for which $Z_P + Z_T < 170$): The experimental observation of the peak in such systems may rule out the nuclear molecule hypothesis.

b) Experiments which probe higher Z_P+Z_T systems: The availability of californium and einsteinium targets makes it feasible to study U+Cf ($Z_P+Z_T = 190$) and U+Es ($Z_P+Z_T = 191$) systems. Such studies would at least contribute to a better understanding of the nature of positron peaks.

c) Coincidence experiments: Positron measurements, in coincidence with heavy ions both in and out of the heavy-ion reaction plane would unambiguously determine the reference frame from which the positrons are emitted.

If either a new particle is produced, or if there is the pair conversion of a nuclear state, it should be possible to observe electrons in coincidence with positrons and heavy ions. In such an experiment, the yield as a function of invariant mass would give us information on the process of formation. In either case, the energy distribution of the electrons measured in coincidence with the positrons should give a peak similar to the positron yield.

d) Electron-positron collisions: It should be possible to unambiguously test the particle hypothesis by observing such particles directly in electron-positron collisions. Since such experiments are in progress,[18] an article including a detailed study of such processes is in preparation.[19]

ACKNOWLEDGEMENTS

A. B. Balantekin is a Eugene P. Wigner Fellow. This research was sponsored by the U.S. Department of Energy under contract DE-AC05-840R21400 with Martin Marietta Energy Systems, Inc., and under contract DE-AC02-ER03074 with Yale University.

REFERENCES

1. J. Rafelski, L. P. Fulcher, and W. Greiner, Phys. Rev. Lett. 27, 958 (1971); W. Pieper and W. Greiner, Z. Phys. 218, 327 (1969); S. S. Gershtein and Y. B. Zeldovich, Lett. Nuovo Cim. 1, 835 (1969); V. S. Popov, Sov. Phys. JETP 32, 526 (1971); see also W. Greiner, ed., Quantum Electrodynamics of Strong Fields (Plenum, N.Y., 1983).
2. J. Schweppe et al., Phys. Rev. Lett. 51, 2261 (1983); M. Clemente et al., Phys. Lett. 137B, 41 (1984).
3. C. Bottcher and M. R. Strayer, Phys. Rev. Lett. 54, 669 (1985); C. Bottcher and M. R. Strayer, contribution to these proceedings.
4. U. Heinz et al., Ann. Phys. (N.Y.) 151, 227 (1983).
5. T. Cowan et al., Phys. Rev. Lett. 54, 1761 (1985).
6. A. B. Balantekin, C. Bottcher, M. R. Strayer, and S. J. Lee, Phys. Rev. Lett. 55, 461 (1985).
7. See J. R. Sapirstein, contribution to these proceedings.
8. R. D. Peccei and H. R. Quinn, Phys. Rev. Lett. 38, 1440 (1977).
9. S. Weinberg, Phys. Rev. Lett. 40, 223 (1978); F. Wilczek, ibid. 40, 279 (1978).
10. F. P. Calaprice et al., Phys. Rev. D20, 2708 (1979).
11. H. Faissner et al., Phys. Lett. 96B, 201 (1980).
12. A. Barroso and G. C. Branco, Phys. Lett. 116B, 247 (1982).
13. F. Wilczek, Phys. Rev. Lett. 39, 1305 (1977).
14. C. Edwards et al., Phys. Rev. Lett. 48, 903 (1982).
15. M. Sivertz et al., Phys. Rev. D26, 717 (1982).
16. W. Lichten and A. Robationo, Phys. Rev. Lett. 54, 781 (1985).
17. D. A. Bromley, J. A. Kuehner, and E. Almqvist, Phys. Rev. Lett. 4, 365 (1960); K. A. Erb and D. A. Bromley, in D. A. Bromley, ed., Heavy Ion Science, Vol. 3 (Plenum, N.Y., 1985), p. 201.
18. K. A. Erb, private communication.
19. A. B. Balantekin, M. R. Strayer, C. Bottcher, and S. J. Lee, in preparation.

FERMION BOUND STATES IN THE NEGATIVE ENERGY CONTINUUM: A QED FOR Z $\gtrsim$ 170 NUCLEI

William B. Campbell
Department of Physics and Astronomy, University of Nebraska-Lincoln
Lincoln, NE 68588-0111

SUMMARY

Certain negative energy continuum solutions for the Dirac Equation with strong external gauge fields have stress tensors which are zero everywhere on a closed surface. These solutions correspond to bound states which have moved into the negative continuum. Such solutions can be truncated at this stress-free surface without having to introduce new external forces. The truncated wave functions are normalized and interpreted as bound state wave functions for fermions confined to the interior of the stress-free surface.

In this paper these wave functions are used to construct a viable Furry picture QED for electrons moving near a nucleus with Z$\gtrsim$ 170. For Z's this large the K-shell levels have moved into the negative continuum. The QED I propose allows one to make systematic calculations such as the spontaneous pair production near a bare Z $\gtrsim$ 170 nucleus.

I. INTRODUCTION

As the importance of strong coupling gauge theories has become clear over the past twenty years, interest in solutions for relativistic wave equations with strong external gauge fields has grown. This work has been motivated in part by the belief that such studies will shed light on mechanisms for confinement of quarks, and gluons. Not only has work been done in Yang-Mills theories[1] but also on the older problem of the electron moving near a high Z nucleus.[2]

A particular phenomenon common to these studies is the onset of instabilities if the coupling gets too strong. J. Mandula discovered such an instability for the Coulomb solution in Yang-Mills theories.[3] For Z $\gtrsim$ 170 the Coulomb field for a bare nucleus becomes unstable to spontaneous pair production.[4] In much earlier work L. Schiff, H. Snyder and J. Weinberg[5] studied similar instabilities for scalar fields.

For particles moving in strong Coulomb fields these instabilities can be related to the appearance of bound states in the negative energy continuum. Indeed for Z $\gtrsim$ 170 the K shell has a binding energy of greater than $2m_ec^2$ (or an $E < -m_ec^2$) and thus two pairs could be produced, the electrons captured in the K shell and the positrons escaping to infinity.

Although the Mandula instability is associated with the classical field of a point source it has a similar interpretation. I have been able to show that only above the Mandula threshold can there be Yang-Mills solutions with energy lower than the Coulomb Energy.[6] Thus glue binding is associated with exceeding the threshold. A quantum interpretation would be that there are bound negative energy glueons.

While we have a good physical picture of bound states moving into the negative energy continuum our relativistic wave equations fail us in that we can no longer calculate normalizable wave functions, precisely because the states are in the negative energy continuum. It has been traditional to argue that this indicates the breakdown of the single particle interpretation for relativistic wave equations.

In the case of the Dirac equation with an external Coulomb field the single particle interpretation has persisted anyway. Certainly the successes of Furry picture QED calculations for high Z atoms has encouraged workers to use the Dirac equation even when $Z \gtrsim 170$.[2,4] While this has not allowed the calculation of the electron wave function for $E < -m_e c^2$ it has allowed for the calculation of positron functions, interpreting as usual the negative energy continuum states as positron states.

The Furry picture QED for $Z \gtrsim 170$ is carried out using the notion of the charged vacuum. That is, all the negative energy continuum electron bound states are assumed filled in the vacuum state, a reasonable assumption since a bare nucleus would be unstable to pair production.[2,4] Some authors extend this charged vacuum by filling all electronic levels for $E < 0$ and reinterpreting them as positron bound states.[7] Then as these levels move into the positron continuum they behave like ordinary resonances.

I have two problems with the charged vacuum interpretation. First if the $E < -m_e c^2$ (or $E < 0$) electron bound states are filled then that will effect the charge distribution near the nucleus and using the Coulomb field of the nucleus alone as the external field can only be an approximation, although a very good one for such high Z. One still has no way of finding the $E < -m_e c^2$ bound state wave functions for the electrons.

The second problem has to do with the nature of the $E < 0$ bound state wave functions as $E \to -m_e c^2$ from above. As Popov has shown,[8] the Dirac K shell wave functions become more localized as $Z \to 170$ (so that $E \to -m_e c^2$). This continues as E goes below $-m_e c^2$ in that the core of the resonant wave function keeps getting more localized as E gets more negative. This has a reasonable physical interpretation for electrons which are becoming more tightly bound but is difficult to interpret for positrons.

What I propose to study here is an alternative based on a different interpretation of certain negative energy solutions for the Dirac equation with an external field. For precisely the circumstance where bound state energies have moved into the negative continuum there exist stationary solutions with $E < -mc^2$ for which the Dirac spinor contribution to the stress tensor is zero everywhere on the surface defined by $E = V - mc^2$, with V the potential energy associated with the external field. The value of E associated with these solutions lies right in the middle of the negative energy resonance which has replaced the bound state. Thus one can truncate the Dirac spinor at this stress-free surface, setting it to zero outside, without having to add any compensating surface force and thereby obtain a normalizable wave function. One can think of the Dirac particle as being confined inside a finite volume by the supercritical external field.

I show in Section II of this paper that these truncated spinors can be used to construct a Furry picture QED which includes the bound electrons with $E < -m_e c^2$.

II. A NEW QED FOR $Z \gtrsim 170$ ATOMS

As mentioned in the Introduction the successes of Furry picture QED for high Z atoms are impressive.[2] For example, the calculated K shell energy level in $_{100}Fm$ agrees with experiment to within 0.02%. But the binding energy for these electrons is already 28% of $m_e c^2$. This suggests that one should take the Furry picture seriously for even higher Z. An essential part of this assumption is the observation that QED corrections to quantities obtained by solving the Dirac equation, while relatively larger than those for low Z, still remain small.

Thus I will assume (as have other authors[2,4]) that one can start with solutions to the Dirac equation for an electron moving in the nuclear Coulomb field alone. This would be an appropriate starting point for the inner electrons in a neutral atom or for an almost fully ionized atom. The ideas I will express here could also be applied in the Hartree-Fock calculations for atoms, but I will assume the atom is almost fully ionized.

This is consistent with the present experimental situation. Highly ionized high Z ions are being collided with the hope of forming, momentarily, combined nuclei with Z > 170 in order to see the spontaneous pair production. Indeed, such pair production may have already been seen.[9]

Using the standard spherical decomposition for Dirac spinors we have:[10]

$$\psi_\kappa^\mu(\vec{r}) \equiv \begin{pmatrix} \dfrac{u_1(r)}{r}\,\chi^\mu_\kappa \\ \\ i\dfrac{u_2(r)}{r}\,\chi^\mu_{-\kappa} \end{pmatrix} , \qquad (1)$$

with χ_κ^μ the appropriate two component spinor angular functions, $-\kappa$ the eigenvalue of $\vec{\sigma} \cdot \vec{L} + 1$ and μ the eigenvalue of j_z. The quantum number κ can take on any integer value except zero.

The stationary state Dirac equation now reduces to a coupled set of radial equations:

$$
\begin{aligned}
-\frac{du_2}{dr} + \frac{\kappa}{r} u_2 &= (E - V - m)u_1 \quad , \\
\frac{du_1}{dr} + \frac{\kappa}{r} u_1 &= (E - V + m)u_2 \quad ,
\end{aligned}
\tag{2}
$$

where we use units such that $\hbar = c = 1$, $m = m_e$ is the mass of the electron, and $V = -eA^0$ is the electrostatic potential energy of an electron moving in the nuclear Coulomb potential $A^0(r)$.

It is easy to verify that near the surface defined by $E = V(r_o) - m$ one of the two independent solutions to Eqs. (2) behaves like:

$$
\begin{aligned}
u_1 &\sim (r - r_o)^2 \quad , \\
u_2 &\sim \text{constant.}
\end{aligned}
\tag{3}
$$

Because $V < 0$ we see that this can only happen for $E < -m$. In general this solution will be badly behaved at $r = 0$ and unacceptable physically.

In order to understand when these solutions are physically important it is useful to obtain them in another way. If we define v_1 by[10]

$$
u_1 \equiv \sqrt{E - V + m}\; v_1 \quad , \tag{4}
$$

then the second order equation for u_1 becomes

$$
\begin{aligned}
&\frac{d^2 v_1}{dr^2} + k^2 v_1 = 0, \\
&k^2 \equiv (E - V)^2 - m^2 - \frac{\kappa(\kappa+1)}{r^2} + \frac{\kappa}{r}\,\frac{V'}{E - V + m} \\
&\qquad - \frac{3}{4}\left(\frac{V'}{E - V + m}\right)^2 - \frac{1}{2}\,\frac{V''}{E - V + m} \quad .
\end{aligned}
\tag{5}
$$

For $-m < E < m$ and r ranging roughly from its value on the surface defined by $E = V + m$ to $r \to \infty$, $k^2 < 0$ and the bound state solutions are found by keeping only the decaying exponential solution. For $E < -m$ there is a "barrier" for r ranging roughly from its value on the surface $E = V + m$ to that on $E = V - m$ (roughly means ignoring angular momentum and other effects). Beyond $E = V - m$ as $r \to \infty$ the solutions are now oscillating in r.

In the barrier region (with either $E > -m$ or $E < -m$) we can always write the decaying exponential solution in the form

$$v_1 = \frac{N}{\sqrt{p}} \exp\left\{ - \int_a^r p \, dr \right\} , \tag{6}$$

where $p > 0$ and p must also satisfy

$$k^2 = \frac{1}{2} \frac{p''}{p} - \frac{3}{4} \left(\frac{p'}{p}\right)^2 - p^2 \tag{7}$$

in order to obtain an exact solution. (The WKB approximation is obtained by neglecting the p" and p' terms.) The constant N is for normalization and a is the value of r at the beginning of the barrier.

For $E > -m$ the v_1 of Eq. (6) must be matched onto that oscillating solution for $r < a$ which is well behaved at $r = 0$. This leads of course to the bound state eigenvalues for E. As Z increases to about 170 and beyond the lowest eigenvalue for E moves below $-m$ and it will now be possible to match solution (6) onto a well behaved oscillating solution for $0 \leqq r < a$. For r somewhat greater than r_0 (the value of r where $E = V - m$) the barrier ends and (6) must be matched onto an oscillating solution that extends to infinity. Clearly this gives us the wave function for the center of the resonance which has replaced the lowest bound state.

Now r_0 is slightly smaller than the value of r where the barrier ends so that the surface $E = V - m$ lies just inside the barrier region. In the immediate neighborhood of r_0 k^2 is dominated by its most singular term and is given by

$$k^2 \cong - \frac{3}{4} \frac{1}{(r-r_0)^2} , \tag{8}$$

where V has been expanded around $r = r_0$. It is easy to verify that a solution to Eq. (7) for the k^2 of (8) is

$$p \cong \frac{1}{|r - r_0|} \tag{9}$$

and for $r \lesssim r_0$

$$v_1 \cong N \sqrt{|r-r_0|} \exp \{-\int_a^{r_0} p\, dr\} \times$$

$$\times \exp \{+ \int_r^{r_0} \frac{1}{r_0-r} dr\}$$

$$\propto |r - r_0|^{3/2} \qquad (10)$$

and since $E - V + m \propto |r-r_0|$ this leads to the u_1 of solution (3).

Thus for $E < -m$ if we match solution (3) in the neighborhood of r_0 with the well behaved solution at $r = 0$ we will obtain the energy and a wave function for the middle of a negative energy resonance. This particular wave function has an interesting physical property.

The contribution to the stress tensor from the Dirac spinors is given by

$$T_{ij} = \frac{i}{2} [\bar{\psi} \gamma_i (\partial_j \psi) - (\partial_j \bar{\psi}) \gamma_i \psi] + (i \leftrightarrow j) \quad . \qquad (11)$$

Here I am using the standard Dirac representation for the γ_i's which means the stress tensor only has terms which pair u_1 with u_2. Thus for solution (3) all components of the stress tensor vanish on the sphere at $r = r_0$. It is essential that u_1 go to zero quadratically at r_0 for this to be true, making this particular solution in the middle of a resonance unique.[11]

It is therefore possible to truncate this solution at $r = r_0$, setting $\psi \equiv 0$ for $r > r_0$, without having to introduce any new forces. Since the Dirac field exerts no stress of any kind on this surface there is no need for rigid walls to hold it in. One can think of such electrons as being confined to the interior of a sphere of radius r_0.

Of course these truncated spinors are no longer solutions to the homogeneous Dirac equation. In the truncated form u_2 will be discontinuous as one can see from Eqs. (3). It is true that u_2 will be relatively small at $r = r_0$, because of the diminishing of both u_1 and u_2 as we move into the barrier region, and thus the truncated ψ may be a good approximation for many purposes. Nonetheless it will not be orthogonal to the stationary solutions of the homogeneous Dirac equation.

The important question is whether these truncated wave functions can be used to construct a meaningful QED. The four-current density operator must be divergenceless and the divergence of the spinor contribution to the energy-momentum four-tensor operator must give the correct volume force.

The construction is quite simple. The levels right in the middle of the negative energy resonances are reinterpreted as electron bound states. These states are treated as discrete levels with the creation and annihilation operators satisfying anticommutation relations appropriate for discrete levels rather than continuous levels. The truncated spinors are normalized to unity just as bound state spinors are, with the consequent interpretation that the charge is located entirely inside r_0.

The current and energy-momentum operators are constructed by using the truncated spinors for diagonal terms but the full untruncated spinors for off-diagonal terms. The untruncated spinors are normalized to unity by integrating only over the finite volume of the r_0 sphere. The untruncated spinors must be used in the off-diagonal terms in order to satisfy the conservation laws. For example, $\partial_\mu j^\mu$ would not be zero but have sources and sinks at $r = r_0$ if the truncated spinors were used in all terms.

As an example of this construction consider the charge density j_0. Suppose that only the K shell level has moved into the negative energy continuum, i.e., only $E_K < -m$. Let ψ_K^t stand for the truncated K level spinor and ψ_K for the full spinor. The electron creation and annihilation operators are $a_K^\dagger$ and a_K and satisfy

$$\{a_K, a_K^\dagger\} = 1 \quad , \tag{12}$$

as they would for any discrete level. In addition all bound states with $-m < E < m$ will be interpreted as electron levels.

All continuum states with $E < -m$ <u>except</u> the one labeled K above will be interpreted as positron states and $b_j^\dagger$ and b_j will be the respective creation and annihilation operators. Here I will use discrete labels to simplify notation.

For the charge density we would write

$$\begin{aligned} j_0 &= \bar{\psi}_K^t \gamma^0 \psi_K^t \, a_K^\dagger \, a_K + :\bar{\psi}_K \, a_K^\dagger \, \gamma^0 \psi': \\ &\quad + :\bar{\psi}' \gamma^0 \psi_K a_K: + :\bar{\psi}' \, \gamma^0 \, \psi': \quad , \end{aligned} \tag{13}$$

where the prime on ψ means the field operator with the K state contribution removed. All other physical operators bilinear in ψ are constructed in a similar way.

The charge operator for (13) is

$$\begin{aligned} Q &\equiv \int j^0 \, d^3r \\ &= a_K^\dagger \, a_K + \sum_i{}' \, a_i^\dagger \, a_i + \sum_j b_j^\dagger \, b_j \quad , \end{aligned} \tag{14}$$

as one would want. Similarly the energy operator is

$$H_0 = E_K a_K^\dagger a_K + \sum_i{}' E_i a_i^\dagger a_i + \sum_j E_j b_j^\dagger b_j \quad . \tag{15}$$

From this point on the Furry picture QED is handled in the usual way. I don't foresee any problems with this approach that differ from those of any such QED. The electron propagators will be difficult to work with as usual.

There are several interesting calculations which can be done without need of the electron propagators. For example, the decay rate for the fully ionized nucleus with $Z > 170$ can be calculated as a pair production process, with the photon in the final state instead of the initial, i.e.,

$$N \rightarrow N + e^- + e^+ + \gamma$$

with the e^- captured in the K shell.

The author appreciates the hospitality of the Aspen Center for Physics where much of this work was done.

BIBLIOGRAPHY

1. The literature on Yang-Mills solutions in external fields is extensive. The following are relevant to the present work but this list is far from exhaustive: J. Mandula, Phys. Rev. D 14, 3497 (1976); R. Jackiw, L. Jacobs and C. Rebbi, Phys. Rev. D 20, 474 (1979); P. Sikivie and N. Weiss, Phys. Rev. Lett. 40, 1411 (1978) and Phys. Rev. D 20, 487 (1979); W. Campbell, D. Joseph and T. Morgan, Nuc. Phys. B 192, 315 (1981); (W. Campbell and T. Morgan, Phys. Rev. D 26, 532 (1982); W. Campbell, Phys. Lett. 103B, 455 (1981) and Phys. Rev. D 27, 1391 (1983).
2. For an extensive review of the high Z electron problem see S. Brodsky and P. Mohr, "Structure and Collisions of Ions and Atoms," I. Sellin, ed.,(Springer-Verlag, Berlin, 1978), pp. 3-67.
3. J. Mandula, Phys. Lett. 67B, 175 (1977). See also M. Magg, Phys. Lett. 74B, 246 (1978).
4. For a thorough review of this instability which also includes a discussion of the Yang-Mills case see M. Soffel, B. Muller and W. Greiner, Phys. Reports 85, 51 (1982).
5. L. Schiff, H. Snyder and J. Weinberg, Phys. Rev. 57, 315 (1940).
6. W. Campbell, Phys. Lett. 122B, 293 (1983) and a related paper in Phys. Rev. D 27, 470 (1983).

7. The charged vacuum is discussed in Refs. 2 and 4 and by L. Fulcher and A. Klein, Phys. Rev. D 8, 2455 (1973) and Ann. of Phys. 84, 335 (1974).
8. V. Popov, Sov. Phys.-JETP 33, 665 (1971).
9. J. Schweppe et al., Phys. Rev. Lett. 51, 2261 (1983) and T. Cowan et al., Phys. Rev. Lett. 54, 1761 (1985).
10. See M. Rose, Relativistic Electron Theory, (John Wiley and Sons, Inc., NY, 1961) Chapter V, Section 26 and Chapter VI, Section 35.
11. There is also a stress-free solution at the surface defined by $E = V + m$ which when matched onto a well behaved solution at $r = 0$ gives an antiresonance, a smaller amplitude inside the barrier than outside. This solution is important in the positron sector but not for electron bound states.

WHAT CAUSES THE SHARP POSITRON SPECTRUM IN HEAVY ATOM COLLISIONS? THE ATOMIC HYPOTHESIS

William Lichten*
Joint Institute for Laboratory Astrophysics
National Bureau of Standards and University of Colorado
Boulder, CO 80309

What causes the remarkable structure in positron spectra found in the experiments carried on at G.S.I.? (See Fig. 1.) In the words of Jack Greenberg at this conference, the "current central issue (is): What is the source of the narrow positron peaks?" Could the cause be

1. Atomic Physics?[1]
2. Nuclear Physics?[3]
3. A new particle? Axion? Higgs Boson? "X^0"? Something else?[4,5]

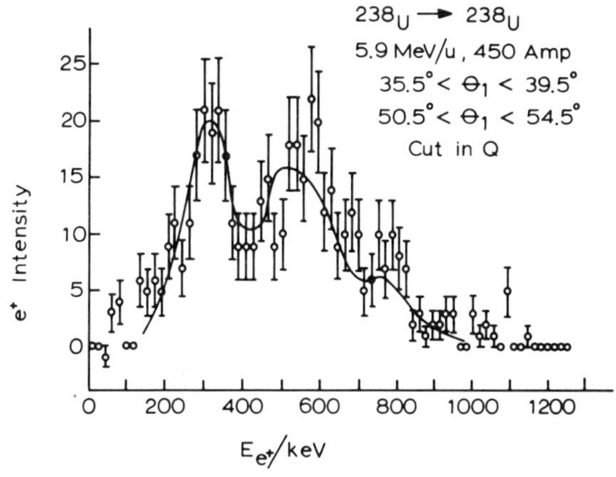

Fig. 1. Positron energy spectrum in U-U collisions. Experimental points: Ref. 2. Theoretical curve: Ref. 1.

I shall not attempt to judge the correctness of hypotheses (2) and (3), but shall discuss some ideas that make (1) a plausible possibility.

The history of atomic physics has shown similar phenomena. Figure 2 shows oscillations in the excitation cross sections for He radiation in He^+-He collisions as a function of bombarding

*1984-1985 JILA Visiting Fellow. Permanent address: Yale University Physics Department, 260 Whitney Avenue, New Haven, CT 06511

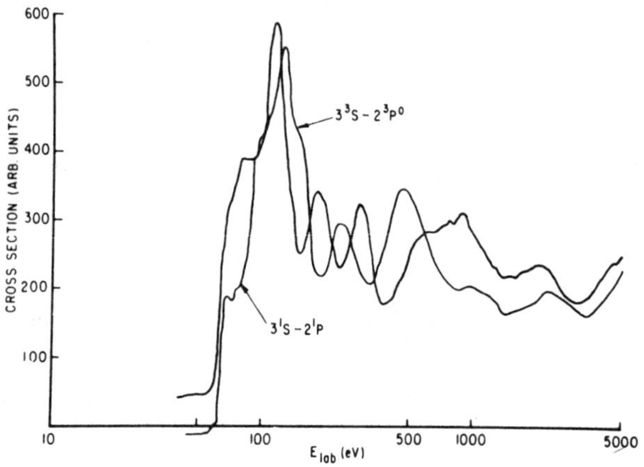

Fig. 2. Excitation cross sections for He radiation in He^+-He collisions as a function of projectile energy. Source: Ref. 7.

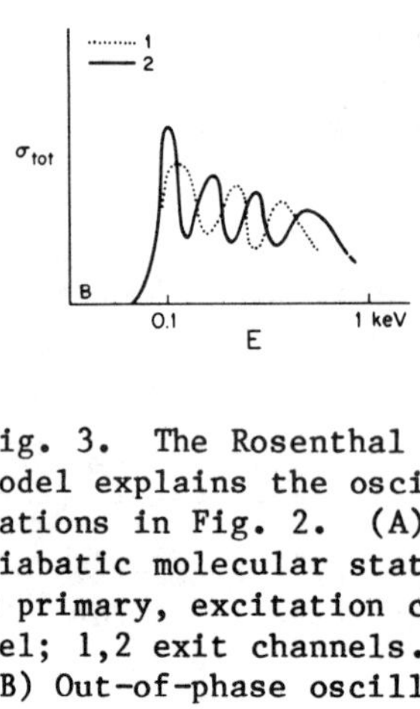

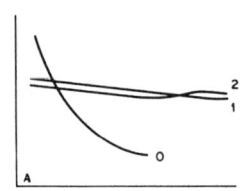

Fig. 3. The Rosenthal model explains the oscillations in Fig. 2. (A) Diabatic molecular states: 0 primary, excitation channel; 1,2 exit channels. (B) Out-of-phase oscillations in total cross sections for states 1 and 2. Source: Ref. 7.

energy.[6,7] The oscillations shown result from out of phase 3^3S and 3^1S excitations, explained by the Rosenthal model.[7] Figure 3 illustrates this model. Panel A shows curves of potential energy versus internuclear separation R for the molecular states. The system enters via the state "0" as the nuclei approach each other (the "primary excitation channel"). A coherent mixture of wave functions of states 1 and 2 occurs at the crossing of curve 0. This mixture reasserts itself at the crossing of states 1 and 2 at larger internuclear distance. The net result is shown in the lower panel of Fig. 3: antiphase oscillations between states 1 and 2. The essential point of the Rosenthal model is an oscillatory interference caused by two interactions, each well localized in time.

A somewhat closer analogy with the sharp positron spectra is the structure in the spectrum of autoionizing electrons emitted during He-He^+ collisions.[8] In this case, one observes sharp peaks (see Fig. 4) in the electron spectrum which cannot correspond to any known autoionizing

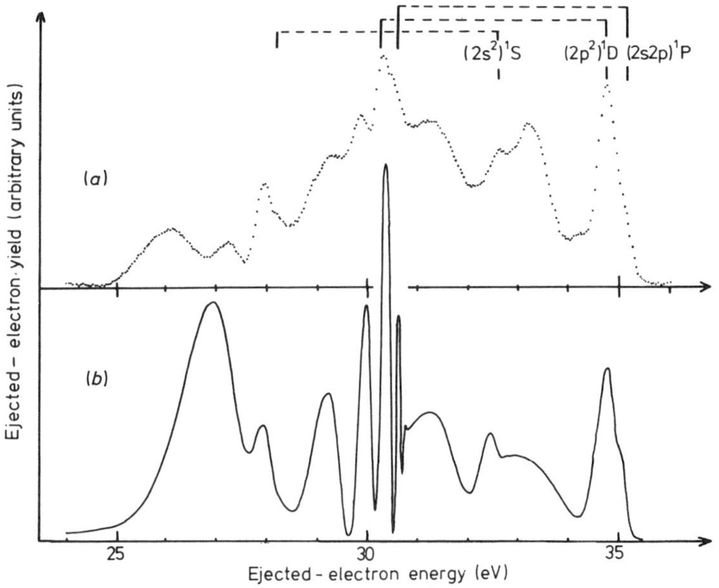

Fig. 4. Electron energy spectra from He^+-He collisions at 1400 eV in the forward and backward directions. (a) Experiment; (b) calculation. With the exception of the large, central peak, and the peak on the extreme right, none of the peaks corresponds to any known, autoionizing level. All the other peaks are produced by interferences in the molecular, autoionizing levels. Source: Ref. 8.

state of the separated atoms. One must conclude that these peaks arise from interferences between molecular states formed during the collision. Since positron emission during atomic collisions is analogous in many ways to electron autoionization, it is plausible that a similar mechanism is operating in both cases.

Figure 5 illustrates another aspect of purely atomic processes that could be relevant to positron emission in a sharply peaked spectrum. It shows the angular distribution of electrons emitted in He^+-He collisions. The marked angular anisotropy comes from the nature of the interference between two widely separated atomic centers. If the mechanism of positron emission were purely atomic, this anisotropy could result and, in turn, help account for the narrow widths observed by the experimentalists (see Fig. 1), due to the reduced Doppler effect.

Next, we ask the question, "What could be the nature of the atomic process that would produce these positron peaks?" We have done some model calculations to try out the various possibilities. The first possibility is a simple, first-order process, involving a single vacancy, the type of process discussed by Gerhard Soff in his

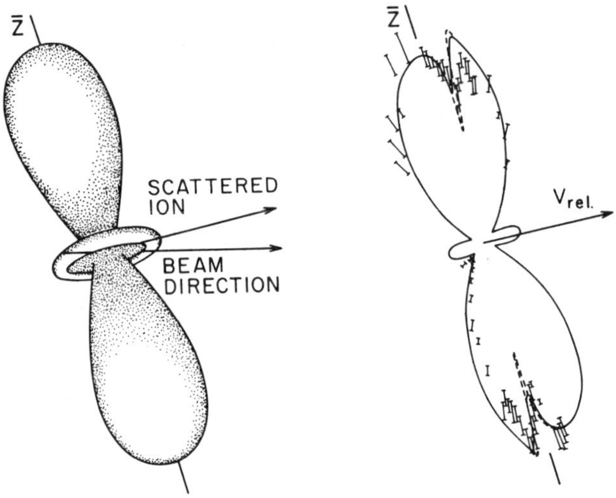

Fig. 5. (Left) Artist's rendition of the angular distribution of electrons emitted from the He^+-He collision complex. Note the narrowness in angular width, which is a consequence of the interference between two separated, atomic centers. This feature would also occur in positron emission via the atomic mechanism, and may account for the smallness of the Doppler width of the spectra. (Right) Verification of the molecular model of autoionization electrons. The curve is a theoretical calculation; the points are experimental values. Source: Ref. 8.

talk. In agreement with the conclusions of the Greiner group, we find that this process could not produce the sharply peaked spectrum. Such a spectrum must arise from an interference between processes widely separated in time. This could arise if the interfering processes occurred in the vicinity of the "strong molecular coupling near 500 f," mentioned by Soff. Such a process could occur in higher order in quantum mechanical perturbation theory:

$$P = \left(\frac{2\Pi}{\hbar}\right) \left(\frac{H_{12}}{\Delta E}\right)^2 V^2 \rho \quad ,$$

where ρ is the density of states, P is the probability of positron emission, H_{12} is a perturbation between two molecular states, separated by an energy ΔE, where state 1 is populated during the collision, but is not allowed to emit positrons, but state 2 emits positrons with a matrix element V. As a possible example, consider the following diagram:

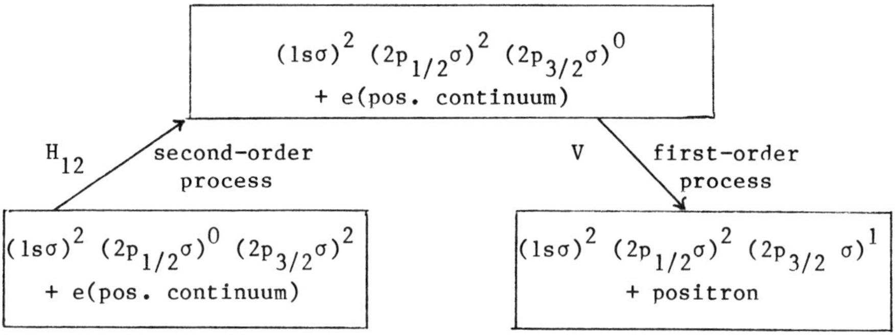

The question is what could be the perturbation H_{12}? This has been considered by the Greiner group,[9] who ruled out the possibility that the process went by electron correlation, which is very weak ($H_{12} \sim 1$ eV). We have considered an alternative perturbation, caused by the nuclear motion, which is large enough in order of magnitude to account for the second-order processes. We conducted a model calculation, which is described in Ref. 1, and obtained the curve shown in Fig. 1. The fit to the experimental data is good enough to lend plausibility to our approach and encourage us to try more rigorous calculations.

Reinhardt, Müller and Greiner[10] have taken these nuclear motion effects into account in their calculations and finds no oscillatory or peaked behavior in the positron spectrum. However, their calculations are within the one-electron approximation only. From the times involved, these higher-order processes need to occur at internuclear distances of $R \sim 500$ f. This is exactly in the region of strong molecular coupling, referred to by Soff. It is just in this region that the one-electron approximation breaks down.[1] Until very thorough investigations are made of many-electron effects, one cannot rule out the atomic explanation definitively.

To summarize, we have found that the resemblance between the positron emission spectra and those found in autoionization during atomic collisions is strong enough to make an atomic explanation a plausible, if unproven, alternative to the nuclear and elementary particle explanations.[1] The objections raised to this explanation are founded on calculations which are too incomplete to be conclusive. Therefore, we continue to offer the hypothesis that the peaks in the positron spectra could be caused by purely atomic phenomena.

REFERENCES

1. W. Lichten and A. Robatino, Phys. Rev. Lett. 54, 669 (1985). See also a comment, ibid., in press.

2. H. Bokemeyer et al., in Quantum Electrodynamics of Strong Fields, edited by Walter Greiner (Plenum, New York, 1981).

3. J. Rafelski, B. Müller, and W. Greiner, Z. Phys. A 285, 49 (1978).

4. T. Cowan et al., Phys. Rev. Lett. 54, 1761 (1985).

5. A. B. Balentekin, C. Bottcher, M. Strayer, and S. J. Lee (unpublished).

6. M. Lipeles, R. Novick and N. Tolk, Phys. Rev. Lett. 15, 815 (1965); S. Dworetsky, R. Novick, W. W. Smith, and N. Tolk, ibid., 18, 939 (1967). See also Ref. 7.

7. H. Rosenthal, Phys. Rev. A 4, 1030 (1971).

8. R. Morgenstern, A. Niehaus, and U. Thielmann, Phys. Rev. Lett. 37, 199 (1976); G. Gerber and A. Niehaus, J. Phys. B: Atom. Molec. Phys. 9, 123 (1976); Q. C. Kessel, R. Morgenstern, B. Müller, and A. Niehaus, Phys. Rev. A 20, 804 (1979). See also, D. Liesen, A. N. Zinoviev, and F. W. Saris, Phys. Rev. Lett. 47, 1392 (1981) and G.S.I. Scientific Report 1981, p. 145 for similar, oscillatory, autoionization spectra.

9. T. De Reus, U. Müller, B. Müller, and W. Greiner, G.S.I. Report No. 82-1, 1981, p. 174.

10. J. Reinhardt, B. Müller, and W. Greiner, Phys. Rev. Lett., Comment (in press).

SESSION ON SUPERHEAVY ATOMS

CHAIRMAN'S SUMMARY

Jack S. Greenberg
A.W. Wright Nuclear Structure Laboratory
Yale University, New Haven, CT 06511

INTRODUCTION

Because relativistic effects play such an essential role in the structure of superheavy atomic systems by dominating both the binding energies and wave functions of the inner-shell electrons, the study of such systems has particular relevance to this workshop. The discussions in this session especially singled out the features of special interest which evolve from the presence of extreme electric field intensities in superheavy atoms. Foremost among these is the achievement of supercritical binding. With the crossing of this boundary we have the unique opportunity to observe, for the first time, the change in the ground state of QED from the neutral to a charged vacuum through the non-perturbative and spontaneous creation of positrons. This topic has motivated much of the work in this field for more than a decade. However, unlike the luxury afforded to theory, the unfortunate reality is that the only experimental access to superheavy atomic systems is through collisions. We are, therefore, led to the necessity of describing and understanding complex collision systems. This brings an added burden to the fundamental problem both theoretically and experimentally. On the other hand, it has the virtue of providing access to new and interesting phenomena.

Our discussions in this session addressed some of the problems and opportunities associated with exploiting the original quasiatomic concept and with pursuing the predicted, as well as the unexpected, experimental consequences. Some of these central issues can be summarized in posing the following questions.

1) To what extent is the superheavy collision system a source of information on the relativistic structure of superheavy atomic species? Do we form quasiatoms at collision energies near the Coulomb barrier? (Clarifying the effects of dynamics and singling out the sources of information are, clearly, salient points to be addressed in this respect.)

2) Is there a significant signal to signature supercritical binding and the spontaneous decay of the QED vacuum in collision systems such as U+U and U+Cm where supercritical binding can be achieved? Is the interplay between the atomic and nuclear degrees of freedom providing such a signature? (i.e. is the QED vacuum being sparked by the formation of metastable, giant nuclear systems?)

3) More generally, what is the source of the anomalous narrow positron peak structures observed in a number of collision systems?

4) What can we learn from collisions at relativistic velocities?

The papers presented have demonstrated that some of these questions have been penetrated with some success. But also, some questions remain unanswered, and, in fact, attempts to answer them have evolved into new issues which are particularly stimulating since they may involve new consequences with fundamental significance. The main catalyst has been the unexpected discovery of sharp positron lines in superheavy collision systems with Coulomb barrier collision energies.

Some of the discussion has been punctuated by theoretical extrapolation of the data possibly beyond its real content and by some interesting speculations. In this respect it may be useful to reiterate the experimental basis of some of the discussions apart from the speculations that follow.

OBSERVATIONAL BASE

In connection with the first question, it should be noted that a large body of experimental information now supports the concept that quasimolecule or quasiatom formation plays an essential role in the physics of superheavy collision systems. Sources of such data include x-ray emission from K-vacancy formation, delta-electron emission, MO x-ray radiation, and positron creation. Dr. Soff discussed some of the diverse set of measurements that all support the concept that quasiatoms are formed in collisions such as U+U, U+Cm, Th+Cm etc. near the Coulomb barrier with deeply bound $1s\sigma$ states. In K-vacancy production the telling features are the large cross sections and the confinement of the cross section to small impact parameter collisions. Both these features attest to the rapid relativistic build-up of the electron density at small internuclear separations and the rapid increase in the binding energy. The strong relativistic effects also reflect in the large cross sections for high energy delta-electron production, in the high energy MO x-rays emitted, and in the large probability for dynamically producing positrons. It is particularly important to note that the fact that dynamic positron production is at a level that can be observed at all is testimony to the coherent effects produced by the formation of the superheavy quasiatom.

Therefore the original suggestions by W. Greiner and, independently, by Gershstein and Zel'dovitch that superheavy quasiatoms can be a source of intense electric fields has proven to be very fruitful. However, the access to the properties and spectroscopy of superheavy atoms through this vehicle is clearly not the conventional one usually associated with stable atoms. The measurements do not provide an energy level diagram. A mapping of the spectroscopy connected with the two-center Dirac equation is obviously not directly available since a time dependence is an inherent feature of the energy level structure. The answer to the first question, therefore, is that the excitation mechanism in the superheavy collision systems provide information on the bound state and continuum wave functions tempered by the dynamics. It also bears emphasis that these type of measurements reflect the binding energies of inner-shell states only in a very approximate way through a qualitative deduction of the energy transferred.

One can say after a number of years of both experimental and theoretical work that the main features of inner-shell vacancy formation, delta-electron emission, and dynamic positron production in superheavy collision systems are well understood theoretically. Both the molecular orbital basis approach, used by the Frankfurt group of W. Greiner, B. Müller, G. Soff, J. Reinhardt, U. Müller, U. Heinz and coworkers and also by T. Tomoda and H. Weidenmüller, and the direct numerical approach of solving the time-dependent Dirac equation discussed by Bottcher and Strayer have been successful in describing the gross features of the positron energy spectra and the production cross sections as a function of the angular distribution of the scattered ions. Given the complexities involved, unknown charge states etc., the state of quantitative agreement between theory and experiment shown by Soff has to be considered good. But probably the ultimate quantitative results that can be achieved do not represent the main impact of this work which has now reached the stage of scrutinizing many details. Rather its main significance may rest with the first qualitative demonstrations of the extreme relativistic effects that are essential ingredients to achieve supercritical binding and to observe the decay of the QED vacuum in overcritical fields. Of course, understanding the vacancy and dynamic positron production processes is a necessary and important step towards any successful search for spontaneous positron emission, if only to serve to determine the background.

In fact, it has become evident from these earlier studies that any practical opportunity to take advantage of the supercritical fields generated in quasiatoms to observe spontaneous positron emission rests with the possibility of extending the time scale for supercritical binding beyond that available in Rutherford scattering. Nuclear reactions are an obvious vehicle for prolonging nuclear contact times. Soff reviewed the schematic models proposed and worked out by J.Rafelski, B. Müller, W. Greiner, J. Reinhardt, U. Müller, G. Soff and coworkers which consider the effects of time delay on the positron spectrum. If we are to take these possibilities seriously, there are, of course, several open questions to be answered: Given the limited experimental information available on nuclear reactions with superheavy collision system, it is not known whether sufficiently long delay times of $\sim 10^{-20}$ to 10^{-19} sec. are achievable. Conventional wisdom argues against this. We may also ask what reactions should be explored. Presently, nuclear theory is not in the position to provide predictions that can guide us unambiguously on these questions.

Therefore, the discovery of narrow positron peaks emitted in supercritical collision systems at bombarding energies near the Coulomb barrier is a provocative development and has had a catalytic influence on the search for spontaneous positron emission. (J. Schweppe **et al.**, Phys. Rev. Lett. **51**, 2261 (1983); M. Clemente **et al.**, Phys. Lett. **137B**, 41 (1984); T. Cowan **et al.**, Phys. Rev. Lett. **54**, 1761 (1985). As occurs in many discoveries, the appearance of the peaks was completely unexpected given our understanding then of the positron production processes and the fact that the experiments were carried out at bombarding energies hovering around the Coulomb barrier where Rutherford scattering dominates the positron production events. Selecting bombarding energies near the Coulomb barrier turns out to be one of those fortuitous choices, dictated in part by the accelerator characteristics, that one can only justify in hindsight. It is not surprising that the source of these positron lines has become the central issue. Indeed, the answers to the questions, posed above, regarding the existence of supercritical binding, the possibility of employing superheavy collision systems to observe spontaneous positron emission, and the possibility of exploiting the positron spectrum as a probe for the formation and properties of superheavy nuclear species all center on understanding the nature and origin of the positron lines.

As a supplement to the mainly theoretical discussions, the experimental observations regarding the positron lines can be summarized as follows:

1) A prominent narrow positron peak has been observed in five supercritical collision systems consisting of combinations of Th and U projectiles with targets of Th, U, and Cm (see Fig. 18 in paper by Soff). The variety of collision partners excludes the possibility that a common collision partner is responsible for a trivial nuclear source of common structure in the positron spectrum. The combined Z_u varies from 180 to 188. For the systems with Z_u=182 and 184, peaks have been observed in two independent experiments (see Figs. 17,18 by Soff) using substantially different detection techniques, kinematic selection of events and analysis procedures. These differences may account for apparent systematic discrepancies in peak energies quoted, although the possibility that the peaks are not really equivalent cannot be discounted. These differences should probably not be taken too seriously presently.

In addition to these dominant reproducible peaks, some of the data may suggest the presence of other smaller structures in the positron spectra. However, the evidence for such structures presently lacks both adequate statistical significance and reproducibility.

2) An interesting feature associated with the peaks is their common small widths of ~80 keV which is narrow on the time scale associated with Rutherford scattering during which supercritical binding can occur; $\Delta E = h/\Delta\tau_R$, $\Delta\tau_R \simeq 10^{-21}$ s. In fact, for the peaks shown in Fig. 18 (Soff), Doppler broadening by a source moving with about the velocity of the center-of-mass can account for the entire peak width. Therefore, it seems that the intrinsic width can, indeed, be significantly narrower and, therefore, the emitting source may be very long-lived, $\tau > 10^{-19}$ (In this connection it should be noted that the fit to the data in Fig. 15 of the paper by Soff incorrectly does not incorporate the effect of Doppler broadening and assumes a lifetime for the nuclear configuration commensurate with an intrinsic width of ~80 keV).

3) The detailed analysis of widths and line shapes provide an effective source of further information. As was mentioned by Soff (re Fig. 16) using the data and analysis presented by Cowan **et al.**, (Phys. Rev. Lett. **54**, 1761 (1985)), the line shapes can be utilized to argue that the velocity of the source emitting the peaks is consistent with the center-of-mass velocity. Moreover, emission by sources moving with the asymptotic velocities of both the projectile and recoil target nuclei at selected scattering angles is not compatible with the measured line widths. However, the word **consistent** has to be emphasized here since these conclusions are based on specifying the direction and velocity of the emitter. These quantities are determined in the experiment only for the scattered ions and the center-of-mass. However, generalizing to the possibility that emission can be by other sources not detected obviates the conclusions of this analysis. For example, a source isotropically emitted from the center-of-mass with a constant distribution in velocity phase space can simulate the line shape measured. With this generalizaton the origin of the line remains an open question.

4) A particularly significant observational point not emphasized in the discussions concerns the beam energy dependence associated with the production of the peaks. This aspect almost escaped detection due to the difficulty of maintaining target stability under bombardment of the very-heavy ions. It was found in the experiments with the EPOS (Schweppe **et al.**, Cowan **et al.**) apparatus that the appearance of the peaks is apparently

associated with narrow intervals of bombarding energies which correspond in each system to a marginal overlap of nuclear surfaces in a head-on collision. The width of this interval appears to be less than the equivalent energy loss in the target by ionization, indicating a resonance-like process. A different energy behavior is reported by Clemente **et al.** for the peaks they detect in U+U collisions. However, the resonance-like behavior could have been disguised here as an effective weak dependence on the projectile energy due to the effect of a target thickness which straddles the energy range explored. It would appear that the production of the peak is associated with nuclear contact.

5) As illustrated in Fig. 15 (Schweppe **et al.**) in Soff's paper, the imposition of kinematic constraints is responsible for revealing the peak in U+CM prominently above the dynamic background. The utilization of good kinematic resolution, in fact, provided the experimental edge which revealed the sharp positron structures in the EPOS experiments. (H. Bokemeyer **et al.**, QED of Strong Fields, W. Greiner ed., Plenum Press, 1983). The selection imposed by kinematic constraints appears to originate from an apparent shift in the kinematic correlations of the scattering events associated with the peak relative to the dominant Rutherford scattering events. The peak events thus seem to have this special status which distinguishes them from the dynamic background.

6) Since their discovery, the important question has been posed whether the origin of the peak positrons can be traced to a trivial source either connected with the apparatus or with common nuclear causes. No evidence to date has been found that the peaks are generated due to properties of the apparatus such as for example a pathological efficiency response function. Moreover, it is extremely unlikely that a similar apparatus effect would be reproduced simultaneously in the EPOS and Orange spectrometer systems which use such very different detection arrangements. It is also very unlikely that known nuclear effects can be responsible for the positron peaks. Internal conversion processes are ruled out by careful study of gamma-ray and electron spectra unless unanticipated very exotic conditions exist in these collision systems. Therefore, gathering all the experimental evidence together argues very strongly against the possibility that the positron peaks are a trivial effect.

7) Besides the mere appearance of the peaks, the most striking feature of the data is the similar peak energies and peak widths found in the supercritical collision systems. This property is present within each set of data in the two independent measurements. A simple average of the peak energies obtained from the EPOS measurement (Cowan **et al.**) for the five systems with Z_u=180-188 is 336±10 keV. The Orange spectrometer (Clemente **et al.**) yields somewhat smaller energies for Z_u= 182 and 184 which appear to depend on scattering angle and bombarding energy. As I mentioned previously, the differences between measurements is something to be noted, but, at the present stage of measurements and theory, it does not present the significant issue implied in Soff's paper especially since the predictive power of theory falls very short of making any meaningful statements about energy differences of the order of the discrepancies found in the measurements. Clearly the overriding qualitative issue confronting theoretical explanation is the near constancy in peak energy, which should not be diminished by the possible existence of experimental inconsistencies presently.

SPECULATIONS

With these experimental facts available the speculations on the origin of the peaks begin. In seeking an explanation for the peaks, it bears keeping in mind that the mere observation of the positron peaks provokes considerable interest on general grounds in addition to their possible connection to spontaneous positron emission. Narrow, low energy positron line spectra are a rare commodity in nature. Their occurrence is anomalous and speaks for an unusual and unorthodox explanation, possibly involving a previously undetected source. This source, of course, has to have a fairly long lifetime in order to account for the width of the peaks.

It is natural to attempt to associate the positrons peaks with spontaneous positron emission particularly since the experiments were carried out to search for this process. Greiner's **et al.**, proposal for the formation of a giant, dinuclear, metastable systems near the Coulomb barrier, of course, was motivated by this possibility and provides a particularly attractive phenomenological framework to explain the earliest data, particularly on the U+Cm collision system. The measured peak energy in this case appears to be consistent with the assumption that the vacuum decays in the supercritical field generated by the composite nucleus with Z_u=188. In addition, the resonance-like dependence of the peak's production on projectile energy and the appearance that the peak may be associated with center-of-mass emission both suggest the formation of such a giant system.

However, this apparent accommodation of experimental observations may only be circumstantial since this proposal encounters serious difficulties when the data base is broadened. Measurements by Cowan **et al.** (Phys. Rev. Lett. **54**, 1761 (1985)) were specifically carried out to test the distinguishing signature of the model that predicts a remarkable Z_u^{20} scaling of the peak energies for systems with similar nuclear charge distributions and states of ionization. As I noted above, they covered a range of supercritical charges Z_u from 180 to 188 and found a near degeneracy in the peak energies in contrast to the factor of ~5 which is predicted by the theoretical calculations shown by Soff in his presentation. These results can only be incorporated in our scenario based on spontaneous positron emission from the giant system if radically different nuclear charge configurations and ionization states are invoked for the compound systems. Indeed, we have seen from Soff's presentation (see primary data in Cowan **et al.**, Phys. Rev. Lett. **54**, 1761 (1985)) that the shapes would have to track fortuitously with Z_u from highly deformed to spherical configurations so as to maintain a constant binding energy for the 1s state. Theory would then have to explain how the giant systems are produced in such diverse shapes and then decay into two fragments that closely mirror elastic scattering. It should also be mentioned that theory presently greatly underestimates the cross sections needed for the formation of the giant system.

To accommodate the near constancy in peaks energies we could alternatively postulate that a particularly stable, spherical nuclear complex is being formed in all the systems with $Z_u \leq 180$. In addition to the fact that this assumption has neither experimental nor theoretical justification, it is also excluded by the kinematic constraints imposed by the good scattering angle resolution achieved in the measurements (Bokemeyer **et al.**; Cowan **et al.**).

Reaching the required charge of $Z_u \leq 180$ involves evaporating a sufficient number of nucleons from some of the systems so as to be readily identified in the kinematic correlations.

A more mundane explanation for the peaks would be internal pair conversion processes in the combined nuclear system. In this case the narrow peak widths do not exclude internal pair conversion as they do for conversion of transitions in the individual nuclei since the Z is now large enough to sufficiently contract the positron energy distribution. Soff considered this possibility and pointed out many questionable features. But in addition, in order for this explanation to be viable, we would have to understand why the transition energy happens to be so similar for the 5 different systems, and why there is no evidence for spontaneous positron emission if the combined system lives long enough to allow the internal conversion process to be observed.

All these consideration raise the obvious question whether the peaks also appear in systems where supercritical binding is not expected. It was pointed out in the discussion by Greenberg that, indeed, this question has been explored by his group in the Th+Ta system with Z_u=163, which is well below the spontaneous positron emission threshold of Z_u=173 for normal nuclear density (GSI Scientific Report 1984 pp. 177, and J. Schweppe ICPEAC, Palo Alto 1985). Again using bombarding conditions corresponding to a marginal overlap of nuclear surfaces, a peak is observed with approximately the same characteristics as those found in the supercritical systems. From the present state of the analysis, it appears that a large fraction of the peak's intensity cannot be associated with a nuclear transition and nor can its width. Soff also mentioned the possibility of a peak appearing in very recent data on the U+Ta system (Kienle **et al.**). Of course, it has not been established that the peaks in the subcritical systems, if substantiated by further analysis, and in the supercritical systems represent the same phenomenon, but spontaneous positron production as the source for **all** the peaks observed in all the systems is clearly excluded.

In another approach to the puzzle Bill Lichten proposed the possibility that an atomic process could produce the positron peaks. In essence as presented it involves interference between processes widely separated in time coupled with many electron effects. The idea has not been worked out in sufficient detail so that its relevance can be judge. Its potential to produce structure has been challenged by Reinhardt **et al.** (Phys. Rev. Lett. comments **55**, 134 (1985); see also counter arguments by Lichten in same issue). Even in the simple form presented, this mechanism cannot produce sharp lines. In this connection, it should be noted that the fit shown in Lichten's contribution to the conference omits Doppler broadening which, if included, would considerably worsen the comparison with experiment. Of course, it would be of interest to see if Lichten's mechanism can produce structure.

We appear to have run out the string of viable known mechanisms without a probable explanation for the origin of the positron peaks. As was previously emphasized, besides the unexpected appearance of the peaks, the striking feature that begs explanation is the near degeneracy in peak energies. Confronting this observation directly, Cowan **et al.**, in Phys. Rev. Lett. **54**, 1761 (1985), speculated that a common source may be responsible for the monoenergetic electrons such as the two-body decay of a previously undetected neutral boson. This particle could be produced either directly or indirectly in the collision. They proposed an experiment to look for the e^+e^-

decay of this particle, pointing out that a clear signal would not only be provided by the sharp energy of the electron equal to the positron peak energy, but that the sum energy of the pair would be free of Doppler broadening and, therefore, produce a very sharp sum-energy line spectrum. Since the conference, this group has carried out such a measurement at GSI, Darmstadt. They have shown that carrying out e^+e^- coincidence measurements is feasible even in the face of a large delta-electron flux from the target. Because of its significance, the results of this experiment is obviously being awaited with great interest.

If such a particle exists, what is it and how is it produced? Following the speculations by Cowan **et al.**, Balantekin **et al.**, have considered possibilities based on a pseudoscalar particle (see this conference). Soff has insisted that he cannot reconcile the production of a pseudoscalar particle with the measured production yield and the energy distribution of the positrons. Naively we may ask if the strong electromagnetic fields participate in the production of such a particle? Is the formation of giant nuclei and long-lived strong electric fields relevant to the production of particles? If the experiment mentioned above yields positive results these questions will have to be addressed.

CONCLUSIONS

Our discussions have shown that the venture into the realm of superheavy atomic systems has proved to be very rewarding. The fact that we have to deal with collision systems is a complicating feature, but the relativistic features are clearly demonstrated in the wave functions and binding energies of the quasiatom. We have seen that relativistic effects completely dominate the description of the electron excitation processes and are responsible for the possibility of observing positron production from atomic processes. In searching for spontaneous positron production we have discovered narrow positron lines. They present us with an interesting puzzle. Resolving their origin is particularly provocative since it may have fundamental consequences. The possiblity that the production of the peaks may involve the strong electric and magnetic fields generated in the superheavy collisions or the formation of giant nuclear complexes emphasizes the general interest in this puzzle involving the interplay of several fields of physics.

List of Participants

Lloyd Armstrong, Jr.
John Hopkins University
Baltimore, MD 21218

C.K.E. Au
Dept. of Physics & Astronomy
University of South Carolina
Columbia, SC 29208

A. Baha Balantekin
Oak Ridge National Lab.
Oak Ridge, TN 37830

William E. Baylis
Department of Physics
University of Windsor
Windsor, Ontario N9B 3P4
CANADA

Adam Bechler
Dept. of Physics & Astronomy
University of Pittsburgh
Pittsburgh, PA 15260

H. Gordon Berry
Argonne National Laboratory
9700 S. Cass Avenue
Argonne, IL 60439

Chris Bottcher
Oak Ridge National Lab.
P. O. Box X
Bldg. 5500
Oak Ridge, TN 37831

Jean-Pierre Briand
LPAN-Institut Curie
11 Rue Pierre & Marie Curie
75231 Paris Cedex 05
FRANCE

William B. Campbell
University of Nebraska
Lincoln, NE 68588-0111

Mau Hsiung Chen
Lawrence Livermore Natl. Lab.
P. O. box 808, L-296
Livermore, CA 94550

Kwok-tsang Cheng
Lawrence Livermore Natl. Lab.
P. O. Box 808, L-17
Livermore, CA 94550

Charles W. Clark
Radiation Physics Division
Physics Bldg., Rm. A251
National Bureau of Standards
Gaithersburg, MD 20899

Hollis E. Dalhed, Jr.
Lawrence Livermore Natl. Lab.
P. O. Box 808, L-17
Livermore, CA 94550

Jean-Paul Desclaux
Centre D'Etudes Nucléaire
de Grenoble
DRF/Service de Physique
85 X - 38041 Grenoble Cedex
FRANCE

Richard D. Deslattes
Physics Bldg., Rm. A251
National Bureau of Standards
Gaithersburg, MD 20899

Klaus Dietz
Physikalisches Institut
der Universität Bonn
Nussallee 12
D-5300 Bonn 1
WEST GERMANY

Richard J. Drachman
Code 680.1
Goddard Space Flight Center
Greenbelt, MD 20771

Gordon W.F. Drake
Dept. of Physics
University of Windsor
Windsor, Ontario N9B 3P4
CANADA

Walter B. England
Department of Chemistry
Univ. of Wisconsin-Milwaukee
Milwaukee, WI 53201

Alex M. Ermolaev
University of Durham Science Lab.
Durham, DH1 3LE
ENGLAND

Burkhard Fricke
Department of Physics
University of Kassel
Heinrich-Plett-Strasse 40
Kassel
WEST GERMANY

Thomas Fulton
Department of Physics & Astronomy
John Hopkins University
34th and Charles Street
Baltimore, MD 21218

Harvey Gould
Lawrence Berkeley Lab.
71-259 Lawrence Berkeley Lab.
Berkeley, CA 94720

Ian P. Grant
Dept. of Theoretical Chemistry
University of Oxford
1 South Parks Road
Oxford OX1 3TG
ENGLAND

Jack S. Greenberg
Yale University
P. O. Box 6666
New Haven, CT 06511

Peter L. Hagelstein
Lawrence Livermore National Lab.
P. O. Box 808
Livermore, CA 94550

Yasuyuki Ishikawa
Department of Chemistry
University of Puerto Rico
Rio Piedras
PUERTO RICO 00931

Verne L. Jacobs
Center for Space Research
Naval Research Laboratory
Washington, DC 20375-5000

Walter R. Johnson
Department of Physics
University of Notre Dame
Notre Dame, IN 46556

Brian R. Judd
Johns Hopkins University
34th and Charles Street
Baltimore, MD 21218

Takashi Kagawa
Department of Physics
Nara Women's University
Nara 630
JAPAN

Hugh P. Kelly
Department of Physics
University of Virginia
Charlottesville, VA 22901

Ernest G. Kessler, Jr.
Quantum Metrology Group
Physics Bldg., Room A141
National Bureau of Standards
Gaithersburg, MD 20899

Longhuan Kim
Dept. of Physics and Astron.
University of Pittsburgh
Pittsburgh, PA 15260

Yong-Ki Kim
Atomic & Plasma Radiation Div.
Physics Bldg., Room A167
National Bureau of Standards
Gaithersburg, MD 20899

William Lichten
Department of Physics
Yale University
New Haven, CT 06520

Kwang-Zu Lu
Atomic and Plasma Radiation Div.
Physics Bldg., Room A167
National Bureau of Standards
Gaithersburg, MD 20899

Georgia A. Martin
Atomic and Plasma Radiation Div.
Physics Bldg. Room A267
National Bureau of Standards
Gaithersburg, MD 20899

William C. Martin
Atomic and Plasma Radiation Div.
NBS Physics Bldg., Room A167
Gaithersburg, MD 20899

L. D. Miller
Army Foreign Sci. & Tech. Center &
Inst. for Nuclear Particle Physics
University of Virginia
220 7th Street, NE
Charlottesville, VA 22901

Marvin Mittleman
Department of Physics
City College of NY
New York, NY 10031

Peter J. Mohr
National Science Foundation
1800 G Street, N.W.
Washington, DC 20550

John D. Morgan, III
Department of Physics
University of Delaware
Newark, DE 19711

Kazem Omidvar
Code 614, NASA
Goddard Space Flight Center
Greenbelt, MD 20771

James Peek
Sandia Laboratory
Albuquerque, NM 87115

Michael S. Pindzola
Department of Physics
Auburn University
Auburn, AL 36849

Greg Pollak
Lawrence Livermore Natl. Lab.
P. O. Box 808, L-18
Livermore, CA 94550

Richard H. Pratt
Dept. of Physics and Astron.
University of Pittsburgh
Pittsburgh, PA 15260

Rocio Jáuregui Renaud
Instituto De Fisica, UNAM
CD Univ., Circuito Exterior
(Apto. Postal 20-364)
MEXICO DF 0100

Andre Robatino
Yale University
34 Livingston Street
New Haven, CT 06511

Sten Salomonson
Department of Physics
University of Virginia
McCormick Road
Charlottesville, VA 22901

Frank C. Sanders
Department of Physics
Southern Illinois University
Carbondale, IL 62901

Jonathan R. Sapirstein
Department of Physics
University of Notre Dame
Notre Dame, IN 46556

James H. Scofield
Lawrence Livermore Natl. Lab.
P. O. Box 808 L-296
Livermore, CA 94550

Gerhard Soff
GSI
Planckstrasse 1
Postfach 110541
6100 Darmstadt 11
WEST GERMANY

Joseph Sucher
Department of Physics and Astron.
University of Maryland
College Park, MD 20742

Andrew W. Weiss
Atomic and Plasma Radiation Div.
Physics Bldg., Room A267
National Bureau of Standards
Gaithersburg, MD 20899

Wolfgang L. Wiese
Atomic and Plasma Radiation Div.
Physics Bldg., Room A267
National Bureau of Standards
Gaithersburg, MD 20899

Ru Y. Yin
Department of Physics and Astron.
University of Pittsburgh
Pittsburgh, PA 15260

Stephen M. Younger
Lawrence Livermore National Lab.
P. O. Box 808, L-17
Livermore, CA 94550

Bernard Zygelman
Harvard College Observatory
60 Garden Street
Cambridge, MA 02138

Author Index

AIP Conference Proceedings

		L.C. Number	ISBN
No. 1	Feedback and Dynamic Control of Plasmas – 1970	70-141596	0-88318-100-2
No. 2	Particles and Fields – 1971 (Rochester)	71-184662	0-88318-101-0
No. 3	Thermal Expansion – 1971 (Corning)	72-76970	0-88318-102-9
No. 4	Superconductivity in *d*- and *f*-Band Metals (Rochester, 1971)	74-18879	0-88318-103-7
No. 5	Magnetism and Magnetic Materials – 1971 (2 parts) (Chicago)	59-2468	0-88318-104-5
No. 6	Particle Physics (Irvine, 1971)	72-81239	0-88318-105-3
No. 7	Exploring the History of Nuclear Physics – 1972	72-81883	0-88318-106-1
No. 8	Experimental Meson Spectroscopy –1972	72-88226	0-88318-107-X
No. 9	Cyclotrons – 1972 (Vancouver)	72-92798	0-88318-108-8
No. 10	Magnetism and Magnetic Materials – 1972	72-623469	0-88318-109-6
No. 11	Transport Phenomena – 1973 (Brown University Conference)	73-80682	0-88318-110-X
No. 12	Experiments on High Energy Particle Collisions – 1973 (Vanderbilt Conference)	73-81705	0-88318-111–8
No. 13	π-π Scattering – 1973 (Tallahassee Conference)	73-81704	0-88318-112-6
No. 14	Particles and Fields – 1973 (APS/DPF Berkeley)	73-91923	0-88318-113-4
No. 15	High Energy Collisions – 1973 (Stony Brook)	73-92324	0-88318-114-2
No. 16	Causality and Physical Theories (Wayne State University, 1973)	73-93420	0-88318-115-0
No. 17	Thermal Expansion – 1973 (Lake of the Ozarks)	73-94415	0-88318-116-9
No. 18	Magnetism and Magnetic Materials – 1973 (2 parts) (Boston)	59-2468	0-88318-117-7
No. 19	Physics and the Energy Problem – 1974 (APS Chicago)	73-94416	0-88318-118-5
No. 20	Tetrahedrally Bonded Amorphous Semiconductors (Yorktown Heights, 1974)	74-80145	0-88318-119-3
No. 21	Experimental Meson Spectroscopy – 1974 (Boston)	74-82628	0-88318-120-7
No. 22	Neutrinos – 1974 (Philadelphia)	74-82413	0-88318-121-5
No. 23	Particles and Fields – 1974 (APS/DPF Williamsburg)	74-27575	0-88318-122-3
No. 24	Magnetism and Magnetic Materials – 1974 (20th Annual Conference, San Francisco)	75-2647	0-88318-123-1

No. 25	Efficient Use of Energy (The APS Studies on the Technical Aspects of the More Efficient Use of Energy)	75-18227	0-88318-124-X
No. 26	High-Energy Physics and Nuclear Structure – 1975 (Santa Fe and Los Alamos)	75-26411	0-88318-125-8
No. 27	Topics in Statistical Mechanics and Biophysics: A Memorial to Julius L. Jackson (Wayne State University, 1975)	75-36309	0-88318-126-6
No. 28	Physics and Our World: A Symposium in Honor of Victor F. Weisskopf (M.I.T., 1974)	76-7207	0-88318-127-4
No. 29	Magnetism and Magnetic Materials – 1975 (21st Annual Conference, Philadelphia)	76-10931	0-88318-128-2
No. 30	Particle Searches and Discoveries – 1976 (Vanderbilt Conference)	76-19949	0-88318-129-0
No. 31	Structure and Excitations of Amorphous Solids (Williamsburg, VA, 1976)	76-22279	0-88318-130-4
No. 32	Materials Technology – 1976 (APS New York Meeting)	76-27967	0-88318-131-2
No. 33	Meson-Nuclear Physics – 1976 (Carnegie-Mellon Conference)	76-26811	0-88318-132-0
No. 34	Magnetism and Magnetic Materials – 1976 (Joint MMM-Intermag Conference, Pittsburgh)	76-47106	0-88318-133-9
No. 35	High Energy Physics with Polarized Beams and Targets (Argonne, 1976)	76-50181	0-88318-134-7
No. 36	Momentum Wave Functions – 1976 (Indiana University)	77-82145	0-88318-135-5
No. 37	Weak Interaction Physics – 1977 (Indiana University)	77-83344	0-88318-136-3
No. 38	Workshop on New Directions in Mossbauer Spectroscopy (Argonne, 1977)	77-90635	0-88318-137-1
No. 39	Physics Careers, Employment and Education (Penn State, 1977)	77-94053	0-88318-138-X
No. 40	Electrical Transport and Optical Properties of Inhomogeneous Media (Ohio State University, 1977)	78-54319	0-88318-139-8
No. 41	Nucleon-Nucleon Interactions – 1977 (Vancouver)	78-54249	0-88318-140-1
No. 42	Higher Energy Polarized Proton Beams (Ann Arbor, 1977)	78-55682	0-88318-141-X
No. 43	Particles and Fields – 1977 (APS/DPF, Argonne)	78-55683	0-88318-142-8
No. 44	Future Trends in Superconductive Electronics (Charlottesville, 1978)	77-9240	0-88318-143-6
No. 45	New Results in High Energy Physics – 1978 (Vanderbilt Conference)	78-67196	0-88318-144-4
No. 46	Topics in Nonlinear Dynamics (La Jolla Institute)	78-57870	0-88318-145-2

No. 47	Clustering Aspects of Nuclear Structure and Nuclear Reactions (Winnepeg, 1978)	78-64942	0-88318-146-0
No. 48	Current Trends in the Theory of Fields (Tallahassee, 1978)	78-72948	0-88318-147-9
No. 49	Cosmic Rays and Particle Physics – 1978 (Bartol Conference)	79-50489	0-88318-148-7
No. 50	Laser-Solid Interactions and Laser Processing – 1978 (Boston)	79-51564	0-88318-149-5
No. 51	High Energy Physics with Polarized Beams and Polarized Targets (Argonne, 1978)	79-64565	0-88318-150-9
No. 52	Long-Distance Neutrino Detection – 1978 (C.L. Cowan Memorial Symposium)	79-52078	0-88318-151-7
No. 53	Modulated Structures – 1979 (Kailua Kona, Hawaii)	79-53846	0-88318-152-5
No. 54	Meson-Nuclear Physics – 1979 (Houston)	79-53978	0-88318-153-3
No. 55	Quantum Chromodynamics (La Jolla, 1978)	79-54969	0-88318-154-1
No. 56	Particle Acceleration Mechanisms in Astrophysics (La Jolla, 1979)	79-55844	0-88318-155-X
No. 57	Nonlinear Dynamics and the Beam-Beam Interaction (Brookhaven, 1979)	79-57341	0-88318-156-8
No. 58	Inhomogeneous Superconductors – 1979 (Berkeley Springs, W.V.)	79-57620	0-88318-157-6
No. 59	Particles and Fields – 1979 (APS/DPF Montreal)	80-66631	0-88318-158-4
No. 60	History of the ZGS (Argonne, 1979)	80-67694	0-88318-159-2
No. 61	Aspects of the Kinetics and Dynamics of Surface Reactions (La Jolla Institute, 1979)	80-68004	0-88318-160-6
No. 62	High Energy e^+e^- Interactions (Vanderbilt, 1980)	80-53377	0-88318-161-4
No. 63	Supernovae Spectra (La Jolla, 1980)	80-70019	0-88318-162-2
No. 64	Laboratory EXAFS Facilities – 1980 (Univ. of Washington)	80-70579	0-88318-163-0
No. 65	Optics in Four Dimensions – 1980 (ICO, Ensenada)	80-70771	0-88318-164-9
No. 66	Physics in the Automotive Industry – 1980 (APS/AAPT Topical Conference)	80-70987	0-88318-165-7
No. 67	Experimental Meson Spectroscopy – 1980 (Sixth International Conference, Brookhaven)	80-71123	0-88318-166-5
No. 68	High Energy Physics – 1980 (XX International Conference, Madison)	81-65032	0-88318-167-3
No. 69	Polarization Phenomena in Nuclear Physics – 1980 (Fifth International Symposium, Santa Fe)	81-65107	0-88318-168-1
No. 70	Chemistry and Physics of Coal Utilization – 1980 (APS, Morgantown)	81-65106	0-88318-169-X

No. 71	Group Theory and its Applications in Physics – 1980 (Latin American School of Physics, Mexico City)	81-66132	0-88318-170-3
No. 72	Weak Interactions as a Probe of Unification (Virginia Polytechnic Institute – 1980)	81-67184	0-88318-171-1
No. 73	Tetrahedrally Bonded Amorphous Semiconductors (Carefree, Arizona, 1981)	81-67419	0-88318-172-X
No. 74	Perturbative Quantum Chromodynamics (Tallahassee, 1981)	81-70372	0-88318-173-8
No. 75	Low Energy X-Ray Diagnostics – 1981 (Monterey)	81-69841	0-88318-174-6
No. 76	Nonlinear Properties of Internal Waves (La Jolla Institute, 1981)	81-71062	0-88318-175-4
No. 77	Gamma Ray Transients and Related Astrophysical Phenomena (La Jolla Institute, 1981)	81-71543	0-88318-176-2
No. 78	Shock Waves in Condensed Mater – 1981 (Menlo Park)	82-70014	0-88318-177-0
No. 79	Pion Production and Absorption in Nuclei – 1981 (Indiana University Cyclotron Facility)	82-70678	0-88318-178-9
No. 80	Polarized Proton Ion Sources (Ann Arbor, 1981)	82-71025	0-88318-179-7
No. 81	Particles and Fields –1981: Testing the Standard Model (APS/DPF, Santa Cruz)	82-71156	0-88318-180-0
No. 82	Interpretation of Climate and Photochemical Models, Ozone and Temperature Measurements (La Jolla Institute, 1981)	82-71345	0-88318-181-9
No. 83	The Galactic Center (Cal. Inst. of Tech., 1982)	82-71635	0-88318-182-7
No. 84	Physics in the Steel Industry (APS/AISI, Lehigh University, 1981)	82-72033	0-88318-183-5
No. 85	Proton-Antiproton Collider Physics –1981 (Madison, Wisconsin)	82-72141	0-88318-184-3
No. 86	Momentum Wave Functions – 1982 (Adelaide, Australia)	82-72375	0-88318-185-1
No. 87	Physics of High Energy Particle Accelerators (Fermilab Summer School, 1981)	82-72421	0-88318-186-X
No. 88	Mathematical Methods in Hydrodynamics and Integrability in Dynamical Systems (La Jolla Institute, 1981)	82-72462	0-88318-187-8
No. 89	Neutron Scattering – 1981 (Argonne National Laboratory)	82-73094	0-88318-188-6
No. 90	Laser Techniques for Extreme Ultraviolt Spectroscopy (Boulder, 1982)	82-73205	0-88318-189-4
No. 91	Laser Acceleration of Particles (Los Alamos, 1982)	82-73361	0-88318-190-8
No. 92	The State of Particle Accelerators and High Energy Physics (Fermilab, 1981)	82-73861	0-88318-191-6

No. 93	Novel Results in Particle Physics (Vanderbilt, 1982)	82-73954	0-88318-192-4
No. 94	X-Ray and Atomic Inner-Shell Physics – 1982 (International Conference, U. of Oregon)	82-74075	0-88318-193-2
No. 95	High Energy Spin Physics – 1982 (Brookhaven National Laboratory)	83-70154	0-88318-194-0
No. 96	Science Underground (Los Alamos, 1982)	83-70377	0-88318-195-9
No. 97	The Interaction Between Medium Energy Nucleons in Nuclei – 1982 (Indiana University)	83-70649	0-88318-196-7
No. 98	Particles and Fields – 1982 (APS/DPF University of Maryland)	83-70807	0-88318-197-5
No. 99	Neutrino Mass and Gauge Structure of Weak Interactions (Telemark, 1982)	83-71072	0-88318-198-3
No. 100	Excimer Lasers – 1983 (OSA, Lake Tahoe, Nevada)	83-71437	0-88318-199-1
No. 101	Positron-Electron Pairs in Astrophysics (Goddard Space Flight Center, 1983)	83-71926	0-88318-200-9
No. 102	Intense Medium Energy Sources of Strangeness (UC-Sant Cruz, 1983)	83-72261	0-88318-201-7
No. 103	Quantum Fluids and Solids – 1983 (Sanibel Island, Florida)	83-72440	0-88318-202-5
No. 104	Physics, Technology and the Nuclear Arms Race (APS Baltimore –1983)	83-72533	0-88318-203-3
No. 105	Physics of High Energy Particle Accelerators (SLAC Summer School, 1982)	83-72986	0-88318-304-8
No. 106	Predictability of Fluid Motions (La Jolla Institute, 1983)	83-73641	0-88318-305-6
No. 107	Physics and Chemistry of Porous Media (Schlumberger-Doll Research, 1983)	83-73640	0-88318-306-4
No. 108	The Time Projection Chamber (TRIUMF, Vancouver, 1983)	83-83445	0-88318-307-2
No. 109	Random Walks and Their Applications in the Physical and Biological Sciences (NBS/La Jolla Institute, 1982)	84-70208	0-88318-308-0
No. 110	Hadron Substructure in Nuclear Physics (Indiana University, 1983)	84-70165	0-88318-309-9
No. 111	Production and Neutralization of Negative Ions and Beams (3rd Int'l Symposium, Brookhaven, 1983)	84-70379	0-88318-310-2
No. 112	Particles and Fields – 1983 (APS/DPF, Blacksburg, VA)	84-70378	0-88318-311-0
No. 113	Experimental Meson Spectroscopy – 1983 (Seventh International Conference, Brookhaven)	84-70910	0-88318-312-9

No. 114	Low Energy Tests of Conservation Laws in Particle Physics (Blacksburg, VA, 1983)	84-71157	0-88318-313-7
No. 115	High Energy Transients in Astrophysics (Santa Cruz, CA, 1983)	84-71205	0-88318-314-5
No. 116	Problems in Unification and Supergravity (La Jolla Institute, 1983)	84-71246	0-88318-315-3
No. 117	Polarized Proton Ion Sources (TRIUMF, Vancouver, 1983)	84-71235	0-88318-316-1
No. 118	Free Electron Generation of Extreme Ultraviolet Coherent Radiation (Brookhaven/OSA, 1983)	84-71539	0-88318-317-X
No. 119	Laser Techniques in the Extreme Ultraviolet (OSA, Boulder, Colorado, 1984)	84-72128	0-88318-318-8
No. 120	Optical Effects in Amorphous Semiconductors (Snowbird, Utah, 1984)	84-72419	0-88318-319-6
No. 121	High Energy e^+e^- Interactions (Vanderbilt, 1984)	84-72632	0-88318-320-X
No. 122	The Physics of VLSI (Xerox, Palo Alto, 1984)	84-72729	0-88318-321-8
No. 123	Intersections Between Particle and Nuclear Physics (Steamboat Springs, 1984)	84-72790	0-88318-322-6
No. 124	Neutron-Nucleus Collisions – A Probe of Nuclear Structure (Burr Oak State Park - 1984)	84-73216	0-88318-323-4
No. 125	Capture Gamma-Ray Spectroscopy and Related Topics – 1984 (Internat. Symposium, Knoxville)	84-73303	0-88318-324-2
No. 126	Solar Neutrinos and Neutrino Astronomy (Homestake, 1984)	84-63143	0-88318-325-0
No. 127	Physics of High Energy Particle Accelerators (BNL/SUNY Summer School, 1983)	85-70057	0-88318-326-9
No. 128	Nuclear Physics with Stored, Cooled Beams (McCormick's Creek State Park, Indiana, 1984)	85-71167	0-88318-327-7
No. 129	Radiofrequency Plasma Heating (Sixth Topical Conference, Callaway Gardens, GA, 1985)	85-48027	0-88318-328-5
No. 130	Laser Acceleration of Particles (Malibu, California, 1985)	85-48028	0-88318-329-3
No. 131	Workshop on Polarized ^{3}He Beams and Targets (Princeton, New Jersey, 1984)	85-48026	0-88318-330-7
No. 132	Hadron Spectroscopy–1985 (International Conference, Univ. of Maryland)	85-72537	0-88318-331-5
No. 133	Hadronic Probes and Nuclear Interactions (Arizona State University, 1985)	85-72638	0-88318-332-3
No. 134	The State of High Energy Physics (BNL/SUNY Summer School, 1983)	85-73170	0-88318-333-1
No. 135	Energy Sources: Conservation and Renewables (APS, Washington, DC, 1985)	85-73019	0-88318-334-X